MULTILEVEL METHODS IN LUBRICATION

TRIBOLOGY SERIES
Editor
D. Dowson (Gt. Britain)

Advisory Board
W.J. Bartz (Germany)
R. Bassani (Italy)
B. Briscoe (Gt. Britain)
H. Czichos (Germany)
K. Friedrich (Germany)
N. Gane (Australia)
W.A. Glaeser (U.S.A.)
H.E. Hintermann (Switzerland)
K.C. Ludema (U.S.A.)
W.O. Winer (U.S.A.)

Vol. 6	Friction and Wear of Polymers (Bartenev and Lavrentev)
Vol. 10	Microstructure and Wear of Materials (Zum Gahr)
Vol. 11	Fluid Film Lubrication – Osborne Reynolds Centenary (Dowson et al., Editors)
Vol. 12	Interface Dynamics (Dowson et al., Editors)
Vol. 13	Tribology of Miniature Systems (Rymuza)
Vol. 14	Tribological Design of Machine Elements (Dowson et al., Editors)
Vol. 15	Encyclopedia of Tribology (Kajdas et al.)
Vol. 16	Tribology of Plastic Materials (Yamaguchi)
Vol. 17	Mechanics of Coatings (Dowson et al., Editors)
Vol. 18	Vehicle Tribology (Dowson et al., Editors)
Vol. 19	Rheology and Elastohydrodynamic Lubrication (Jacobson)
Vol. 20	Materials for Tribology (Glaeser)
Vol. 21	Wear Particles: From the Cradle to the Grave (Dowson et al., Editors)
Vol. 22	Hydrostatic Lubrication (Bassani and Piccigallo)
Vol. 23	Lubricants and Special Fluids (Stepina and Vesely)
Vol. 24	Engineering Tribology (Stachowiak and Batchelor)
Vol. 25	Thin Films in Tribology (Dowson et al., Editors)
Vol. 26	Engine Tribology (Taylor, Editor)
Vol. 27	Dissipative Processes in Tribology (Dowson et al., Editors)
Vol. 28	Coatings Tribology – Properties, Techniques and Applications in Surface Engineering (Holmberg and Matthews)
Vol. 29	Friction Surface Phenomena (Shpenkov)
Vol. 30	Lubricants and Lubrication (Dowson et al., Editors)
Vol. 31	The Third Body Concept: Interpretation of Tribological Phenomena (Dowson et al., Editors)
Vol. 32	Elastohydrodynamics – '96: Fundamentals and Applications in Lubrication and Traction (Dowson et al., Editors)
Vol. 33	Hydrodynamic Lubrication – Bearings and Thrust Bearings (Frêne et al.)
Vol. 34	Tribology For Energy Conservation (Dowson et al., Editors)
Vol. 35	Molybdenum Disulphide Lubrication (Lansdown)
Vol. 36	Lubrication at the Frontier – The Role of the Interface and Surface Layers in the Thin Film and Boundary Regime (Dowson et al., Editors)

TRIBOLOGY SERIES, 37
EDITOR: D. DOWSON

MULTILEVEL METHODS IN LUBRICATION

C.H. Venner
Faculty of Mechanical Engineering
University of Twente
The Netherlands

A.A. Lubrecht
Laboratoire de Mécanique des Contacts
UMR CNRS 5514, INSA de Lyon
France

2000

ELSEVIER
Amsterdam – Lausanne – New York – Oxford – Shannon – Singapore – Tokyo

ELSEVIER SCIENCE B.V.
Sara Burgerhartstraat 25
P.O. Box 211, 1000 AE Amsterdam, The Netherlands

First edition 2000

Library of Congress Cataloging in Publication Data
A catalog record from the Library of Congress has been applied for.

ISBN: 0 444 50503 2

∞ The paper used in this publication meets the requirements of ANSI/NISO Z39.48-1992 (Permanence of Paper).
Printed in The Netherlands.

To our families

Foreword

The problem of an efficient numerical solution of a realistic and therefore complex system of equations occupies many researchers in many disciplines. For various reasons, but mainly to approximate reality, one wants a very large number of unknowns. Using classical solution techniques, the solution of such a system of equations would take too long, and sometimes MultiLevel techniques are used to accelerate convergence. Over the last one and a half decade, the authors have studied the problem of ElastoHydrodynamic Lubrication, governed by a complex integro-differential equation. Their work has resulted in a very efficient and stable solver. In this book they describe the different intermediate problems analyzed and solved, and how those ingredients finally come together in the EHL solver. A number of these intermediate problems, like Hydrodynamic Lubrication and Dry Contact, are useful in their own right. In the Appendix the full codes[1] of the Poisson problem, the Hydrodynamic Lubrication problem, the Dry Contact problem and the ElastoHydrodynamic Lubrication problem are given. These codes are all written in the language 'C', based on the 'ANSI-C' version as described in the second edition of Kernighan and Ritchie's 'The C programming language' [70]. The authors have tried to find a compromise between readibility for the non-C-initiated and simplicity of the code. The book is intended to be an 'intermediate' level MultiGrid book, somewhere between an introduction to MultiGrid as 'A MultiGrid Tutorial' by W.L. Briggs [25] and the advanced 'MultiGrid Techniques: 1984 Guide with Applications to Fluid Dynamics' by A. Brandt [15]. This book can be read in several ways; as a cookbook of how to write and **debug** an efficient EHL code, as an in depth analysis of the difficulties of the system of equations describing EHL or as an example of how MultiLevel methods can be applied to a complex system of equations. Whatever approach of the topic is chosen by the reader, the authors hope that the book will teach them something new and valuable, concerning the possibilities and efficiency of MultiLevel techniques or the intricacies of EHL. The authors are aware of their debt to Prof. Achi Brandt who has taught them the basics of MultiGrid techniques and all the more advanced techniques described in this book.

The authors would like to thank Mr.'s J. Ståhl and F. Pubilier for their contribution to the hydrodynamic problem in Chapter 4 and the line relaxation in that chapter respectively. They also thank Prof. D. Dowson, Prof. J. Frêne, Dr. F.P.H. van Beckum, Ir. H. Moes, Drs. D.E.A. van Odyck and Mr. B.C. Jacod, MSc, for proofreading different chapters and for their suggestions and corrections that improved the readibility of the book. Special thanks are due to Ms. V. Merchat for insisting on clarity and compati-

[1]These codes are all running codes, used in the results sections. They were lifted straight from their working environment, to minimize errors.

bility and her many corrections of the text and the codes. Dr. F. Colin has contributed in rendering the codes machine independent, by performing tests on different platforms. Finally the authors would like to thank Mr. J.-M. Dumont, (a true LaTeX wizzard), and Mr. W. Lette, who as system administrators took care of regular back-ups, and thereby limited the consequences of human errors.

The authors have taken much care to ensure that the algorithms described and the codes given are correct. However, the authors can not be held liable for any errors in the codes, nor for the concequences resulting from such errors. The codes provided in this book are intended for personal use by the reader. Under no condition is the reader allowed to use these codes for commercial purposes. If the reader is interested in a commercial use of the codes, he/she is requested to contact the authors.

March, 2000

Enschede,
C.H. Venner

Lyon,
A.A. Lubrecht

Notation

a	radius of Hertzian contact $[m]$ $a = \sqrt[3]{(3\text{w}R_x)/(2E')}$
$a_{i,j}$	element of matrix A (Chapter 2)
aen	approximate error norm
A	matrix containing the discrete operator $A\underline{u} = \underline{f}$ (Chapter 2)
B	matrix containing part of the discrete operator (Chapter 2)
c	clearance between journal and shaft $[m]$ (Chapter 4)
d	dimesion of the problem
D, E, F	matrix containing parts of A (Chapter 2)
D	bearing diameter $[m]$ (Chapter 4)
e	excentricity between journal and shaft $[m]$ (Chapter 4)
e_i^h	discretisation error $e_i^h = u_i - u_i^h$
en	error norm
E_1, E_2	modulus of elasticity of bodies 1 and 2 $[Pa]$
E'	reduced modulus of elasticity $[Pa]$ $2/E' = (1-\nu_1^2)/E_1 + (1-\nu_2^2)/E_2$
f	general right hand side function
$\underline{f}^h$	discretized right hand side (vector)
$\underline{f}$	right hand side vector
$_Pf$	right hand side function of pressure equation
$_Hf$	right hand side function of film thickness equation
$_Wf$	right hand side function of force balance equation
g	general function describing the boundary conditions
G	dimensionless material parameter $G = \alpha E'$ (Chapter 6)
h	film thickness $[m]$ (Chapter 4, 6)
h	gap height $[m]$ (Chapter 5)
h_0	rigid body dispacement $[m]$
h	fine grid mesh size
h_x, h_y	fine grid mesh size in x and y direction
h_θ	fine grid mesh size in θ direction (Chapter 4)
H	dimensionless film thickness $H = h/c$ (Chapter 4)
H	dimensionless film thickness $H = hR_x/a^2$ (Chapter 5, 6)
H^{HD}	Hamrock & Dowson dimensionless film thickness $H^{HD} = h/R_x$ (Chapter 6)
H^M	Moes dimensionless film thickness $H^M = (h/R_x)/\sqrt{2U}$ (Chapter 6)
H_0	dimensionless rigid body dispacement $H_0 = h_0R_x/a^2$

H	coarse grid mesh size
H_x, H_y	coarse grid mesh size in x and y direction
$\imath$	imaginary number $\imath = \sqrt{-1}$
I	identity matrix $I_{i,j} = \delta_{ij}$ (Chapter 2)
I_H^h	inter-grid operator (interpolation) from grid H to grid h
$\mathbb{I}_H^h$	inter-grid operator (high order interpolation) from grid H to grid h
I_h^H	inter-grid operator (restriction) from grid h to grid H
i, j	fine grid indices in x and y direction
I, J	coarse grid indices in x and y direction
k	ratio length to radius of a journal bearing (Chapter 4)
k, l	grid level
K	continuous deformation kernel
K^{hh}	discrete deformation kernel on grid h
$\tilde{K}^{hh}$	approximat discrete deformation kernel on grid h
L	length of journal $[m]$ (Chapter 4)
L	dimensionless material parameter (Moes) $L = G(U)^{0.25}$ (Chapter 6)
L	differential operator
L^h	linear discrete fine grid operator
L^H	linear discrete coarse grid operator
$L^h\langle \underline{u}^h \rangle$	non-linear discrete fine grid operator working on $\underline{u}^h$
$L^H\langle \underline{u}^H \rangle$	non-linear discrete coarse grid operator working on $\underline{u}^H$
$\mathcal{M}$	iteration matrix (Chapter 2)
M	2d dimensionless load parameter (Moes) $M = W(U)^{-0.75}$ (Chapter 6)
n	number of (grid)points in a direction
n_x	number of (grid)points in x direction
n_y	number of (grid)points in y direction
N	number of (grid)points $N \approx n^d$
p	order of the discretisation error (Chapter 2, 5)
p	pressure $[Pa]$
p_0	coefficient in the Roelands viscosity pressure relation $[Pa]$
p_h	Hertzian pressure $[Pa]$ $p_h = \sqrt[3]{(3\mathrm{w}E'^2)/(2\pi^3 R_x^2)}$
P	dimensionless pressure $P = pc^2/(12\eta u_m R)$ (Chapter 4)
P	dimensionless pressure $P = p/p_h$ (Chapter 5, 6)
r	residual
$\underline{r}^h$	discrete residual (vector)
rn	residual norm
wr	residual of the dimensionless force balance equation
R	average radius of shaft and journal $[m]$ (Chapter 4)
R_{1x}, R_{2x}	radius of curvature in x direction of bodies 1 and 2 $[m]$
R_x	reduced radius of curvature in x direction $[m]$ $1/R_x = 1/R_{1x} + 1/R_{2x}$

R_{1y}, R_{2y}	radius of curvature in y direction of bodies 1 and 2 $[m]$
R_y	reduced radius of curvature in y direction $[m]$
	$1/R_y = 1/R_{1y} + 1/R_{2y}$
s	number of relaxation sweeps
S	Sommerfeld number $S = (\mathrm{w}/\eta L u_s)(c^2/R^2)$
t	time $[s]$
T	dimensionless time $T = t u_m/a$
u	continuous solution of the differential equation
u_i	value of u in x_i, $u_i = u(x_i)$
$u_{i,j}$	value of u in (x_i, y_j)
u_i^h	discrete solution of the discrete set of equations in x_i
$u_{i,j}^h$	discrete solution of the discrete set of equations in (x_i, y_j)
$\underline{u}^h$	vector containing the solution of the discrete set of equations
$\bar{u}_i^h, \tilde{u}_i^h$	approximations to u_i^h
u_x	first derivative $u_x = \partial u/\partial x$
u_{xx}	second derivative $u_{xx} = \partial^2 u/\partial x^2$
u_1	velocity of lower surface $[m/s]$
u_2	velocity of upper surface $[m/s]$
u_m	mean velocity $[m/s]$ $u_m = (u_1 + u_2)/2$
u_s	sum velocity $[m/s]$ $u_s = u_1 + u_2$
U	dimensionless speed parameter $U = (\eta_0 u_m)/(E' R_x)$ (Chapter 6)
$\underline{v}^h$	numerical error $v_i^h = u_i^h - \tilde{u}_i^h$
$\underline{v}^m$	error vector after m iterations $\underline{v}^h = \underline{u}^h - \underline{u}^m$
x, y, z	coordinates $[m]$
X, Y, Z	dimensionless coordinates with respect to a: $X = x/a$
w	elastic deformation $[m]$
$\underline{w}_k$	k^{th} eigenvector of matrix $\mathcal{M}$ with eigenvalue μ_k (Chapter 2)
w	load $[N]$
$\mathrm{w_x}$	load in x direction $[N]$ (Chapter 4)
$\mathrm{w_y}$	load in y direction $[N]$ (Chapter 4)
W	dimensionless load (Chapter 4)
W	dimensionless load parameter $W = \mathrm{w}/(E' R_x^2)$ (Chapter 6)
WU	work unit (1 wu is equavalent of one fine grid relaxation)
z	pressure viscosity index, Roelands equation $[-]$

α	pressure viscosity index $[Pa^{-1}]$
α	ratio of mesh size $\alpha = h_y/h_x$ (Chapter 2)
α	ratio of mesh size $\alpha = k h_y/h_x$ (Chapter 4)
$\bar{\alpha}$	dimensionless parameter $\bar{\alpha} = \alpha p_h$ (Chapter 6)
δ	maximum approach/deformation $[m]$
δ_i^h	change due to relaxation (Chapter 2)
$\delta_{i,j}^h$	change due to relaxation (Chapter 5)
ϵ	dimensionless excentricity $\epsilon = e/c$ (Chapter 4)

η	viscosity $[Pa\ s]$
η_0	viscosity at ambient pressure $[Pa\ s]$
$\bar{\eta}$	dimensionless viscosity $\bar{\eta} = \eta/\eta_0$
θ	angular frequency (Chapter 2, 3)
θ	dimensionless coordinate $\theta = \zeta/R$ (Chapter 4)
θ	oil film thickness to gap height ratio (Chapter 6)
λ	waviness wavelength $[m]$
$\bar{\lambda}$	dimensionless parameter $\bar{\lambda} = (12u_m\eta_0R_x^2)/(a^3p_h)$ (Chapter 6)
μ	asymptotic error amplification factor $\mu = \rho(\mathcal{M})$ (Chapter 2)
$\bar{\mu}$	asymptotic smoothing factor (Chapter 3,4,5,6)
$\mu(\theta)$	amplitude amplification factor
ν_0, ν_1, ν_2	number of relaxation sweeps, on coarsest grid, before V-cycle and after V-cycle respectively
ν_1, ν_2	Poisson ratio of bodies 1 and 2 $[-]$
ξ	coefficient in the Reynolds equation
ρ	density $[kg/m^3]$
ρ_0	density at atmospheric pressure $[kg/m^3]$
$\bar{\rho}$	dimensionless density $\bar{\rho} = \rho/\rho_0$
$\tau^h(x)$	truncation error in discrete operators
Ψ	attitude angle (Chapter 4)
ω	general underrelaxation factor
ω_1	underrelaxation factor in pressure equation
ω_2	underrelaxation factor in force balance equation
Ω	computational domain
Ω^h	discretized domain
Ω_1, Ω_2	computational sub-domain

Teaching

The authors have taught the Multigrid techniques described in this book for over a decade to students (fourth and fifth year and PhD) and to various colleagues. They have come to the conclusion that the only way to *fully understand* a certain problem is to have programmed and debugged a code that performs the required Multigrid task. Only then the students' claim that he/she understands the problem/method can be taken seriously. As with most things, the proof is in doing! The most thorough way of teaching Multigrid would be to ask the student to write the program from scratch, but this approach has two different drawbacks. First of all it would take the student a lot of time, and secondly a wide variety of programs would emerge, which are difficult and time consuming to correct/debug for the supervisor. As a compromise we tend to give the student a complete program in which the Multigrid routines are 'emptied', and comment lines indicate where active code has to be added. This approach has several advantages. It allows step by step programming and debugging of the different rountines, as the code can be compiled and run at every instant (even though it does not perform any useful tasks). Furthermore, it ensures that the structure and house-keeping routines are fixed and bug-free. A final advantage is that the Multigrid stucture and administrative routines can remain virtually unchanged when different types of problems are dealt with. They can thus be written and debugged once and used many times. Most of the time it is only necessary to add one or more (matrices of) variables. Consequently, the complex task of building a working Multigrid code can be broken down into a series of less complex sub-tasks, each of which requires one or few subroutines to be completed:

- build a one level relaxation routine that can perform the relaxation on any required level, and analyse the discretization error and the speed of convergence as a function of the number of grid points.

- create a two level routine, by completing the inter-grid routines, using the Correction Scheme, then perform many level tests and analyse the convergence speed.

- extend the routines to obtain a Full Approximation Scheme and test its convergence rate at many levels.

- extend the driving routine from V-cycles to a Full MultiGrid (FMG) routine and analyse the discretization and numerical errors, using the difference of converged solutions on different grids.

In general the one or two dimensional Poisson problem is best suited to start with. Beginning with an emptied program, students (even those with little programming experience) can normally write a working Multigrid program in five four hour sessions. These five sessions can be complemented by five more theoretical lectures of one or two hours, explaining relaxation and interpolation in detail, and introducing them to Smoothing Rate Analysis and other Multigrid concepts explained in this book.

Only after the student has completely understood the Poisson solver (and preferably also the hydrodynamic program, including cavitation), it is possible to tackle the dry contact problem. The approach suggested for teaching the MultiLevel Multi-Integration is slightly different from the one used for the Poisson problem. Once again the authors would like to stress that this problem should only be tackled after the Poisson solver has been built successfully. Even though the fast integration uses different techniques then those used for the solution of differential equations, the grid structure is identical. Furthermore, the fast solver described in the last step, requires a previous experience with the Poisson problem:

- disable most of the dry contact solver **DRY2d.c** in *main()* and keep only one film thickness calculation. Then build the classical integration *calcku()*.
- build the second order fast integration by completing *sto2k1()*, *coarsen2x()*, *coarsen2y()*, *refine2x()* and *refine2y()*.
- compute the integrals using *deep* = *1* and correct over the entire domain. As a result the integrals should be equal (up to machine precision) to the ones calculated classically, which simplifies debugging. Then use more than one level deep. Finally, test with small numbers of correction points.
- repeat for the fourth order routines and introduce the concept of 'ghost points'.
- repeat for the sixth order routines, etc.
- build the fast solver using distributed relaxation.

The advantage of this approach is that the fast integration can be introduced without the necessity to include the 'ghost points' immediately. Once the principle is understood, more accurate, higher order routines can be built, requiring the use of the 'ghost points'. When the fast integration routines work correctly, the fast contact solver can be built.

As an example the emptied subroutine *relax()* from the Poisson solver is given. Three lines representing the relaxation, the residual norm calculation and the work unit calculation have to be added. The lines are indicated by comment statements:

```
void relax(Stack *U, int k)
{
/* perform Gauss-Seidel relaxation on gridlevel k */

Level *L;
double rn;
int    i,j;

L =U->Lk+k;
```

```
for (i=1;i<=L->ii-1;i++)
  for (j=1;j<=L->jj-1;j++)
    /*  one point Gauss-Seidel relaxation */

rn=0.0;
for (i=1;i<=L->ii-1;i++)
  for (j=1;j<=L->jj-1;j++)
    /* resnorm update */

/* wu update */
printf("\nLevel %d Residual %8.5e Wu %8.5e",k,
        rn/((L->ii-1)*(L->jj-1)),U->wu);
}
```

Note that the syntax of the subroutine is correct, and thus it can be compiled and run, even though it doesnot perform any useful task in its 'emptied' state.

The degree to which the programs are 'emptied', can be varied according to the students' level, and the amount of time available. Using the emptied program, students can concentrate on writing the essential parts of the Multigrid code. After having completed the Poisson program, they can then be asked to extend the program to solve the Hydrodynamic Lubrication problem or a related problem like the Stokes problem outlined in the 1984 Guide [15]. Having mastered the basics of Multigrid they can now progress in a more autonomous way. During this process of advanced Multigrid they will often discover that they have not yet understood the basics completely, or that they now see the basic aspects in a very different way. Another common observation is that **debugging** a Multigrid program is often non-trivial and that the order outlined above is useful even for complex problems and/or 'advanced' multigridders. Another useful strategy to adopt when confronted with less-than-optimal convergence is to simplify the problem at hand as much as possible and then to try to study this simpler problem. Whenever the simpler problem gives satisfactory convergence, the changes are transplanted to the full problem and a new analysis cycle starts.

Contents

Chapter 1

Introduction

This book is about the steps necessary for the construction of an efficient and stable numerical solver for the ElastoHydrodynamic Lubrication problem. Therefore it is necessary to have a detailed understanding of the equations describing the ElastoHydrodynamic Lubrication problem with respect to their computational complexity and their stability. Thus this chapter is devoted to the analysis of the full set of equations describing the ElastoHydrodynamic Lubrication (abbreviated to EHL) problem. The problems related to the system of equations, the differential equation, the integral equation and the global equation will be highlighted. In particular the differential equation is very non linear and thus stability problems occur for certain sets of operating conditions.

This non linearity is caused by two different phenomena; coefficients in the Reynolds equation that depend on the pressure, and the cavitation condition. Furthermore, under some high load conditions the equation suffers from weak coupling in the y direction, in certain parts of the solution domain. Finally, it is shown that the differential equation suffers from slow convergence using classical solution techniques for large numbers of unknowns.

The integral part of the equations is characterised by a non-local character which will require a modification of the classical one point relaxation methods. Furthermore, when using fine grids with many points, the number of operations required to calculate this integral becomes very large, making a more advanced calculation necessary.

Finally, the global equation describing the load balance (or force equilibrium) also requires special attention before a stable and efficient solution algorithm can be obtained.

Since the system of equations describing the EHL problem is complicated, we will start our study on a much simpler equation, and gradually work towards the full EHL system, adding one difficulty at the time. After each step we will require that the algorithm performs with a similar efficiency as the one at the end of the previous step.

1.1 Justification

The regime of ElastoHydrodynamic Lubrication (EHL) is mainly found in lubricated non conforming contacts of machine elements. Common examples are: rolling element bearings (see Figure 1.1), gears, cams and tappets, etc.

Figure 1.1: *Example of ElastoHydrodynamically Lubricated contacts, the contacts between the rolling element and the inner and outer raceway in rolling element bearings.*

As the contact area is limited and the forces are generally large, causing high pressures, elastic deformation of the contacting bodies takes place. As the contacting surfaces are generally moving with respect to one another, a lubricant (generally oil or grease) is used to limit friction and wear. The high pressure inside the contact compresses the oil and causes a dramatic increase in its viscosity, thus the lubricant properties like density and especially viscosity are far from constant.

These two characteristics of the EHL contact are essential for the physical comprehension as well as for the efficient numerical solution, and they play a central role in building an efficient solver. As such an EHL contact is characterised by:

- important elastic deformations, and
- important piezo viscous effects.

Apart from being governed by a complex system of equations, the EHL problem also requires a very detailed solution. One reason is physical: to optimize contact performance, a full EHL film completely separating the two surfaces is required. Such a film is very thin; its thickness is generally of the order of 0.1 to 1 μm. This means that the oil film thickness is smaller or even much smaller than the surface roughness of the two surfaces, and that this surface roughness will influence or even determine the way in which the contact will operate. In order to model the influence of surface roughness on EHL, a very fine grid with many points describing the roughness geometry will be needed, which results in the requirement for an efficient solver. Furthermore, the extreme non linearity of the system of equations requires a very stable solution method.

A second reason that requires a very fine grid is purely numerical: the highly loaded cases have very localised film thickness minima, and can only be computed on very fine grids. Otherwise, the precision in the film thickness calculated becomes very small, and in extreme cases it can result in locally negative film thicknesses. For both reasons very fine grids are required, and thus an efficient solver capable of generating such a detailed solution in a reasonable time is essential. Other reasons for requiring a very efficient and stable solution algorithm include the necessity to solve transient problems, which involves the solution of a problem similar to the stationary problem for many (thousands) time steps.

1.2 History

The key equations used in the study of ElastoHydrodynamic Lubrication have been known for more than a century. In 1881, H. Hertz [60] published his study of the contact between two spherical bodies. The contact area and the pressure distribution under dry contact conditions are still frequently used these days for instance to approximate the sub-surface stress field. Furthermore, the dry contact solution, which carries his name, is used as an asymptotic solution to make the EHL equations dimensionless. In 1886, O. Reynolds [92] published the equation describing the slow fluid flow in narrow gaps, in order to explain the pressure generation in journal bearings, experimentally observed by B. Tower. This equation forms the basis of the lubrication theory, and carries his name. A third publication complements the two previous equations; in 1893 Barus [5] published how the viscosity of oils increases as a function of pressure.

For a detailed account of the history of EHL the reader is referred to Dowson [35], [37] and Gohar [45]. Here only a brief résumé is given. In the beginning of this century, Martin [80] and Gümbel [51] tried to apply the Reynolds equation to the lubrication of gear teeth. However, the film thicknesses that they predicted were much too small compared to the surface roughness, to explain successful long term operation. It was not until the work of Ertel [42] and Grubin [50] that the Reynolds equation was combined with the elastic deformation equation to yield realistic film thickness predictions for line contacts. In 1951, the first 'numerical' solution of the EHL problem was published by Petrusevich [90], including a strange singularity in the pressure distribution: the pressure spike. The foundation of modern numerical solutions of the EHL problem was laid in the early 60's by Dowson and Higginson [35], solving the line contact problem for a variety of operating conditions, and providing a film thickness equation based on these calculations. In the mid seventies the increased capacity of computers allowed Hamrock and Dowson [55] to solve the circular contact problem. Again a film thickness equation based on their numerical results was published. Since then numerical work has been extended to study more complex lubricant rheologies, thermal and dynamic effects, as well as the influence of sub-contact size features.

The history of Multigrid methods is much more recent, the first to describe an algorithm using more than two grids (levels) was Fedorenko [43], however, he did not realise the potential efficiency of the method. Its full efficiency was demonstrated by Brandt in the early 70's [12, 13], and local grid refinements were introduced some ten years later by Bai and Brandt [4]. Since then Multigrid methods have been applied to many different problems at the Weizmann Institute of Science [16] and elsewhere. Of the many researchers active in this field we only name a few: Hackbusch [52], Hemker [58], McCormick [81] and Wesseling [108]. For more information the reader is referred to the abundant literature in this rapidly expanding field.

1.3 Description of the EHL Problem

In its simplest form, the complete EHL problem is described by a number of equations:

- the Reynolds equation describes the flow of a Newtonian fluid inside a narrow gap and the cavitation condition states that the pressure should remain positive
- the elastic deformation equation describes the deformed geometry of the gap
- the viscosity-pressure relation gives the viscosity as a function of the pressure
- the density-pressure relation gives the density as a function of the pressure
- the force balance equation states that the integral of the pressure in the film should balance the applied load.

The interaction of all these equations results in a very non linear system of equations. Furthermore, this non linear character of the system of equations changes as a function of the operating conditions, which makes the solution process difficult to optimize. In the next sections the different equations will be outlined and discussed.

It should be noted that this system describes the EHL problem in its simplest form. The problem can be made more realistic, but also more complicated, by the introduction of:

- non Newtonian fluid rheology
- non linear elastic, or plastic behaviour of the contacting bodies
- important transient effects including the dependence of load, speed and geometry on time.

1.3.1 Reynolds Equation

The Reynolds equation [92] is derived from the Navier-Stokes equation for a slow viscous flow. This means that both inertia forces and external forces are neglected with respect to viscous forces. The second simplification is due to the narrow gap: the dimensions in the z direction are much smaller than those in both x and y direction, see Figure 1.2. Using the condition of no-slip at the wall boundary, one obtains the velocity profile as a function of z. Continuity of mass flow in a cell stretching from one surface to the other finally gives:

$$\frac{\partial}{\partial x}(\frac{\rho h^3}{12\eta}\frac{\partial p}{\partial x}) + \frac{\partial}{\partial y}(\frac{\rho h^3}{12\eta}\frac{\partial p}{\partial y}) - \frac{\partial(u_m \rho h)}{\partial x} - \frac{\partial(\rho h)}{\partial t} = 0 \tag{1.1}$$

where p is the pressure, $h = h_1 + h_2$ is the film thickness, or gap height, u_m is the mean surface velocity $u_m = (u_1 + u_2)/2$, η is the viscosity, ρ the density of the lubricant and t represents time. The x axis is aligned with the direction of the mean velocity u_m, and the derivation has to be extended to include pivoting. For the case where u_m is constant, and

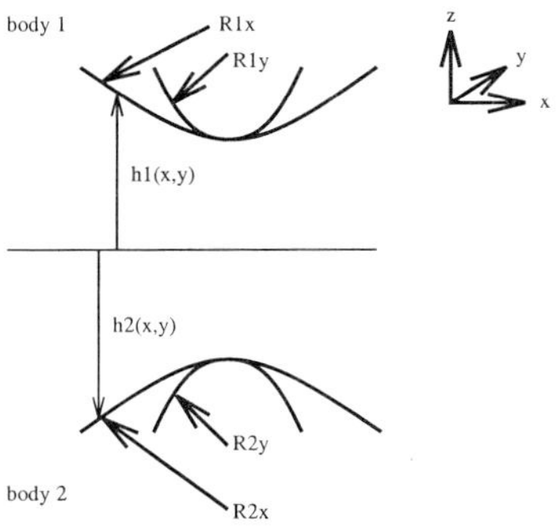

Figure 1.2: *Coordinate system in the contact area.*

neglecting density variations which are at maximum of the order of 30% (see Figure 1.5), this equation reduces to:

$$\frac{\partial}{\partial x}(\frac{h^3}{12\eta}\frac{\partial p}{\partial x}) + \frac{\partial}{\partial y}(\frac{h^3}{12\eta}\frac{\partial p}{\partial y}) - u_m\frac{\partial h}{\partial x} - \frac{\partial h}{\partial t} = 0 \qquad (1.2)$$

The first two terms in this equation describe the flow due to the pressure gradient and are called the *Poiseuille* terms. The second term describes the flow due to the mean surface velocity and is called the *Couette* term or *wedge* term. . The last term represents the flow due to the squeeze effects and is referred to as the *squeeze* term. . If transient effects are absent one obtains the the stationary Reynolds equation:

$$\frac{\partial}{\partial x}(\frac{h^3}{12\eta}\frac{\partial p}{\partial x}) + \frac{\partial}{\partial y}(\frac{h^3}{12\eta}\frac{\partial p}{\partial y}) - u_m\frac{\partial h}{\partial x} = 0 \qquad (1.3)$$

Finally, in some cases the contact dimensions in the y direction are very large compared to those in the x dimension. This situation is referred to as a line contact situation. Under these conditions the equation further simplifies to:

$$\frac{\partial}{\partial x}(\frac{h^3}{12\eta}\frac{\partial p}{\partial x}) - u_m\frac{\partial h}{\partial x} = 0 \qquad (1.4)$$

This one dimensional equation can be integrated once to yield a first order differential equation. However, this integrated form has no advantages for the numerical solution and is not used here. In fact, because the two dimensional equation can not be integrated, the present form is even preferred. As a second order differential equation the one dimensional problem can be studied as a prelude to the two dimensional problem, or, the other way around, numerical techniques developed for the two dimensional problem can be translated relatively easily to the one dimensional problem.

1.3.2 Cavitation

The pressures obtained from the solution of the Reynolds equation in the previous section are generally positive, however, nothing prevents the theoretical solutions for pressure from taking on negative values. An example from Hydrodynamic Lubrication is the Sommerfeld solution (see Chapter 4), which predicts positive pressures in the converging gap where $\partial h/\partial x < 0$ and negative pressures in the diverging gap where $\partial h/\partial x > 0$. Such negative pressures are not physically acceptable, since a fluid can not sustain important negative pressures (tension). In such cases, the fluid will evaporate (boil) and the pressure is limited by the vapour pressure of the fluid. This process is called cavitation, and the vapour pressure is generally small compared to the pressure generated in an EHL contact. As such, the vapour pressure can be neglected, and the pressure is limited from below by the zero or ambient pressure. The combination of the Reynolds equation with the cavitation condition changes the problem to a complementarity problem: the complete solution domain Ω is divided into two distinct domains Ω_1 and Ω_2. The Reynolds equation with $p > 0$ is valid in part of the computational domain called Ω_1, and the cavitation condition is valid in the rest of the domain Ω_2. This complementarity equation will be denoted as:

$$\begin{aligned} &\frac{\partial}{\partial x}\left(\frac{h^3}{12\eta}\frac{\partial p}{\partial x}\right) + \frac{\partial}{\partial y}\left(\frac{h^3}{12\eta}\frac{\partial p}{\partial y}\right) - u_m\frac{\partial h}{\partial x} = 0 \qquad && (x,y) \in \Omega_1 \\ &p(x,y) = 0 && (x,y) \in \Omega_2 \end{aligned} \tag{1.5}$$

The sub division of the computational domain Ω into two sub domains Ω_1 and Ω_2 is unknown, and has to be established numerically. Furthermore, the boundary between the two domains depends on the operating conditions, and has thus to be established each time from scratch. In Figure 1.3 this boundary is represented by a solid line. To the left of this line the Reynolds equation is valid: Ω_1, to the right cavitation occurs: Ω_2.

Since the pressures in an EHL contact are generally very large, the atmospheric pressure is neglected and the boundary conditions imposed on the pressure are:

$$p(x,y) = 0 \qquad (x,y) \in \partial\Omega \tag{1.6}$$

It can be pointed out that strictly speaking these conditions are only required on the boundary $\partial\Omega_1$. However, as part of this boundary is not known in advance, it is more interesting to prescribe the conditions on the boundary of the entire domain $\partial\Omega$. Furthermore, it should be pointed out that the boundary conditions are compatible with the cavitation conditions, since both the vapour pressure and the atmospheric pressure have been neglected.

1.3.3 Viscosity-Pressure

The simplest viscosity pressure relationship is exponential, under the name Barus [5]. It reads:

$$\eta(p) = \eta_0 \exp(\alpha p) \tag{1.7}$$

where η_0 is the atmospheric viscosity and α is called the pressure viscosity coefficient. For mineral oils this coefficient varies between $1 \cdot 10^{-8}$ and $2 \cdot 10^{-8}$ Pa^{-1}. For pressures of 1 GPa a viscosity increase of $\exp(15) \approx 3 \cdot 10^6$ is predicted, which is fairly large. For

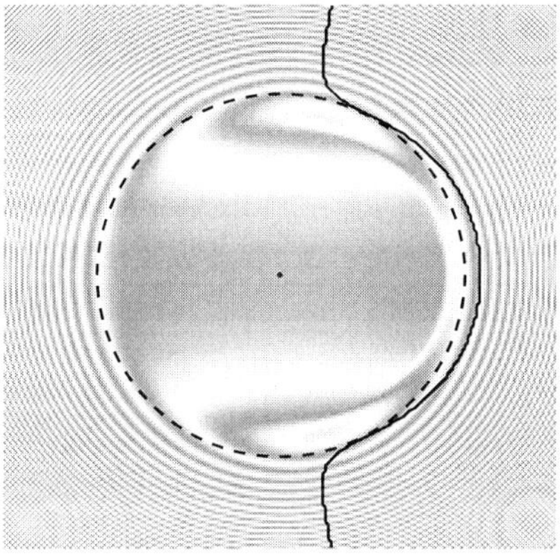

Figure 1.3: *Film thickness map with the cavitation boundary (drawn line).*

pressures of 3 GPa a viscosity increase of $\exp(45) \approx 3 \cdot 10^{19}$ is predicted, which is truly enormous. More realistic and complicated viscosity pressure relations such as the one proposed by Roelands [93] can be used:

$$\eta(p) = \eta_0 \ \exp(\ln(\eta_0) + 9.67)(-1 + (1 + \frac{p}{p_0})^z)) \tag{1.8}$$

where z is the pressure viscosity index, typically $z = 0.6$, and $p_0 = 1.96 \cdot 10^8$ [Pa] is a constant.

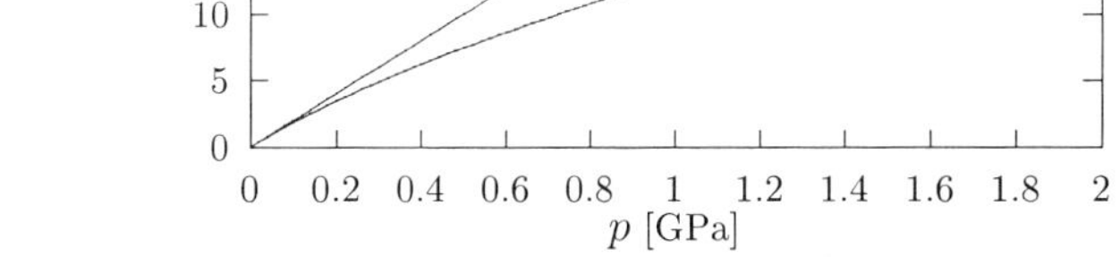

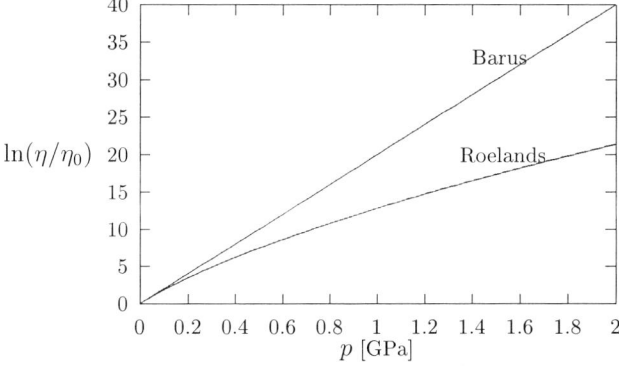

Figure 1.4: *Relative viscosity η/η_0 as a function of the pressure p in [GPa].*

A more precise viscosity pressure relation at high pressures, is not really necessary to accurately predict the film thicknesses. At these pressure levels, the viscosity has become so high anyway, that the (large) pressure gradients do not influence the flow anymore. In other words, due to the viscosity, the Poiseuille term in (1.4) tends to zero, and leaves the Couette term to dominate the Reynolds equation.

1.3.4 Density-Pressure

A simple density pressure relation is given by the Dowson and Higginson relation [35]. It reads:

$$\rho(p) = \rho_0 \frac{5.9 \cdot 10^8 + 1.34\, p}{5.9 \cdot 10^8 + p}$$

where ρ_0 is the atmospheric density and p is given in Pa. Please note the two asymptotes: for small p ($p \to 0$), $\rho = \rho_0$, for very large p ($p \to \infty$), $\rho = 1.34\, \rho_0$.

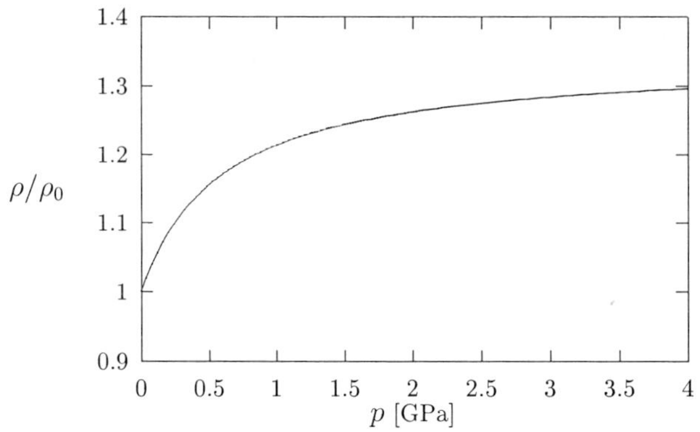

Figure 1.5: *Relative density ρ/ρ_0 as a function of the pressure p in $[GPa]$.*

1.3.5 Film Thickness Equations

As stated in the introduction, the elastic deformation of the two contacting bodies plays an important role in EHL. In order to approximate the elastic deformations of real bodies, two hypotheses are used:

- the deformation is linear elastic, and the two contacting bodies have uniform and isotropic properties.
- the contact dimensions a are small compared to the size of the bodies ($a << R_x$) allowing the approximation of the bodies by two semi-infinite half spaces.

Both hypotheses are generally valid and the approximations obtained agree very well with experimental results. In this case, the elastic deformation $w(x, y)$ due to the pressure distribution $p(x, y)$ can be approximated by:

$$w(x,y) = \frac{2}{\pi E'} \int_{-\infty}^{+\infty} \int_{-\infty}^{+\infty} \frac{p(x',y')\, dx'\, dy'}{\sqrt{(x-x')^2 + (y-y')^2}} \qquad (1.9)$$

where

$$\frac{2}{E'} = \frac{1-\nu_1^2}{E_1} + \frac{1-\nu_2^2}{E_2} \tag{1.10}$$

and ν_1 and ν_2 represent the Poisson ratio of body 1 and 2. E_1 and E_2 are the elastic moduli of the two bodies 1 and 2 respectively. E' is called the reduced elastic modulus.

Approximating the gap between the two elliptical bodies of reduced radii of curvature R_x and R_y by a parabola, finally gives the gap height $h(x,y)$ as

$$h(x,y) = h_0 + \frac{x^2}{2R_x} + \frac{y^2}{2R_y} + \frac{2}{\pi E'} \int_{-\infty}^{+\infty} \int_{-\infty}^{+\infty} \frac{p(x',y')\,dx'\,dy'}{\sqrt{(x-x')^2+(y-y')^2}} \tag{1.11}$$

where h_0 represents the rigid body approach.
In a line contact (contact dimensions in the y direction much larger than those in x), p is assumed to be constant in the y direction and the above equation should be replaced by:

$$h(x) = h_0 + \frac{x^2}{2R_x} - \frac{2}{\pi E'} \int_{-\infty}^{+\infty} p(x') \ln\left(\frac{x-x'}{x_0}\right)^2 dx' \tag{1.12}$$

where x_0 is a reference distance where the deformation is taken to be zero.

1.3.6 Force Balance

The integral of the pressure distribution obtained from the Reynolds equation should balance the externally applied load w, in order to have an equilibrium of forces. For the two dimensional problem this condition reads:

$$\mathrm{w} = \int_{-\infty}^{+\infty} \int_{-\infty}^{+\infty} p(x',y')\,dx'\,dy' \tag{1.13}$$

where w is the applied load in the two dimensional case.
For the line contact case, the force balance equation reads:

$$\mathrm{w}_1 = \int_{-\infty}^{+\infty} p(x')\,dx' \tag{1.14}$$

where w_1 is the applied load per unit length.

In both the one and two dimensional problem, the **local** equations describing the pressure and film thickness distributions $p(x,y)$ and $h(x,y)$, are complemented by a single **global** Equation (1.13) and (1.14), that determines the value of the h_0 constant in the film thickness Equation (1.11) and (1.12).

1.4 Simplification

The system of equations describing the EHL problem, being far from simple, our goal is double: to obtain a stable and efficient (close to $O(N)$) solution method. This means that if N is the number of points in which the solution is to be calculated, the solution will involve a number of operations (multiplications and divisions) which is proportional to N and thus that the total computing time will be proportional to the number of points N. In order to achieve these two aims we will have to start our study with the analysis of a much simpler problem and try to achieve the required properties: efficiency and stability on this simpler problem. Once these **two** criteria are fulfilled, we will extend the problem and add one more difficulty at a time. The step by step process is then repeated.

From the full EHL problem, which is treated in Chapter 6, two major simplifications can be made in a straightforward manner: neglecting the viscosity pressure relation and studying the isoviscous case, and neglecting the elastic deformations and studying the rigid case. A combination of both simplifications leads to the problem of Hydrodynamic Lubrication (HL) with an isoviscous lubricant and a rigid geometry, and this problem is addressed in Chapter 4.

A second limit for the EHL problem is found for the case $h \to 0$, the low speed or high load asymptote. In that situation the Reynolds equation reduces to the contact equation: $h \geq 0$. This equation is the so called dry contact problem that will be addressed in Chapter 5. However, in order to reduce the complexity of the integral equation to close to $O(N)$, we need to introduce a fast integration technique: MultiLevel Multi Integration (MLMI). This is done in the second part of Chapter 5.

Finally the HL problem can be simplified by neglecting cavitation and by taking the film thickness to be constant: $h(x,y) = constant$. The Reynolds equation simplifies to the Poisson equation (Laplace equation):

$$\frac{\partial^2 p}{\partial x^2} + \frac{\partial^2 p}{\partial y^2} = f(x,y) \tag{1.15}$$

where $f(x,y)$ represents a right hand side forcing term. Without this forcing term, only trivial solutions are possible if the boundary conditions prescribe constant values. In order to adhere to the conventional notation the variable p will be replaced by the variable u:

$$\frac{\partial^2 u}{\partial x^2} + \frac{\partial^2 u}{\partial y^2} = f(x,y) \tag{1.16}$$

A final simplification reduces the problem to a single dimension:

$$\frac{\partial^2 u}{\partial x^2} = f(x) \tag{1.17}$$

This problem is classical and has an analytic solution for simple right hand side functions and boundary conditions. However, the numerical analysis will focus on the reduction of the complexity to $O(N)$ operations (Chapter 3), and will introduce the analysis of the speed of convergence, the intergrid transfers and other routines which will prove very useful and will remain largely unchanged in the next chapters. As such the Poisson equation is very well suited to start our study of Multilevel or MultiGrid techniques.

As all these chapters will rely heavily on standard, and some not so standard, numerical techniques, the next chapter will give an introduction to numerical methods, introducing discretization, discretization error, relaxation and convergence speed analysis.

1.5 Model Problems

In the previous section the Reynolds equation was simplified to obtain the Poisson equation, using a number of intermediate steps. This section describes the equation(s) obtained at each step, but now starting from the simplest one (Poisson) and working towards the most complex one (Reynolds). Each section describes the intermediate problem, the dimensionless equations and the boundary conditions, thereby sketching the lay-out of the chapters of the book.

1.5.1 Poisson Problem

Even though, a major simplification is required to change the Reynolds equation to the Poisson equation, this much simpler equation can still teach us an important lesson concerning the speed of convergence of an iterative solver, or more precisely, the slowness of convergence, since the number of relaxation sweeps necessary will increase as the number of discrete unknowns N increases. As such the humble Poisson equation will be the subject of Chapter 3. The variables u, x and y can be interpreted as dimensionless variables, eliminating the need of a dimensionless equation.

$$\frac{\partial^2 u}{\partial x^2} + \frac{\partial^2 u}{\partial y^2} = f(x,y) \qquad (x,y) \in \Omega \tag{1.18}$$

with a set of Dirichlet boundary conditions: $u(x,y) = 0, \ (x,y) \in \partial\Omega$.

1.5.2 Hydrodynamic Lubrication

We will then continue our study by analysing the case of stationary Hydrodynamic Lubrication, assuming a constant viscosity η, a constant density ρ, a constant mean velocity u_m and a rigid geometry. The general Reynolds equation (1.1) simplifies to:

$$\frac{\partial}{\partial x}(h^3 \frac{\partial p}{\partial x}) + \frac{\partial}{\partial y}(h^3 \frac{\partial p}{\partial y}) - 12\eta u_m \frac{\partial h}{\partial x} = 0 \tag{1.19}$$

This equation describes for instance the pressure generation in a journal bearing, in which the radius of the shaft and the journal differ by the clearance c which is (very) small compared to the average radius $R = D/2$. The bearing has a length L, and lubricant is supplied at the top of the bearing ($x = 0$) through a groove at ambient pressure $p = 0$. For $c << R$ the radius of the journal and shaft can be approximated by R and the gap height $h(x)$ can be approximated by $h(x) = c(1 + \epsilon \cos(x/R))$, where ϵ is the relative eccentricity.

In general capitals are used for dimensionless parameters (exceptions are R, L and E'). Dimensionless Greek characters are indicated by a bar over the original symbol.

Introducing the dimensionless variables $H = h/c$, $P = pc^2/(12\eta u_m R)$, $\theta = x/R$, $Y = y/L$, and adding the complementarity condition of cavitation, the dimensionless Reynolds equation reads:

$$\begin{aligned} &\frac{\partial}{\partial\theta}(H^3\frac{\partial P}{\partial\theta}) + \frac{R^2}{L^2}\frac{\partial}{\partial Y}(H^3\frac{\partial P}{\partial Y}) - \frac{\partial H}{\partial\theta} = 0 && (\theta, Y) \in \Omega_1 \\ &P(\theta, Y) = 0 && (\theta, Y) \in \Omega_2 \end{aligned} \tag{1.20}$$

with

$$H(\theta) = 1 + \epsilon\cos(\theta) \tag{1.21}$$

This equation is valid on the computational domain $\Omega = \Omega_1 + \Omega_2$ which is a rectangle of $0 \le \theta \le 2\pi$, $0 \le Y \le 1$. The boundary conditions on the domain read $P(\theta, Y) = 0$, $(\theta, Y) \in \partial\Omega$.

1.5.3 Dry Contact

When the lubricant film thickness is very small compared to the elastic deformation of the contacting bodies, the pressure distribution and the deformation can be approximated by the dry contact problem. When two bodies are loaded together, the gap between them should become zero (contact, positive pressure: domain Ω_1), or remain positive (no contact, zero local pressure: domain Ω_2), In mathematical terms this is a complementarity problem. The two equations are valid on the sub domains Ω_1 and Ω_2 respectively, but the division of the domain Ω into the two sub domains is a priori unknown. The complementarity problem can be expressed as:

$$\begin{aligned} &h(x,y) = 0,\ p(x,y) > 0 && (x,y) \in \Omega_1 \\ &h(x,y) > 0,\ p(x,y) = 0 && (x,y) \in \Omega_2 \end{aligned} \tag{1.22}$$

$$h(x,y) = h_0 + \frac{x^2}{2R_x} + \frac{y^2}{2R_y} + \frac{2}{\pi E'}\int_{-\infty}^{+\infty}\int_{-\infty}^{+\infty}\frac{p(x',y')\,dx'\,dy'}{\sqrt{(x-x')^2 + (y-y')^2}} \tag{1.23}$$

where the applied load w should be balanced by the integral over the pressure distribution. This results in the force balance equation:

$$\mathrm{w} = \int_{-\infty}^{+\infty}\int_{-\infty}^{+\infty} p(x,y)\,dx\,dy \tag{1.24}$$

Assuming $R_y = R_x$, using the dimensionless variables $P = p/p_h$, $H = hR_x/a^2$, $X = x/a$ and $Y = y/a$ the contact problem can be written as:

$$\begin{aligned} &H(X,Y) = 0,\ P(X,Y) > 0 && (X,Y) \in \bar{\Omega}_1 \\ &H(X,Y) > 0,\ P(X,Y) = 0 && (X,Y) \in \bar{\Omega}_2 \end{aligned} \tag{1.25}$$

where the symbol $\bar{\Omega}$ is used to denote the *dimensionless* domain.
From the equations in Chapter 5, the relation giving the Hertzian parameters p_h and a allow the gap Equation (1.23) to be written as:

$$H(X,Y) = H_0 + \frac{X^2}{2} + \frac{Y^2}{2} + \frac{2}{\pi^2}\int_{-\infty}^{+\infty}\int_{-\infty}^{+\infty}\frac{P(X',Y')\,dX'\,dY'}{\sqrt{(X-X')^2+(Y-Y')^2}} \tag{1.26}$$

The same relations allow the dimensionless force balance equation to be written as:

$$\int_{-\infty}^{+\infty}\int_{-\infty}^{+\infty} P(X,Y)\,dX\,dY = 2\pi/3 \tag{1.27}$$

Since the pressure is only larger than zero in the contact domain $\bar{\Omega}_1$ the integrals can be limited to:

$$\boxed{H(X,Y) = H_0 + \frac{X^2}{2} + \frac{Y^2}{2} + \frac{2}{\pi^2}\iint_{\bar{\Omega}_1}\frac{P(X',Y')\,dX'\,dY'}{\sqrt{(X-X')^2+(Y-Y')^2}}} \tag{1.28}$$

$$\boxed{\iint_{\bar{\Omega}_1} P(X,Y)\,dX\,dY = 2\pi/3} \tag{1.29}$$

If we study the contact zone $\bar{\Omega}_1$ separately for a moment, the equation $H = 0$ is in reality an integral equation in P. Standard, one point relaxation will lead to divergence of the solution process on fine grids. Furthermore, the calculation of the deformation will lead to very long computing times on these fine grids. These problems, together with an efficient treatment of the complementarity condition and the global equation will have to be resolved in order to obtain an overall efficient solver for the dry contact problem.

1.5.4 ElastoHydrodynamic Lubrication

Since the highly loaded EHL pressure distribution and elastic deformation are very close to the Hertzian dry contact solution, it seems reasonable to introduce dimensionless parameters based on these asymptotic dry contact solutions.

Its application has two different advantages, first of all it allows the introduction of two parameters that determine the contact operating conditions. The second advantage is that the dimensionless pressure and film thickness values are of the order of 1.0, resulting is a maximum precision of the numerical calculations.

We will study the dimensionless two dimensional Reynolds equation, based on the circular Hertzian solution with a the radius of the contact circle and p_h the maximum Hertzian contact pressure. The relations between p_h and a and the contact geometry, operating conditions and materials parameters, are outlined in Chapter 5. Again capitals are used for dimensionless parameters and dimensionless Greek characters are indicated by a bar over the original symbol.

Thus we introduce $P = p/p_h$, $X = x/a$, $Y = y/a$ and $H = hR_x/a^2$, based on Hertz and $T = u_m t/a$, $\bar{\eta} = \eta/\eta_0$ and $\bar{\rho} = \rho/\rho_0$.

Using these parameters the incompressible Reynolds equation (1.2) reduces to:

$$\frac{\partial}{\partial X}\left(\frac{a^3 p_h}{12\eta_0 R_x^2}\frac{H^3}{\bar{\eta}}\frac{\partial P}{\partial X}\right) + \frac{\partial}{\partial Y}\left(\frac{a^3 p_h}{12\eta_0 R_x^2}\frac{H^3}{\bar{\eta}}\frac{\partial P}{\partial Y}\right) - u_m\frac{\partial H}{\partial X} - u_m\frac{\partial H}{\partial T} = 0 \tag{1.30}$$

Introducing $\xi = H^3/(\bar{\eta}\bar{\lambda})$, with $\bar{\lambda} = (12\eta_0 u_m R_x^2)/(a^3 p_h)$, we can simplify this equation to:

$$\frac{\partial}{\partial X}\left(\xi\frac{\partial P}{\partial X}\right) + \frac{\partial}{\partial Y}\left(\xi\frac{\partial P}{\partial Y}\right) - \frac{\partial H}{\partial X} - \frac{\partial H}{\partial T} = 0 \tag{1.31}$$

Each individual term in this equation is now dimensionless and ξ can be regarded as a coefficient of X and Y which varies over the domain. As the Reynolds equation is only valid for positive pressures, the equation has to be extended to a complementarity equation:

$$\begin{array}{ll} \frac{\partial}{\partial X}\left(\xi\frac{\partial P}{\partial X}\right) + \frac{\partial}{\partial Y}\left(\xi\frac{\partial P}{\partial Y}\right) - \frac{\partial H}{\partial X} - \frac{\partial H}{\partial T} = 0 & (X,Y) \in \bar{\Omega}_1 \\ P(X,Y) = 0 & (X,Y) \in \bar{\Omega}_2 \end{array} \tag{1.32}$$

Where the precise division into the two sub domains is a priori unknown.
Using the same parameters the stationary one dimensional Reynolds equation (1.4) can be written as:

$$\begin{array}{ll} \frac{\partial}{\partial X}\left(\xi\frac{\partial P}{\partial X}\right) - \frac{\partial H}{\partial X} = 0 & (X,Y) \in \bar{\Omega}_1 \\ P(X,Y) = 0 & (X,Y) \in \bar{\Omega}_2 \end{array} \tag{1.33}$$

The dimensionless film thickness equation including the elastic deformation equation becomes with these parameters:

$$H(X,Y) = H_0 + \frac{X^2}{2} + \frac{Y^2}{2} + \frac{2R_x p_h}{\pi a E'}\int_{-\infty}^{+\infty}\int_{-\infty}^{+\infty}\frac{P(X',Y')\,dX'\,dY'}{\sqrt{(X-X')^2+(Y-Y')^2}} \tag{1.34}$$

As can be found from the equations in Chapter 5, the ratio of the Hertzian parameters p_h and a gives $p_h/a = E'/(\pi R_x)$, thus:

$$H(X,Y) = H_0 + \frac{X^2}{2} + \frac{Y^2}{2} + \frac{2}{\pi^2}\int_{-\infty}^{+\infty}\int_{-\infty}^{+\infty}\frac{P(X',Y')\,dX'\,dY'}{\sqrt{(X-X')^2+(Y-Y')^2}} \tag{1.35}$$

Finally the dimensionless force balance equation becomes:

$$\frac{\mathrm{w}}{p_h a^2} = \int_{-\infty}^{+\infty} \int_{-\infty}^{+\infty} P(X',Y')\,dX'\,dY' \qquad (1.36)$$

Using the fact that the Hertzian pressure distribution is a semi-ellipsoid, and that the lubricated load and the dry contact load are the same, we find with:

$$\mathrm{w} = \frac{2\pi p_h a^2}{3} \qquad (1.37)$$

that:

$$\boxed{\int_{-\infty}^{+\infty} \int_{-\infty}^{+\infty} P(X',Y')\,dX'\,dY' = \frac{2\pi}{3}} \qquad (1.38)$$

1.6 Conclusion

In this chapter the history of ElastoHydrodynamic Lubrication has been sketched, as well as a short history of Multigrid methods. Next the physical processes playing a role in EHL were outlined as well as the equations describing these processes. In the following section the EHL problem was simplified step by step until the Poisson equation was obtained. In the final sections the inverse path was described, starting from the Poisson equation and adding difficulties in a step by step process. A number of these intermediate problems are themselves interesting, and are described seperately. The next chapters will closely follow the outline give in this section, but first a summary of important numerical tools will be given in the next chapter.

1.7 Advanced Topics

1.7.1 Weak Coupling

The Reynolds equation (1.31) describing the flow in thin EHL films shows two very different behaviours at low and high pressure. At low pressure, the viscosity is close to its ambient value η_0, and the deformations are small. The Reynolds equation in this case has a nearly elliptical character, and the influence of changing the pressure in the Reynolds equation on the coefficients ξ can be neglected. For high pressures, meaning those inside the Hertzian zone, the exponential pressure viscosity relation (1.7) causes the coefficients ξ to become very small, indeed negligible. The dimensionless time dependent Reynolds equation (1.31) reduces to:

$$\frac{\partial H}{\partial X} + \frac{\partial H}{\partial T} = 0 \qquad (1.39)$$

This is a one dimensional transport equation. Its solution is: $H(X - T) = constant$. For the stationary case the Reynolds equation in the high pressure zone simplifies to:

$$\frac{\partial H}{\partial X} = 0 \qquad (1.40)$$

with the solution: $H(X) = constant$. This dramatic reduction of the Reynolds equation has two consequences:

- The two dimensional equation becomes a one dimensional equation in the high pressure zone, in other words; one completely looses the coupling in the Y direction. In reality the coupling in the Y direction will not disappear, it will become very weak, and special attention has to be paid to this problem.
- The second consequence is that the equation $\partial H/\partial X = 0$ linked with the high pressures, changes the original partial differential equation with almost elliptical character to an integral equation. For these high pressure conditions, the variation of the film thicknesses H through the changes of P are very important, and have to be accounted for in the solution process.

Between these two extreme cases the character of the system of equations will evolve from almost 'elliptical' to almost 'integral equation', as a function of the operating conditions, but also as a function of the dimensionless position X (and Y) in the contact. A good solver should thus be capable of dealing efficiently with both characters, and with the intermediate cases.

Chapter 2

Numerical Methods: Introduction

In the previous chapter some tribological problems were formulated in terms of systems of partial differential equations. Only in asymptotic cases these problems can be solved analytically. Generally, a numerical approach is required. In this chapter an introduction is given into techniques and tools from the field of numerical mathematics. The one and two dimensional Poisson problem as described at the end of the previous chapter will be used to illustrate the various aspects of the numerical solution of a partial differential problem. In particular, attention is directed to the numerical solution by means of a standard iterative method such as Jacobi and Gauss-Seidel relaxation. The amount of work needed to obtain the solution is expressed in terms of the number of unknowns. The presented results clearly show that such simple iterative methods, due to their local nature, can not lead to efficient solvers for problems with (realistically) large numbers of unknowns. Understanding the nature of the convergence behaviour of these simple iterative schemes however does form the first essential step on the way to obtaining efficient solvers.

As this book is aimed at students in engineering, physics and other technical sciences, mathematically rigorous descriptions and proofs are not given. For such proofs the reader is referred to the numerous books on numerical methods that exist e.g. [33],[54].

2.1 Model Problems

To illustrate the various aspects discussed in this and the following chapters a model problem is used. This is the so-called Poisson equation in one and two dimensions. These problems were selected not only for the purpose of illustration. They are also of interest by themselves, being simplified versions of some real problems that appear in various scientific fields and applications. Understanding how to create an efficient numerical solver for these model problems is an essential first step towards efficient solvers for more complex problems such as the tribological problems outlined in Section 1.5.

The two dimensional Poisson problem is given by:

$$\begin{aligned}\frac{\partial^2 u}{\partial x^2}+\frac{\partial^2 u}{\partial y^2} &= f(x,y) \quad \text{for} \quad (x,y)\in\Omega \\ u(x,y) &= g(x,y) \quad \text{for} \quad (x,y)\in\partial\Omega\end{aligned} \tag{2.1}$$

The domain is taken to be the unit square $\Omega = [0,1]\times[0,1]$. u is the unknown function and $f(x,y)$ is a known function given on Ω. $g(x,y)$ is a known function given on the boundary Ω. A boundary condition of this form is generally referred to as a Dirichlet boundary condition. A further simplification is obtained if the variations in y direction are negligible. In that case the problem reduces to the one dimensional Poisson equation.

The one dimensional problem is given by:

$$\begin{aligned}\frac{\partial^2 u}{\partial x^2} &= f(x) \quad x\in\Omega \\ u(x) &= g(x) \quad x\in\partial\Omega\end{aligned} \tag{2.2}$$

and a domain $\Omega = [0,1]$ will be assumed. Again u is the unknown function and f and g are assumed to be given. Note that here the boundary of Ω consists of only two points, and the information that g carries is limited to two real values. Strictly speaking, the notation with the partial derivative ∂ is not needed for the one dimensional problem, but as the one dimensional problem is considered as a simplification of the two dimensional problem, and studied as a prelude to it, the ∂ is maintained.

2.2 Discretization

The first step in the numerical solution process is generally referred to as *discretization*. The domain on which the problem is to be solved is covered with a grid. A grid is defined as a set of points, which in two dimensions are intersection points of two families of straight lines parallel to the coordinate axes, see Figure 2.1. (In three dimensions the grid is generated by the intersection of planes parallel to the coordinate planes. In one dimension it is just a set of points $\{x_i\}$ on the interval $[0,1] : 0 \equiv x_0 < x_1 < ... < x_n \equiv 1$) Now, rather than solving the unknown u as a continuous function of the coordinates carrying an infinite amount of information, the aim is restricted to solving u at the grid points only, which are finite in number. Finally one settles for even less: the solution of an approximation to the values of u in the grid points. To this end, the continuous problem (partial differential equation(s) + boundary conditions) is transformed to a set of equations in terms of the unknown values at grid points. This set of equations will be denoted as the discretized equations or the discrete system of equations. The system of equations should of course have as many equations as unknowns.

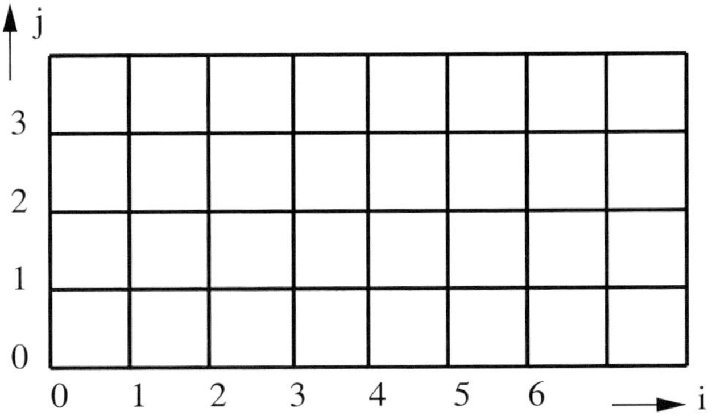

Figure 2.1: *The domain Ω is covered with a grid Ω^h, with mesh size h.*

There are several ways to derive a discrete system of equations from the differential problem each with their own merits for a given class of problems, e.g. Finite Element discretization, Variational Approach, Finite Difference Method, Finite Volume Approach. Below the Finite Difference Method is described. In that case a system of equations from which an approximation to u in each of the grid points can be obtained, is constructed. At each point of the grid a term (sometimes a combination of terms) of the differential equation is replaced by an approximation containing the value in the grid point itself and the values in grid points in its vicinity. The approximate relation is derived by means of Taylor expansions. For example, consider a one dimensional problem. A uniform grid of points x_i is assumed numbered, i.e. $x_i = ih$ for fixed $h = 1/n$ and $i = 0, ..., n$. The approach for non-uniform grids is not essentially different but the expressions become more complex. Let u_i denote the value of the unknown function u at the point of the grid x_i i.e. $u_i = u(x_i)$. Obviously, the values of u for neighbouring points can be related to each other using Taylor expansions. For example u_{i+1} can be expressed as:

$$u_{i+1} = u_i + h\frac{\partial u}{\partial x}\bigg|_{x_i} + \frac{h^2}{2}\frac{\partial^2 u}{\partial x^2}\bigg|_{x_i} + \frac{h^3}{3!}\frac{\partial^3 u}{\partial x^3}\bigg|_{x_i} + \frac{h^4}{4!}\frac{\partial^4 u}{\partial x^4}\bigg|_{x_i} + h.o.t. \tag{2.3}$$

where $h.o.t.$ denotes terms of higher order in h. In the same way u_{i-1} can be expressed as:

$$u_{i-1} = u_i - h\frac{\partial u}{\partial x}\bigg|_{x_i} + \frac{h^2}{2}\frac{\partial^2 u}{\partial x^2}\bigg|_{x_i} - \frac{h^3}{3!}\frac{\partial^3 u}{\partial x^3}\bigg|_{x_i} + \frac{h^4}{4!}\frac{\partial^4 u}{\partial x^4}\bigg|_{x_i} + h.o.t. \tag{2.4}$$

and also:

$$u_{i-2} = u_i - 2h\frac{\partial u}{\partial x}\bigg|_{x_i} + \frac{(2h)^2}{2}\frac{\partial^2 u}{\partial x^2}\bigg|_{x_i} - \frac{(2h)^3}{3!}\frac{\partial^3 u}{\partial x^3}\bigg|_{x_i} + \frac{(2h)^4}{4!}\frac{\partial^4 u}{\partial x^4}\bigg|_{x_i} + h.o.t. \tag{2.5}$$

From Equation (2.3) it follows that:

$$\left.\frac{\partial u}{\partial x}\right|_{x_i} = \frac{u_{i+1} - u_i}{h} - \frac{h}{2}\left.\frac{\partial^2 u}{\partial x^2}\right|_{x_i} - \frac{h^2}{3!}\left.\frac{\partial^3 u}{\partial x^3}\right|_{x_i} - \frac{h^3}{4!}\left.\frac{\partial^4 u}{\partial x^4}\right|_{x_i} + h.o.t. \tag{2.6}$$

which can be written as:

$$\left.\frac{\partial u}{\partial x}\right|_{x_i} = \frac{u_{i+1} - u_i}{h} + \tau^h(x_i) \tag{2.7}$$

Consequently, $(u_{i+1} - u_i)/h$ forms an approximation to the first derivative of u at $x = x_i$ with an error τ^h:

$$\tau^h(x) = -\frac{h}{2}\frac{\partial^2 u}{\partial x^2}(x) - \frac{h^2}{3!}\frac{\partial^3 u}{\partial x^3}(x) - \frac{h^3}{4!}\frac{\partial^4 u}{\partial x^4}(x) + h.o.t. \tag{2.8}$$

As this error results from truncation of the Taylor series it will be referred to as the *truncation error*. τ^h will be used as a symbol for any error of this sort. By definition $\tau^h(x)$ is a power series in h with coefficients containing the local value of derivatives of u. Assuming the series (2.8) is convergent for some $h > 0$ its value will go to zero for h going to zero. For functions u that are smooth on the scale of the mesh size h its behaviour will be dominated by the leading term and thus $(u_{i+1} - u_i)/h$ can be characterized as an $O(h)$ approximation, or, referring only to the power of h in the leading term of the truncation error, as a *first order* approximation. Obviously using Equation (2.3) is not the only way to obtain an approximation to $\partial u/\partial x$ at x_i. An alternative is to use Equation (2.4):

$$\left.\frac{\partial u}{\partial x}\right|_{x_i} = \frac{u_i - u_{i-1}}{h} + \tau^h(x_i) \tag{2.9}$$

with

$$\tau^h(x) = \frac{h}{2}\frac{\partial^2 u}{\partial x^2} - \frac{h^2}{3!}\frac{\partial^3 u}{\partial x^3} + \frac{h^3}{4!}\frac{\partial^4 u}{\partial x^4} + h.o.t. \tag{2.10}$$

Hence, also $(u_i - u_{i-1})/h$ forms an $O(h)$ approximation to $\partial u/\partial x$ at $x = x_i$. The only difference between the two approximations is in the details of the truncation error, compare Equations (2.8) and (2.10).

In the same way, i.e. by combining Taylor Series and eliminating terms, higher order approximations (than first order) can be obtained. For example, using the grid points x_i, x_{i-1}, and x_{i-2} it follows that:

$$\frac{3u_i - 4u_{i-1} + u_{i-2}}{2h} = \left.\frac{\partial u}{\partial x}\right|_{x_i} - \tau^h(x_i) \tag{2.11}$$

with

$$\tau^h(x) = -\frac{h^2}{3}\frac{\partial^3 u}{\partial x^3} - \frac{h^3}{4}\frac{\partial^4 u}{\partial x^4} + h.o.t. \tag{2.12}$$

Thus, the left hand side of Equation (2.11) forms a second order approximation to $\partial u/\partial x$ at x_i. Note that to obtain the higher accuracy an additional grid point is used compared to the two points (i and $i+1$, or i and $i-1$) sufficient for a first order approximation. In general, the higher the required order the larger the number of error terms that should be eliminated and thus the larger the number of discrete points to be used. However, symmetry considerations can help to reduce the number of points needed. For example, if in the present case the points x_{i+1} and x_{i-1} are used one obtains:

$$\frac{u_{i+1} - u_{i-1}}{2h} = \left.\frac{\partial u}{\partial x}\right|_{x_i} - \tau^h(x_i) \tag{2.13}$$

with

$$\tau^h(x) = -\frac{h^2}{3!}\frac{\partial^3 u}{\partial x^3} - \frac{h^4}{5!}\frac{\partial^5 u}{\partial x^5} + h.o.t. \tag{2.14}$$

Hence, the left hand side of Equation (2.13) forms a second order approximation to $\partial u/\partial x$ at x_i using only two points. Note that this central approximation is in fact the average of the one-sided first order approximations Equations (2.7) and (2.9). As the leading $O(h)$ error terms are the same but of opposite sign, see Equations (2.8) and (2.10), they cancel when the average is taken.

Approximations for higher order derivatives can be constructed too. For example adding Equations (2.3) and (2.4) and subtracting two times u_i one obtains:

$$\frac{u_{i+1} - 2u_i + u_{i-1}}{h^2} = \left.\frac{\partial^2 u}{\partial x^2}\right|_{x_i} - \tau^h(x_i) \tag{2.15}$$

with

$$\tau^h(x) = -\frac{h^2}{12}\frac{\partial^4 u}{\partial x^4} + h.o.t. \tag{2.16}$$

which shows that the left hand side of Equation (2.15) is a second order approximation to $\partial^2 u/\partial x^2$ at x_i.

The central idea of *finite difference discretization* is to replace each derivative in the partial differential equation by a discrete expression omitting the truncation errors τ^h. Let the problem to be solved be the one dimensional model problem, i.e. to solve $u(x)$ from (2.2) for $0 \leq x \leq 1$ with $u(0) = g(0)$ and $u(1) = g(1)$ given. This task is then replaced by solving (by computer) an approximation u_i^h to $u(x_i)$ in each grid point $x_i = ih$ from the set of discrete equations:

$$\frac{u_{i+1}^h - 2u_i^h + u_{i-1}^h}{h^2} = f_i^h \tag{2.17}$$

for all points $1 \leq i \leq n-1$ and with the values at the boundary u_0^h and u_n^h given. The right hand side f_i^h is simply the right hand side function f in x_i i.e. $f_i^h = f(x_i)$. However, in special cases a local averaging may be preferred, e.g. $f_i^h = \frac{1}{h}\int_{x_i-h/2}^{x_i+h/2} f(s)ds$. The system of $(n-1)$ equations from which the unknowns $u_1^h ... u_{n-1}^h$ must be solved, can subsequently be written in matrix form, see Section 2.3.

Note that the notation u_i^h with superscript h is introduced to indicate that the solution of Equation (2.17) is a grid-dependent result. As the finite difference on the left is a second order approximation to $\partial^2 u/\partial x^2$, the resulting system of equations forms a second order approximation to the continuous problem posed by Equation (2.2). After all, using Equation (2.15), Equation (2.2) can be written as:

$$\frac{u_{i+1} - 2u_i + u_{i-1}}{h^2} + \tau^h(x_i) = f_i^h \tag{2.18}$$

Thus, if $\lim_{h\to 0} \tau^h(x_i) = 0$ for all x_i the discrete equation becomes an increasingly more accurate approximation to the continuous differential equation with decreasing mesh size h. If this is the case the discretization is said to be *consistent*. Equation (2.16) shows that this is obviously the case here. For a consistent discretization one may expect that also the difference between u and u^h decreases with decreasing mesh size. This is indeed true, if another requirement is satisfied, i.e. if the problem stable. By stability it is meant that the solution depends continuously on the boundary conditions and the right hand side, i.e. a small change in the right hand side f, and/or boundary conditions g only results in small variations in the solution. For further details and general discussions on well posedness of the discrete problem the reader is referred to books on numerical mathematics: see Hackbusch and Kevorkian [53, 69]. For the present chapter the importance of this property becomes clear from Equation (2.18). This equation shows that solving the discrete problem (2.17) with a right hand side f is the same as solving the continuous equation but with slightly different right hand side $f - \tau^h$ (2.18).

The error in the discrete approximation, as a result of the truncation error τ^h in the equation, is defined for each grid point as the difference between the continuous and the discrete solution:

$$e_i^h = u_i - u_i^h \tag{2.19}$$

and is referred to as the *discretization* error. In principle, i.e. if there are no singularities etc., it is of the same order in h as the truncation error τ^h. Note that e^h and τ^h are fundamentally different.

τ^h is an error made in approximating the *equation* and e^h is the error in the *solution* made as a result of solving the approximate system of equations (discretization error).

So far the examples were restricted to the one dimensional case. However, the extension to two dimensions is straightforward. In that case a grid point is denoted by (x_i, y_j) and the value of the function u at this point by $u_{i,j}$. Instead of h there can be two different mesh sizes h_x and h_y. For example, the two dimensional model problem given by Equation (2.1), is replaced by the discrete system:

$$\frac{u_{i+1,j}^h - 2u_{i,j}^h + u_{i-1,j}^h}{h_x^2} + \frac{u_{i,j+1}^h - 2u_{i,j}^h + u_{i,j-1}^h}{h_y^2} = f_{i,j}^h \tag{2.20}$$

for $1 \le i \le (n_x - 1)$ and $1 \le j \le (n_y - 1)$ with $u_{i,j}^h = g(x_i, y_j)$ given for $i = 0$ or $j = 0$ or $i = n_x$ or $j = n_y$.

Equation (2.20) forms a second order approximation to the continuous problem given by Equation (2.1), and it is probably the most widely used discretization for this problem. The error made in approximating the equation, i.e. the truncation error is given by:

$$\tau^h = -\frac{h_x^2}{12}\frac{\partial^4 u}{\partial x^4} - \frac{h_y^2}{12}\frac{\partial^4 u}{\partial y^4} + h.o.t. \tag{2.21}$$

Finally, to facilitate certain types of analysis, or writing a computer program, the system of equations for all i and j can be written in matrix form, see Section 2.3.

Summarizing, to discretize a problem by means of finite differences means that each derivative in the continuous equation is replaced by finite differences, such that the resulting discrete equation approximates the continuous equation with some prescribed accuracy. Obviously for more complex equations, e.g. if mixed derivatives are involved, the derivations are a bit more complicated than for the model problems considered here, however, the approach is not essentially different. Note that in principle one can not only prescribe the order of accuracy but also design a discretization such that specific terms in the error may cancel.

2.3 Systems of Equations

The result of the discretization process can be written as a matrix system of equations. Using Finite Difference discretization one obtains a system of equations for $\underline{u}^h$:

$$A\underline{u}^h = \underline{f}^h \tag{2.22}$$

where A is the matrix, $\underline{u}^h$ a vector containing the unknowns, and $\underline{f}^h$ a right hand side vector. As an example, the matrix equations are given resulting from the finite difference discretization of the model problems, i.e. Equations (2.17) and (2.20). For the one dimensional problem given by Equation (2.17) ordering the equations in order of increasing i (lexicographic) the discrete matrix equation is:

$$\frac{1}{h^2}\begin{pmatrix} -2 & 1 & & & & \\ 1 & -2 & 1 & & & \\ & & . & . & . & \\ & & & 1 & -2 & 1 \\ & & & & 1 & -2 \end{pmatrix}\begin{pmatrix} u_1^h \\ u_2^h \\ . \\ u_{n-2}^h \\ u_{n-1}^h \end{pmatrix} = \begin{pmatrix} f_1^h - u_0^h/h^2 \\ f_2^h \\ . \\ f_{n-2}^h \\ f_{n-1}^h - u_n^h/h^2 \end{pmatrix} \left.\vphantom{\begin{pmatrix} a\\a\\a\\a\\a \end{pmatrix}}\right\} \text{(n}-\text{1) elements} \tag{2.23}$$

where the dots indicate repetition of the pattern and only the non-zero matrix elements have been written. These non-zero terms are identical for each interior point, with respect to the diagonal. Instead of using the full matrix A it is convenient to use only this repetitive pattern, giving the non-zero coefficients of the discrete operator. The central coefficient relates to u_i^h, the one to the left to u_{i-1}^h etc. This notation is called a stencil:

$$\frac{1}{h^2}[\begin{array}{ccc} 1 & -2 & 1 \end{array}] \tag{2.24}$$

Note that the Dirichlet boundary condition is simply incorporated in the right hand side vector. As it turns out the matrix one obtains is a tri-diagonal matrix. This is off course due to the lexicographic order. From a point of view of solving the matrix equation with a so-called direct solver this bandwidth should be as narrow as possible, see Section 2.4. For one dimensional partial differential problems the matrix of the discrete problem will often be a band matrix with the minimum width of the band depending on the width of the discrete operator. In the present case the discrete equation at a point i only uses values of u in the nearest neighbours $i-1$ and $i+1$ which then gives a bandwidth of 1.

Next consider the two dimensional problem given by Equation (2.20). If the equations are organised in the matrix in lexicographic order which for two dimensions implies in order of increasing i and j again a band matrix is obtained. However, the width of the band is now determined by the number of points in x direction. For a given (i,j) the discrete equation contains the points $(i,j+1)$ and $(i,j-1)$ which appear at positions (n_x-1) to the left and (n_x-1) to the right in the matrix. Taking a closer look shows that the matrix is still a tridiagonal matrix if it is regarded as a matrix of $(n_y-1)(n_y-1)$ blocks of size $(n_x-1)(n_x-1)$. As an illustration the system of equations for $n_x=5$, $n_y=4$ and $h_x=h_y=h$ is given by (2.22) with:

$$A = \frac{1}{h^2}\left(\begin{array}{cccc|cccc|cccc}
-4 & 1 & & & 1 & & & & & & & \\
1 & -4 & 1 & & & 1 & & & & & & \\
 & 1 & -4 & 1 & & & 1 & & & & & \\
 & & 1 & -4 & & & & 1 & & & & \\
\hline
1 & & & & -4 & 1 & & & 1 & & & \\
 & 1 & & & 1 & -4 & 1 & & & 1 & & \\
 & & 1 & & & 1 & -4 & 1 & & & 1 & \\
 & & & 1 & & & 1 & -4 & & & & 1 \\
\hline
 & & & & 1 & & & & -4 & 1 & & \\
 & & & & & 1 & & & 1 & -4 & 1 & \\
 & & & & & & 1 & & & 1 & -4 & 1 \\
 & & & & & & & 1 & & & 1 & -4
\end{array}\right) \tag{2.25}$$

$$\underline{u}^h = \begin{pmatrix} u^h_{1,1} \\ u^h_{2,1} \\ u^h_{3,1} \\ u^h_{4,1} \\ u^h_{1,2} \\ u^h_{2,2} \\ u^h_{3,2} \\ u^h_{4,2} \\ u^h_{1,3} \\ u^h_{2,3} \\ u^h_{3,3} \\ u^h_{4,3} \end{pmatrix} \quad \text{and} \quad \underline{f}^h = \begin{pmatrix} f^h_{1,1} - u^h_{1,0}/h^2 - u^h_{0,1}/h^2 \\ f^h_{2,1} - u^h_{2,0}/h^2 \\ f^h_{3,1} - u^h_{3,0}/h^2 \\ f^h_{4,1} - u^h_{4,0}/h^2 - u^h_{5,1}/h^2 \\ f^h_{1,2} - u^h_{0,2}/h^2 \\ f^h_{2,2} \\ f^h_{3,2} \\ f^h_{4,2} - u^h_{5,2}/h^2 \\ f^h_{1,3} - u^h_{1,4}/h^2 - u^h_{0,3}/h^2 \\ f^h_{2,3} - u^h_{2,4}/h^2 \\ f^h_{3,3} - u^h_{3,4}/h^2 \\ f^h_{4,3} - u^h_{4,4}/h^2 - u^h_{5,3}/h^2 \end{pmatrix} \tag{2.26}$$

The block-structure is indicated by the drawn lines. Once again it is more convenient to use only the repetitive part of the matrix A, giving the coefficients of the discrete operator. The central coefficient relates to $u^h_{i,j}$, the one to its left to $u^h_{i-1,j}$, the one above $u^h_{i,j+1}$, etc.

$$\frac{1}{h^2}\begin{bmatrix} & 1 & \\ 1 & -4 & 1 \\ & 1 & \end{bmatrix} \tag{2.27}$$

Using this block structure the system of equations for arbitrary n_x,n_y and h_x, h_y can be written as:

$$\begin{pmatrix} -2I-B & I & & & \\ I & -2I-B & I & & \\ & & \cdot\ \cdot\ \cdot & & \\ & & I & -2I-B & I \\ & & & I & -2I-B \end{pmatrix} \begin{pmatrix} \underline{u}_1^h \\ \underline{u}_2^h \\ \cdot \\ \underline{u}_{n_y-2}^h \\ \underline{u}_{n_y-1}^h \end{pmatrix} = \begin{pmatrix} \underline{f}_1^h \\ \underline{f}_2^h \\ \cdot \\ \underline{f}_{n_y-2}^h \\ \underline{f}_{n_y-1}^h \end{pmatrix} \quad (2.28)$$

where I is the $(n_x-1)(n_x-1)$ identity matrix $I_{i,j} = \delta_{ij}$ (Kronecker delta) multiplied with $1/h_y^2$ and B is the $(n_x-1)(n_x-1)$ matrix:

$$B = \frac{1}{h_x^2} \begin{pmatrix} 2 & -1 & & & \\ -1 & 2 & -1 & & \\ & \cdot & \cdot & \cdot & \\ & & -1 & 2 & -1 \\ & & & -1 & 2 \end{pmatrix} \quad (2.29)$$

$\underline{u}_j^h$ are vectors of length (n_x-1):

$$\underline{u}_j^h = \begin{pmatrix} u_{1,j}^h \\ u_{2,j}^h \\ \cdot \\ u_{n_x-2,j}^h \\ u_{n_x-1,j}^h \end{pmatrix} \quad (2.30)$$

for $1 \leq j \leq (n_y-1)$. $\underline{f}_j^h$ are also vectors of length (n_x-1). For $2 \leq j \leq (n_y-2)$ they are given by:

$$\underline{f}_j^h = \begin{pmatrix} f_{1,j}^h - u_{0,j}^h/h_x^2 \\ f_{2,j}^h \\ \cdot \\ f_{n_x-2,j}^h \\ f_{n_x-1,j}^h - u_{n_x,j}^h/h_x^2 \end{pmatrix} \quad (2.31)$$

and for $j = 1$ and $j = (n_y-1)$ by:

$$\underline{f}_1^h = \begin{pmatrix} f_{1,1}^h - u_{1,0}^h/h_y^2 - u_{0,1}^h/h_x^2 \\ f_{2,1}^h - u_{2,0}^h/h_y^2 \\ \cdot \\ f_{n_x-2,1}^h - u_{n_x-2,0}^h/h_y^2 \\ f_{n_x-1,1}^h - u_{n_x-1,0}^h/h_y^2 - u_{n_x,1}^h/h_x^2 \end{pmatrix} \quad (2.32)$$

and

$$\underline{f}_{n_y-1}^h = \begin{pmatrix} f_{1,n_y-1}^h - u_{1,n_y}^h/h_y^2 - u_{0,n_y-1}^h/h_x^2 \\ f_{2,n_y-1}^h - u_{2,n_y}^h/h_y^2 \\ \cdot \\ f_{n_x-2,n_y-1}^h - u_{n_x-2,n_y}^h/h_y^2 \\ f_{n_x-1,n_y-1}^h - u_{n_x-1,n_y}^h/h_y^2 - u_{n_x,n_y-1}^h/h_x^2 \end{pmatrix} \quad (2.33)$$

Having constructed the discrete system of equations the next task is to obtain the solution. Based on the way the system is solved, different methods can be distinguished, i.e. direct and iterative methods. These methods are discussed in the following paragraphs.

2.4 Direct Solver

Characteristic for a direct method is that the solution is obtained in a fixed number of operations, i.e. the solution process terminates after a finite number of operations with the exact solution (up to rounding errors). In fact a direct method can be seen as effectively inverting the matrix A in (2.22):

$$\underline{u}^h = A^{-1}\underline{f}^h \tag{2.34}$$

Various direct methods exist, each with their own merits for specific types of matrices. For a detailed description of algorithms the reader is referred to e.g. [46]. As a typical example standard Gaussian elimination with backsubstitution is described below. Assume that A is an $N \times N$ matrix and the element at row i and column j will be denoted by $a_{i,j}$. The ith element of the vector $\underline{u}^h$ and $\underline{f}^h$ will be denoted by u_i^h and f_i^h respectively. The ith equation of the system (2.22) can then be written as:

$$\sum_{j=1}^{N} a_{i,j} u_j^h = f_i^h \tag{2.35}$$

Solving the system by means of Gaussian elimination consists of two steps. In the first step the system (2.22) is rewritten into a system of equations:

$$\bar{A}\underline{u}^h = \underline{\bar{f}}^h \tag{2.36}$$

where $\bar{A}$ is an upper triangular matrix. This is a matrix in which $\bar{a}_{i,j} = 0$ for $j < i$. This system is obtained in the following way. Using (2.35) for $i = 1$ the first unknown u_1^h is eliminated from (2.35) for all $i > 1$ in which it appears. This is done by subtracting the equation for $i = 1$ multiplied with $a_{i,1}/a_{1,1}$ from equation i. The new i^{th} equation is then given by:

$$\sum_{j=1}^{N} \tilde{a}_{i,j} u_j^h = \tilde{f}_i^h \tag{2.37}$$

where

$$\tilde{a}_{i,j} = a_{i,j} - \frac{a_{i,1}}{a_{1,1}} a_{1,j} \tag{2.38}$$

and

$$\tilde{f}_i^h = f_i^h - \frac{a_{i,1}}{a_{1,1}} f_1^h \tag{2.39}$$

Obviously $\tilde{a}_{i,1} = 0$ for $i > 1$. Thus a new system has been created in which the first column is zero, except for the first row. Next, from this new system, in the same way as described above for u_1^h, u_2^h is eliminated from all equations $i > 2$ by subtracting from equation i the second equation multiplied with $\tilde{a}_{i,2}/\tilde{a}_{2,2}$. This leads to a system of equations with a matrix that has a zero first column except for the first element (row 1), *and* a zero second column except for the first two elements (rows 1 and 2). This process of elimination is repeated until the system of equations (2.36) is obtained.

The elimination process described above runs into trouble if at a given point the equation used to eliminate a given unknown has a zero (or very small) element on the diagonal, for example the elements $a_{1,1}$ and $\tilde{a}_{2,2}$ used in two steps described above. In general the element used to eliminate is referred to as the "pivot". In the above it was assumed that at each stage the ith equation can be used to eliminate the ith unknown from the equations $j > i$. If this is not possible, one should use another equation, i.e. another pivot, leading to strategies such as "partial" and "complete" pivoting, see [46].

The second phase of the process is to solve u_j^h for all j from (2.36). This phase is referred to as *backsubstitution*. First the value u_N^h is determined. As the last equation of the triangular matrix only contains the diagonal element:

$$u_N^h = \frac{\bar{f}_N^h}{\bar{a}_{N,N}} \tag{2.40}$$

Subsequently for $i = N - 1$ downto $i = 1$ the value of i is computed according to:

$$u_i^h = (\bar{f}_i^h - \sum_{j=i+1}^{N} \bar{a}_{i,j} u_j^h)/\bar{a}_{i,i} \tag{2.41}$$

The two steps of the Gaussian elimination with backsubstitution described above can be summarized and written in a general form referred to as LU factorization. For these and further details concerning direct methods the reader is referred to the literature.

From the description of Gaussian elimination given above it may be clear that direct methods have the advantage that they are simple, straightforward, and often easy to implement. Little knowledge is required of the system of equations. In fact many numerical packages contain routines performing the task of solving (2.22) for given A and $\underline{f}^h$ and programming code is published in various books, e.g. see [89]. These standard routines are ideally suited to quickly solve small and simple systems of equations.

To determine whether such a direct method can lead to an efficient solver for the possibly large systems (large N) of equations resulting from the discretization of partial differential problems, such as the model problems (2.2) and (2.1), the number of operations needed to obtain the solutions must be determined. First assume that the matrix has a band structure with band width b. It can easily be verified that for such cases the first phase of the Gaussian elimination requires $O(b^2N)$ operations where an operation means a multiplication and addition or division. The work count of the second phase, backsubstitution, is $O(bN)$. For large N the amount of work in the first phase will dominate and the total number of operations, and thus the cpu time required to obtain the solution, will be $W = O(b^2N)$ This work count reflects the background of the remark in the previous section that it is beneficial to order the equations in such a way that the band width is minimal. For the one dimensional model problem, Equation (2.23) shows that in this case $b = 1$ and $N = (n-1)$. Consequently this problem can be solved in

$O(N)$ operations, and thus in a cpu time proportional to the number of unknowns which is the ideal work count. For the two dimensional problem the band width of the matrix is $b = (n_x - 1)$, see Equations (2.25) and (2.28), and the total number of unknowns $N = (n_x-1)(n_y-1)$. The amount of work needed to solve this problem with Gaussian elimination is thus: $W = O(n_x-1)^3(n_y-1)$ which for $n_x = n_y = n$ gives $W = O(n^4)$. For a d dimensional problem with $O(n)$ points in each direction, ordering the equations in lexicographic order the band width of the matrix will be $b = O(n^{d-1})$, and $N = O(n^d)$. The amount of work needed to solve the system is then given by:

$$W = O(n^{3d-2}) = O(N^{3-\frac{2}{d}}) \tag{2.42}$$

where $N = O(n^d)$ is the total number of points on the grid. Thus for a two dimensional problem the cpu time needed to solve the problem will be $O(N^2)$.

So far we assumed a banded matrix. However, if the matrix is full the situation is worse. In that case $b = N$ and thus the total amount of work will be $O(N^3)$ for both a one and two dimensional problem. This occurs for example for the EHL problem, and also leads to excessive storage requirements. A band matrix can efficiently be stored e.g. by storing only the $O(bN)$ non zero elements. Even when applying Gaussian Elimination the system does not fully fill up to a triangular matrix. However, if the matrix is full, it needs to be stored completely, requiring $O(N^2)$ memory positions.

Summarizing it is concluded that for small problems a direct solver can be the most efficient way to solve the discrete equations. However, for large N this will obviously lead to large computing times as the required cpu time quickly increases with increasing N. As a result, direct methods are often not suited for practical problems. In that case the need to *quickly* solve the problem with about $n = 100$, and thus $N = 10^4$ for $d = 2$ (and $N = 10^6$ for $d = 3$) is certainly not an exceptional demand. Consequently, the turn around time to obtain the numerical solution will be unacceptably large, and an algorithm based on a direct solver will not be suitable to carry out extensive parameter studies, which in practice are often needed, i.e. where the problem must be solved often to investigate the effects of certain problem parameters on the solution.

2.5 Iterative Solver

An iterative solution method for the system of equations:

$$A\underline{u} = \underline{f} \tag{2.43}$$

generally starts with an initial guess $\underline{u}^0$. Subsequently, a sequence of successive iterates (iterands) $\underline{u}^m$ is computed for $m = 1, 2...$:

$$\underline{u}^0 \to \underline{u}^1 \to \underline{u}^2 \to ... \to \underline{u}^m \to \underline{u}^{m+1} \to ... \tag{2.44}$$

where the arrow indicates the rule of the iterative process. In the following sections it is assumed that the calculation of $\underline{u}^{m+1}$ makes use of $\underline{u}^m$ only, not of the preceding iterates. In that case the iterative method can be characterised by a prescription: $\underline{u}^{m+1} = \Phi(\underline{u}^m)$, for $m \geq 0$, where Φ contains data of the system (2.22), i.e. it depends on the matrix A and on the right hand side $\underline{f}$. This latter dependence is sometimes explicitly included by writing $\underline{u}^{m+1} = \Phi(\underline{u}^m, \underline{f})$. Thus an iterative method is a (linear or non-linear) mapping, which subsequently facilitates its analysis, e.g. see Hackbusch [54]. An iterative method is called linear if $\Phi(\underline{u}, \underline{f})$ is linear in $\underline{u}$ and $\underline{f}$, i.e. if there are matrices $\mathcal{M}$ and $\mathcal{N}$ such that:

$$\Phi(\underline{u}, \underline{f}) = \mathcal{M}\underline{u} + \mathcal{N}\underline{f} \tag{2.45}$$

where the matrix $\mathcal{M}$ is referred to as the *iteration matrix* of the iteration Φ. The iteration should of course be such that for all right hand sides $\underline{f}^h$ any solution satisfying $A\underline{u}^h = \underline{f}^h$ is a stationary point:

$$\Phi(\underline{u}^h, \underline{f}^h) = \underline{u}^h \tag{2.46}$$

This property is referred to as consistency of the process. The error in the approximation $\underline{u}^m$ is now defined as:

$$\underline{v}^m = \underline{u}^h - \underline{u}^m \tag{2.47}$$

Substitution of $\underline{u}^m$ from Equation (2.47) and using Equation (2.46) one can rewrite the iterative process in terms of the influence on the error:

$$\underline{v}^{m+1} = \mathcal{M}\underline{v}^m \tag{2.48}$$

For the approximate solution to become more accurate with increasing iteration number (convergence) one needs:

$$|\underline{v}^{m+1}| = |\mathcal{M}\underline{v}^m| < |\underline{v}^m| \tag{2.49}$$

From this equation it follows that a linear iterative method with iteration matrix $\mathcal{M}$ converges if and only if the spectral radius, i.e. the largest absolute eigenvalue of the matrix $\mathcal{M}$ is smaller than 1. The matrix $\mathcal{M}$ is related to the matrix A, which in its turn is determined by the particularities of the discretization. Therefore, for a given iterative process to converge, the matrix A may have to satisfy certain conditions, e.g. be diagonally dominant. At this point no further details are given. The main point is that the iterative method and the method of discretization can not be chosen independently.

The largest eigenvalue of the iteration matrix does not only determine *if* the process converges, it also determines the asymptotic speed of convergence of the process, i.e. the amount by which one iteration reduces the error in the approximation. For example if the matrix $\mathcal{M}$ has a full set of eigenvectors the error can be written as:

$$\underline{v}^0 = \sum_k \alpha_k \underline{w}_k \tag{2.50}$$

where $\underline{w}_k$ are the eigenvectors of $\mathcal{M}$ with eigenvalues μ_k and α_k are coefficients. The error in the mth iterand can then be written as:

$$\underline{v}^m = \sum_k \alpha_k \mu_k^m \underline{w}_k \tag{2.51}$$

The error components associated with the absolute smallest eigenvalue will thus be reduced very quickly and the asymptotic convergence is determined by the absolute largest eigenvalue. Even if the matrix $\mathcal{M}$ does not have a full set of eigenvectors one can still show that for convergence of the iterative process the absolute value of all eigenvalues of $\mathcal{M}$ must be smaller than unity. In that case this conclusion is achieved by an analysis based on $\mathcal{M}$ transformed into a so-called Jordan form.

There are many different iterative processes, and it goes beyond the scope of this chapter to give a detailed overview. Two very basic processes referred to as Jacobi and Gauss-Seidel relaxation respectively are often used in practice. These processes are useful to illustrate the various aspects of iteratively solving the system of equations. For the one and two dimensional model problems these processes are described in detail in the following sections.

2.6 Relaxation

In the previous paragraph a brief general description of an iterative process was given in terms of matrices and vectors. This notation is convenient if one wishes to analyse the properties of the process by means of eigenvectors and eigenvalues. However, for many iterative processes it is not necessary to use or store the full matrix. After all, the equation at a given point of the grid only relates the unknowns in a limited number of points in its direct vicinity. In this paragraph a simple iterative process is described for the discretized one and two dimensional model problems given by Equation (2.2) and (2.1) respectively.

In words the process can be described as follows. Given an approximation to the solution of the problem in each grid point, a new approximation is computed by scanning the grid points in a prescribed order, changing the value in each grid point such that the local equation in that point is satisfied. The process comes in two flavours. If the new values are computed using only old values in the surrounding grid points it is referred to as Jacobi relaxation (Simultaneous Displacement). If the changes made when relaxing previous grid points are taken into account when relaxing the next grid point it is referred to as Gauss-Seidel relaxation (Successive Displacement). For Jacobi relaxation the order in which the grid points are relaxed is irrelevant. For Gauss-Seidel relaxation it makes a difference. Often the grid points are scanned in order of increasing grid indices in which case one refers to it as Gauss-Seidel relaxation with lexicographic ordering.

2.6.1 Jacobi Relaxation

For convenience the discretized one dimensional model problem described by Equation (2.2) is repeated here. For each grid point $x_i = ih$ with $1 \le i \le (n-1)$ solve the discrete approximation u_i^h to $u(x_i)$ from:

$$\frac{u_{i-1}^h - 2u_i^h + u_{i+1}^h}{h^2} = f_i^h \tag{2.52}$$

with $u_0^h = g(0)$ and $u_n^h = g(1)$ given.

Jacobi relaxation applied to the solution of this model problem is described as follows. Let $\tilde{u}_i^h$ denote the current approximation to u_i^h. For each grid point i a new approximation $\bar{u}_i^h$ to u_i^h is computed according to:

$$\bar{u}_i^h = \tilde{u}_i^h + \omega\delta_i^h \tag{2.53}$$

where ω is the relaxation factor and δ_i^h is given by:

$$\delta_i^h = -\frac{h^2 r_i^h}{2}, \tag{2.54}$$

and r_i^h is the *residual* defined as:

$$r_i^h = f_i^h - \frac{\tilde{u}_{i-1}^h - 2\tilde{u}_i^h + \tilde{u}_{i+1}^h}{h^2} \tag{2.55}$$

As can be shown easily $\omega = 1$ implies that the new approximation to u_i^h is such that given the values $\tilde{u}_{i-1}^h$ and $\tilde{u}_{i+1}^h$ in the neighbouring points, the discrete equation at the point i is satisfied:

$$\frac{\tilde{u}_{i-1}^h - 2\bar{u}_i^h + \tilde{u}_{i+1}^h}{h^2} = f_i^h \tag{2.56}$$

For $\omega < 1$ the process is referred to as *damped* Jacobi, or Jacobi with underrelaxation. In that case only part of the change needed to solve the local equation is applied. On the role of the relaxation factor, see Section 2.7. Notice that the new value $\bar{u}_i^h$ for each i is computed using 'old' values only, i.e. the changes to be made can be computed in parallel and applied simultaneously to all grid points. Therefore Jacobi relaxation is referred to as a Simultaneous Displacement scheme. One step of computing and applying the changes to each of the grid points constitutes one iteration or one relaxation sweep.

Finally, please note that in order to implement the process in a computer program, the complete matrix is never needed. At most three arrays (vectors) are required. One to store the old approximation $\tilde{u}_i^h$ for all grid points, one for the new approximation $\bar{u}_i^h$, and one for the right hand side in all grid points f_i^h.

Next consider the two dimensional equation given by (2.1). This requires the solution of an approximation $u_{i,j}^h$ to $u(x_i = ih_x, y_j = jh_y)$ from:

$$\frac{u_{i-1,j}^h - 2u_{i,j}^h + u_{i+1,j}^h}{h_x^2} + \frac{u_{i,j-1}^h - 2u_{i,j}^h + u_{i,j+1}^h}{h_y^2} = f_{i,j}^h \tag{2.57}$$

for all interior points $1 \le i \le (n_x - 1)$ and $1 \le j \le (n_y - 1)$. On the boundary, e.g. for $i = 0$ and $i = n_x$ and for $j = 0$ and $j = n_y$ the solution is given by $u_{i,j}^h = g(x_i, y_j)$.

Jacobi relaxation applied to the solution of this problem can be described as follows. Let $\tilde{u}_{i,j}^h$ denote the current approximation to $u_{i,j}^h$. For each point a new approximation $\bar{u}_{i,j}^h$ is computed:

$$\bar{u}_{i,j}^h = \tilde{u}_{i,j}^h + \omega\delta_{i,j}^h \tag{2.58}$$

where

$$\delta_{i,j}^h = -\left(\frac{2}{h_x^2} + \frac{2}{h_y^2}\right)^{-1} r_{i,j}^h \tag{2.59}$$

with $r_{i,j}^h$ the residual which is given by:

$$r_{i,j}^h = f_{i,j}^h - \frac{\tilde{u}_{i-1,j}^h - 2\tilde{u}_{i,j}^h + \tilde{u}_{i+1,j}^h}{h_x^2} - \frac{\tilde{u}_{i,j-1}^h - 2\tilde{u}_{i,j}^h + \tilde{u}_{i,j+1}^h}{h_y^2} \tag{2.60}$$

Again, one step of computing and applying the changes to all grid points forms one relaxation sweep or one iteration. The changes can be computed and applied simultaneously, or put in a different way, when the grid points are scanned in some order, the order does not influence the resulting solution.

2.6.2 Gauss-Seidel Relaxation

When computing a new approximation for a given grid point one may already use the new approximation in grid points relaxed previously. This is the case for Gauss-Seidel relaxation. Consider the one dimensional problem first. Let $\tilde{u}_i^h$ again denote the current approximation in each grid point. Subsequently, the grid points are scanned in e.g. lexicographic order. Starting with grid point $i = 1$ a new approximation $\bar{u}_i^h$ to u_i^h is computed according to Equation (2.53) and (2.54). However now r_i^h is the so-called *dynamic* residual which is defined by:

$$r_i^h = f_i^h - \frac{\bar{u}_{i-1}^h - 2\tilde{u}_i^h + \tilde{u}_{i+1}^h}{h^2} \tag{2.61}$$

It is referred to as *dynamic* as it is the residual computed at the point of relaxing the equation, i.e. using the most recent approximation to u_i^h available at the different grid points appearing in the operator. Thus, the new value $\bar{u}_{i-1}^h$ is already used since, in lexicographic order, by the time the point i is scanned a new value has already replaced the old one in the point $i-1$. For $\omega = 1$ the change applied to $\tilde{u}_i^h$ is such that the discrete equation at the grid point i in terms of the most recent values available is satisfied, i.e. $\bar{u}_i^h$ satisfies:

$$\frac{\bar{u}_{i-1}^h - 2\bar{u}_i^h + \tilde{u}_{i+1}^h}{h^2} = f_i^h \tag{2.62}$$

Each pass over the grid scanning all points is referred to as a *relaxation sweep*. Sometimes it is used with $\omega > 1$ in which case it is referred to as Successive Over Relaxation (S.O.R) whereas $\omega < 1$ is referred to as underrelaxation. In this latter case only part of the change needed to solve the local equation is applied. The importance of the relaxation factor is specified in more detail in Section 2.7. To implement Gauss-Seidel relaxation in a computer program only two arrays (vectors) are needed. One for the right hand side f_i^h in all grid points, and one for the approximation to u_i^h. This latter array will contain $\tilde{u}_i^h$ for all i at the start of the sweep. At any point during the relaxation sweep it contains $\tilde{u}_i^h$ in points i not yet relaxed and $\bar{u}_i^h$ in points i already relaxed. At the end of the sweep it contains $\bar{u}_i^h$ for all points i. This is exactly what is meant by successive displacement.

One does not need a separate array for the old approximation. The old approximation is simply overwritten during the relaxation sweep.

Gauss-Seidel relaxation applied to the two dimensional problem implies that the grid points are scanned in some prescribed order, e.g. with increasing i and j, which is the so-called lexicographic order. At each point a new approximation is computed according to Equations (2.58) and (2.59), however, instead of using Equation (2.60) $r^h_{i,j}$ is the *dynamic* residual defined as:

$$r^h_{i,j} = f^h_{i,j} - \frac{\bar{u}^h_{i-1,j} - 2\tilde{u}^h_{i,j} + \tilde{u}^h_{i+1,j}}{h_x^2} - \frac{\bar{u}^h_{i,j-1} - 2\tilde{u}^h_{i,j} + \tilde{u}^h_{i,j+1}}{h_y^2} \tag{2.63}$$

Thus, to compute a new approximation at the point (i, j) the new approximation previously computed at the points $(i-1, j)$ and $(i, j-1)$ is already used. A complete pass, scanning in this manner all the grid points, forms one relaxation sweep, or, one iteration, and to implement the process in a computer program only two arrays are needed, as for the one dimensional problem.

2.7 Performance

The number of iterations needed to solve the problem with an iterative method depends on the asymptotic speed of convergence, and the required accuracy of the approximate solution. In this section the convergence speed is analysed for the two relaxation processes described in the previous section. In particular, estimates are given of the number of relaxation sweeps needed to reduce the error by a fixed amount, and to 'solve' the problem.

For an iterative solution process with an asymptotic error amplification factor $\mu = \rho(\mathcal{M})$ the error after M iterations satisfies:

$$|\underline{v}^M| \leq \mu^M |\underline{v}^0| \tag{2.64}$$

Please note that $\underline{v}^M$ indicates the Mth iterand whereas μ^M indicates μ to the power M. If one requires an error reduction from $|\underline{v}^0|$ to $|\underline{v}^0|/c$ with $c > 1$, M should roughly satisfy:

$$M_c > \frac{\ln(c)}{\ln(1/\mu)} \tag{2.65}$$

To quantify M_c for Jacobi and Gauss-Seidel relaxation applied to the model problems the value of μ should be determined. As explained in Section 2.5 the asymptotic speed of convergence of the relaxation process is determined by the spectral radius: the largest eigenvalue of the iteration matrix $\mu = \rho(\mathcal{M})$. If μ is bounded away from 1, independent of the mesh size, only a few iterations are needed to obtain the required error reduction. However, often $\mu = 1 - O(h^2)$, i.e. grid dependent, and the smaller the mesh size (the finer the grid) the closer it is to unity.

To illustrate this behaviour, the eigenvalues of the Jacobi and Gauss-Seidel iteration matrix for the model problems are analysed below. First the matrix $\mathcal{M}$ is derived. For this purpose the matrix A in Equation (2.22) is written as:

$$A = E + D + F \tag{2.66}$$

where E is the matrix containing all elements from A below the diagonal, i.e. $E_{i,j} = A_{i,j}$ for $j < i$ and $E_{i,j} = 0$ otherwise. In the same way F is the matrix containing all elements from A above the diagonal such that $F_{i,j} = A_{i,j}$ for $j > i$ and $F_{i,j} = 0$ otherwise. Finally, D is a matrix containing only the diagonal elements of A: $D_{i,i} = A_{i,i}$ and $D_{i,j} = 0$ if $i \neq j$.

Let $\underline{\tilde{u}}^h$ denote the vector with the approximation before relaxation and $\underline{\bar{u}}^h$ the vector with the new approximation and assume Jacobi relaxation. The equivalent of (2.53) and (2.58) in vector notation is:

$$\underline{\bar{u}}^h = \underline{\tilde{u}}^h + \omega \underline{\delta}^h \tag{2.67}$$

where $\underline{\delta}^h$ satisfies:

$$A\underline{\tilde{u}}^h + D\underline{\delta}^h = \underline{f}^h \tag{2.68}$$

and

$$\underline{\delta}^h = D^{-1}(\underline{f}^h - A\underline{\tilde{u}}^h) \tag{2.69}$$

The error before relaxation is defined as:

$$\underline{\tilde{v}}^h = \underline{u}^h - \underline{\tilde{u}}^h \tag{2.70}$$

and the error after relaxation as:

$$\underline{\bar{v}}^h = \underline{u}^h - \underline{\bar{u}}^h \tag{2.71}$$

Substitution of $\underline{\tilde{u}}^h$ and $\underline{\bar{u}}^h$ from Equations (2.70) and (2.71) in Equation (2.67) and (2.69) and using the fact that $A\underline{u}^h = \underline{f}^h$ gives:

$$\underline{\bar{v}}^h = (I - \omega D^{-1}A)\underline{\tilde{v}}^h \tag{2.72}$$

so for Jacobi relaxation the iteration matrix $\mathcal{M}$ is defined as:

$$\mathcal{M}_J = I - \omega D^{-1}A \tag{2.73}$$

with A from Equation (2.23) and (2.28) for the one and two dimensional problem respectively, and D obtained from A according to Equation (2.66). For the one dimensional model problem it follows that the eigenvectors of $\mathcal{M}_J$ are given by:

$$w_i^k = \sin(\frac{k\pi i}{n}) \tag{2.74}$$

with $1 \leq k \leq (n-1)$. As an example Figure 2.2 shows a typical eigenvector represented on the grid. Note that the eigenvectors are simply harmonic functions represented on the grid, with a wavelength $\lambda_k = 2/k$ (In general $\lambda_k = 2L/k$ if the domain is $[0, L]$).
The eigenvalues of $\mathcal{M}_J$ are given by:

$$\mu_k = 1 - \omega + \omega \cos(\frac{k\pi}{n}) \tag{2.75}$$

for $1 \leq k \leq (n-1)$. For $\omega = 1$ they form pairs with the same value but opposite sign $\mu_k = \mu_{n-k}$, and possibly one eigenvalue 0.

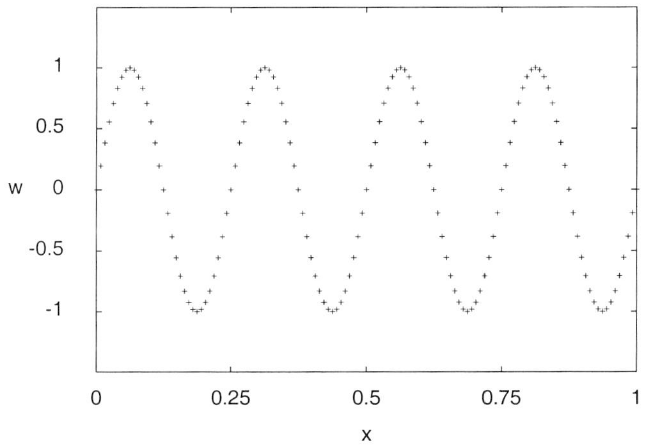

Figure 2.2: *Eigenvector w_k of the Jacobi iteration matrix for for the 1d problem represented on the grid ($n = 128$, $k = 8$).*

For the process to converge one needs $|\mu_k| < 1$ which requires $0 < \omega \leq 1$ and the absolute largest eigenvalue (spectral radius) μ is given by:

$$\mu = \max_{k\in[1,n]}(|(1-\omega) + \omega\cos(\frac{k\pi}{n})|, |(1-\omega) - \omega\cos(\frac{k\pi}{n})|) \tag{2.76}$$

which generally gives:

$$\mu = (1-\omega) + \omega\cos(\frac{\pi}{n}) \approx (1 - \frac{\omega\pi^2}{2n^2}) \tag{2.77}$$

and as $h = 1/n$ one indeed obtains $\mu = 1 - O(h^2)$.

Next consider Gauss-Seidel relaxation in lexicographic order. In that case the iteration matrix $\mathcal{M}$ is given by:

$$\mathcal{M}_G = -(E + \omega^{-1}D)^{-1}((1-\omega^{-1})D + F) \tag{2.78}$$

with D, E, and F obtained from A according to (2.66) and A itself as in Equation (2.23) and (2.28) for the one and two dimensional problem respectively.

The eigenvalues of $\mathcal{M}_G$ can be derived once the eigenvalues of $\mathcal{M}_J$ are known. For $\omega = 1$ each pair of a positive and negative eigenvalue of $\mathcal{M}_J$ gives one eigenvalue of the $\mathcal{M}_G$ that is the square of this Jacobi eigenvalue. The other eigenvalues are all 0.

For the one dimensional model with $\omega = 1$ the eigenvalues of $\mathcal{M}_G$ are thus given by:

$$\mu_k = \cos^2(\frac{k\pi}{n}) \tag{2.79}$$

for $1 \leq k \leq (n-1)/2$ if n is odd and for $1 \leq k \leq n/2 - 1$ if n is even. The remaining eigenvalues are 0. The eigenvectors associated with the non-zero eigenvalues are:

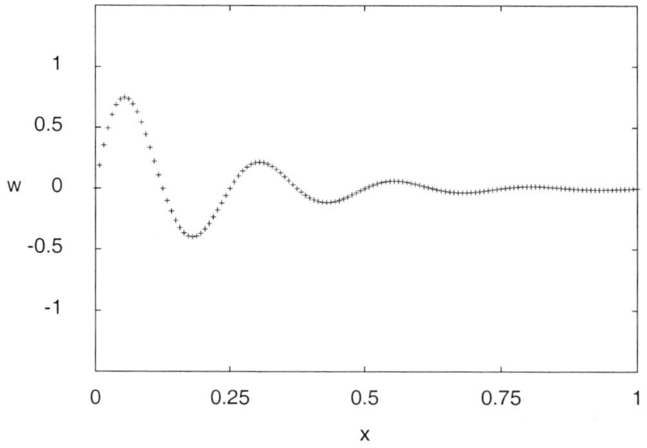

Figure 2.3: *Eigenvector w_k of the Gauss-Seidel iteration matrix for for the 1d problem represented on the grid ($n = 128$, $k = 8$).*

$$w_i^k = \mu_k^{i/2} \sin(\frac{k\pi i}{n}) \tag{2.80}$$

and due to the factor $\mu_k^{i/2}$ they are are not such simple harmonics as for Jacobi. This is illustrated by Figure 2.3.

The largest eigenvalue μ is obtained for $k = 1$:

$$\mu = \cos^2(\frac{\pi}{n}) \tag{2.81}$$

and for $n >> 1$, μ can be approximated by:

$$\boxed{\mu = \cos^2(\frac{\pi}{n}) \approx (1 - \frac{\pi^2}{2n^2})^2 \approx 1 - \frac{\pi^2}{n^2}} \tag{2.82}$$

Consequently, the Gauss-Seidel process converges faster than Jacobi relaxation. However, with $h = 1/n$ one still has $\mu = 1 - O(h^2)$.

Next the two dimensional problem is considered. Using Equations (2.73), and (2.66) with A as defined by (2.23) it can be shown that the eigenvectors of the Jacobi iteration matrix for the 2 dimensional problem are given by:

$$w_{i,j}^{k,l} = \sin(\frac{k\pi i}{n_x}) \sin(\frac{l\pi j}{n_y}) \tag{2.83}$$

Each eigenvector represents a 2d harmonic function represented on the grid, and the largest eigenvalue is associated with the eigenvector containing the harmonic with the largest wavelength: $\lambda_x = 2$ and $\lambda_y = 2$ ($\lambda_x = 2L_x$ and $\lambda_y = 2L_y$ if the domain is $[0, L_x] \times [0, L_y]$ instead of $[0, 1] \times [0, 1]$).

The eigenvalues are:

$$\mu_{k,l} = 1 - \omega + \frac{\omega\alpha^2}{1+\alpha^2}\cos(\frac{k\pi}{n_x}) + \frac{\omega}{1+\alpha^2}\cos(\frac{l\pi}{n_y}) \tag{2.84}$$

where $\alpha = h_y/h_x$. For $h_x = h_y = h$ and $n_x = n_y = n = 1/h$ one obtains:

$$\mu_{k,l} = (1-\omega) + \frac{\omega}{2}\cos(\frac{k\pi}{n}) + \frac{\omega}{2}\cos(\frac{l\pi}{n}) \tag{2.85}$$

The largest eigenvalue is:

$$\boxed{\mu = (1-\omega) + \omega\cos(\frac{\pi}{n}) \approx 1 - \frac{\omega\pi^2}{2n^2}} \tag{2.86}$$

and with $h = 1/n$ again $\mu = 1 - O(h^2)$ follows.

Finally consider Gauss-Seidel relaxation in lexicographic order. The iteration matrix is given by (2.78) with D, E and F obtained from A using Equation (2.66) and A defined by (2.28). As mentioned above the eigenvalues of $\mathcal{M}_G$ can be determined using the eigenvalues of the $\mathcal{M}_J$, and for $\omega = 1$ each pair of eigenvalues of $\mathcal{M}_J$ yields one eigenvalue of the $\mathcal{M}_G$ that is the square of the Jacobi eigenvalue. The other eigenvalues are 0. The non-zero eigenvalues are:

$$\mu_{k,l} = \left(\frac{\alpha^2}{1+\alpha^2}\cos(\frac{k\pi}{n_x}) + \frac{1}{1+\alpha^2}\cos(\frac{l\pi}{n_y})\right)^2 \tag{2.87}$$

for $1 \le k \le (n_x-1)/2$ and $1 \le l \le (n_y-1)/2$ if n_x and n_y are odd, and for $1 \le k \le n_x/2-1$ and $1 \le l \le n_y/2-1$ if n_x and n_y are even. The eigenvectors associated with the non-zero eigenvalues are:

$$w_{i,j}^{k,l} = \mu_{k,l}^{(i+j)/2}\sin(\frac{k\pi}{n_x})\sin(\frac{l\pi}{n_y}) \tag{2.88}$$

and are simply the two dimensional equivalent of (2.80). The asymptotic speed of convergence is determined by the largest eigenvalue which is given by:

$$\mu = \left(\frac{\alpha^2}{1+\alpha^2}\cos(\frac{\pi}{n_x}) + \frac{1}{1+\alpha^2}\cos(\frac{\pi}{n_y})\right)^2 \tag{2.89}$$

which for $h_x = h_y = h$ and $n_x = n_y = n = 1/h$ gives:

$$\boxed{\mu = \cos^2(\frac{\pi}{n}) \approx \left(1 - \frac{\pi^2}{2n^2}\right)^2 \approx 1 - \frac{\pi^2}{n^2}} \tag{2.90}$$

Thus, the largest eigenvalue for the Gauss-Seidel iteration matrix is again smaller than the largest eigenvalue of the Jacobi iteration matrix, which implies that also for the two dimensional model problem Gauss-Seidel relaxation will converge faster. Finally, it should be noted for Jacobi relaxation as well as for Gauss-Seidel relaxation the same values of μ are obtained for the one and two dimensional problem, i.e. assuming $h_x = h_y = h$ and

$n_x = n_y = n$. The asymptotic error reduction per relaxation sweep is $\mu = |\cos(\pi/n)|$ for Jacobi relaxation ($\omega = 1$) and $\mu = \cos^2(\pi/n)$ for Gauss-Seidel relaxation. So, the asymptotic convergence speed apparently only depends on the mesh size of the problem, i.e. the number of points in one (each) dimension of the problem. Qualitatively this can be explained as follows. As the equation in each grid point involves only a few neighbouring grid points, it takes roughly $O(n)$ steps before the information from the boundaries reaches the interior grid points, or before the information from the boundary at $x = 0$ reaches the points near the boundary $x = 1$. Thus it only depends on the number of points in each dimension of the problem, and not on the total number of points. Nevertheless, with decreasing mesh size, the convergence becomes increasingly slow. To illustrate this behaviour Table 2.1 gives μ calculated by Equations (2.77), (2.86) and Equations (2.82), (2.90) as a function of n for Jacobi and Gauss-Seidel relaxation.

Finally, as an illustration μ for Gauss-Seidel relaxation as a function of n is given in Table 2.1. The table confirms that $\mu = 1 - O(n^{-2}) = 1 - O(h^2)$. The table clearly shows that the asymptotic speed of convergence is very slow. Each iteration only reduces the error by a very small amount, and thus it may be expected that the number of iterations to reduce the error by a given factor will be large for large values of n.

	Jacobi		Gauss-Seidel	
n	μ	$1 - \frac{\pi^2}{2n^2}$	μ	$1 - \frac{\pi^2}{n^2}$
16	0.98079	0.98072	0.96194	0.96145
32	0.99518	0.99518	0.99044	0.99036
64	0.99880	0.99880	0.99759	0.99759
128	0.999699	0.999699	0.999398	0.999398
256	0.999925	0.999925	0.999849	0.999849
512	0.9999812	0.9999812	0.9999624	0.9999624
1024	0.99999529	0.99999529	0.999990587	0.999990587

Table 2.1: *Asymptotic error reduction factor μ for Jacobi and Gauss-Seidel relaxation in lexicographic order applied to the one and two dimensional model problem.*

This is shown in Table 2.2 where M_{10} as defined by Equation (2.65) is given, i.e. the required number of iterations to increase the accuracy by one digit, as a function of n. The results show that using Gauss-Seidel relaxation the number of sweeps needed to reduce the error by a factor 10 is half the number of sweeps needed with Jacobi relaxation. However, this is of little comfort as the table shows that each time the mesh size is halved, the required number of iterations becomes 4 times larger. Apparently it is proportional to n^2. Indeed, for Jacobi relaxation, substitution of μ from Equation (2.77) or (2.86) with $\omega = 1$ in Equation (2.65) gives:

$$M_c = \frac{2n^2}{\pi^2} \ln(c) \tag{2.91}$$

Substitution of μ from Equation (2.82) or (2.90) in Equation (2.65) gives:

	Jacobi		Gauss-Seidel	
n	$\ln(1/\mu)$	M_{10}	$\ln(1/\mu)$	M_{10}
16	0.019401721	118	0.03880344	59
32	0.004826904	477	0.009653808	238
64	0.001205269	1910	0.002410539	955
128	0.000301226	7644	0.000602453	3822
256	0.000075300	30578	0.000150601	15289
512	0.000018824	122321	0.000037649	61159
1024	0.000004706	489287	0.000009412	244643

Table 2.2: *Estimated number of relaxation sweeps required to increase the accuracy by 1 digit, as a function of* n *for Jacobi and Gauss-Seidel relaxation with lexicographic ordering applied to the one and two dimensional model problem.*

$$M_c = \frac{n^2}{\pi^2}\ln(c) \tag{2.92}$$

for Gauss-Seidel relaxation. Generally estimates of the amount of work are given in terms of the total number of nodes on the grid. Let N be this total number of nodes on the grid. With $h = 1/n$ in each dimension:

$$N = O(n^d) \tag{2.93}$$

and

$$M_c = O(N^{2/d}\ln(c)) \tag{2.94}$$

So, expressed in the total number of points on the grid, the number of iterations required to reduce the error by a factor of c is $O(N^2)$ for the one dimensional problem and $O(N)$ for the two dimensional problem.

Finally, the amount of work needed to 'solve' the problem is determined, where 'solve' implies reduce the error to the level of the discretization error. After all, our original task is not to exactly solve the discrete problem, but to solve the continuous differential Equation (2.2) or (2.1). By definition even the exact solution to the discrete problem is only an approximation to the desired solution of the continuous problem with an error of magnitude $|\underline{e}^h|$. This implies that it is a waste to solve the discrete problem exactly. Any approximation $\underline{\tilde{u}}^h$ to $\underline{u}^h$ with an error $|\underline{\tilde{v}}^h| = |\underline{u}^h - \underline{\tilde{u}}^h| = O(|\underline{e}^h|)$ is as good an approximation to the differential solution u as the exact solution $\underline{u}^h$ of the discrete problem. If there are no singularities, for a p order discretization $|\underline{e}^h| = O(h^p)$. Thus, assuming an initial error $|\underline{v}^0| = O(1)$ an estimate of the number of iterations M_s required to reduce the error to the level of the discretization error can be obtained from:

$$\mu^{M_s} < h^p \tag{2.95}$$

which gives:

$$M_s > \frac{p\ln(1/h)}{\ln(1/\mu)} \tag{2.96}$$

Using $\mu = 1 - O(n^{-2})$, $p = 2$, for second order discretization, and $h = 1/n$ one obtains:

$$M_s = O(n^2 \ln(n)) \tag{2.97}$$

and expressed in terms of the total number of grid points N:

$$M_s = O(\frac{1}{d}N^{2/d}\ln(N)) \tag{2.98}$$

As each relaxation requires one pass over the grid, i.e. computing an updated value for each point on the grid, the amount of work of one relaxation sweep is $O(N)$ and the total work required to solve the problem is:

$$W = O(M_s N) = O(\frac{1}{d}N^{\frac{2}{d}+1}\ln(N)) \tag{2.99}$$

Thus the total work, and thereby the computing time needed to solve the problem, increases faster than linearly with the number of nodes. For $d = 2$ the cpu time will roughly be proportional to $O(N^2)$ which is the same work count as a direct solver. For $d = 3$ it will be slightly more efficient than a direct solver. However, for the large values of N often required for practical problems the turn around time of the numerical solution algorithm will also be unacceptably large. Summarizing, an algorithm based on these iterative methods alone will not be suitable to efficiently carry out parameter studies for practical problems.

2.8 Local Mode Analysis

In the previous section emphasis was on the asymptotic speed of convergence and its consequences for the number of iterations needed to solve the problem. This was done by determining the iteration matrix and its eigenvalues. For the Jacobi iteration matrix the eigenvectors were harmonic functions represented on the grid, and the largest eigenvalue was associated with the eigenvectors with the largest wavelength on the grid. For the Gauss-Seidel relaxation the situation was a bit more complex and the eigenvectors were no longer such simple harmonics. Moreover, the set of eigenvectors is not complete. It may be clear that in general the computation of the exact eigenvectors and eigenvalues of the iteration matrix may not be simple. However, often the iteration matrices resemble so-called circulant matrices for which the eigenvectors are exactly the harmonics components represented on the grid. As a result to a good approximation the convergence behaviour of iterative processes can be analysed by determining its effect on Fourier components of the error, instead of using the matrix analysis of the preceding paragraphs. The resulting analysis is referred to as "Local Mode Analysis". For a formal justification as to why this analysis gives an accurate prediction the reader is referred to Brandt [17]. Below it is presented as a simple alternative tool that will give essential understanding of the nature of the convergence of the iterative process. Moreover, it lays the foundation for the measures to be taken to overcome the slow convergence.

2.8.1 1d Problem

Consider the discrete one dimensional model problem (2.52). With Jacobi relaxation the new approximation in grid point i is defined by:

$$\bar{u}_i^h = \tilde{u}_i^h + \omega\delta_i^h \tag{2.100}$$

with

$$\delta_i^h = -\frac{h^2}{2}\left[f_i^h - \frac{\tilde{u}_{i-1}^h - 2\tilde{u}_i^h + \tilde{u}_{i+1}^h}{h^2}\right] \tag{2.101}$$

By definition the error before relaxation is given by:

$$\tilde{v}_i^h = u_i^h - \tilde{u}_i^h \tag{2.102}$$

and in the same way the error after the relaxation is defined by:

$$\bar{v}_i^h = u_i^h - \bar{u}_i^h \tag{2.103}$$

Substitution of Equation (2.102) and (2.103) in (2.100) and (2.101), and using the fact that $\underline{u}^h$ is the exact discrete solution, one obtains:

$$\bar{v}_i^h = (1-\omega)\tilde{v}_i^h + \frac{\omega}{2}(\tilde{v}_{i-1}^h + \tilde{v}_{i+1}^h) \tag{2.104}$$

This equation states that the error after relaxation in a given point is a combination of the error before relaxation in the point itself and the error before relaxation in two of its neighbours. For $\omega = 1$ the error after relaxation in point i is the average of the error before relaxation in point $i-1$ and the error before relaxation in the point $i+1$.

The relaxation is a local process, i.e. points several mesh sizes away affect each other exponentially little. Hence for the purpose of analysing the relaxation behaviour, one may regard the grid as embedded in a rectangular domain and expand both $\underline{\tilde{v}}$ and $\underline{\bar{v}}$ in a Fourier series:

$$\tilde{v}_i^h = \sum_{0<|\theta|\leq\pi} \tilde{A}(\theta)e^{\imath\theta i} \tag{2.105}$$

and

$$\bar{v}_i^h = \sum_{0<|\theta|\leq\pi} \bar{A}(\theta)e^{\imath\theta i} \tag{2.106}$$

where θ is the angular frequency, A is the amplitude of the component of angular frequency θ and $\imath = \sqrt{-1}$. The summations are over a subset of $|\theta| \leq \pi$, i.e. only those components that the grid can uniquely represent. Note that only assumptions about the error are made, not about the solution itself. The local nature in terms of the error allows us to disregard the effect of the boundary conditions. These conditions are important for the solution itself. However, their effect on the error is often (for elliptic problems) felt only a few mesh sizes into the domain. If this is not the case a separate analysis can always be done for the effect of relaxation near the boundary. In any case, the present analysis effectively decouples the boundary from the interior. As a result the effect of relaxation on the different Fourier components of the error can be analysed straightforwardly using only the discrete equation for the interior points.

As mentioned earlier, to a good approximation the behaviour of many simple iterative processes can be analysed assuming that relaxation maps a given Fourier component onto itself. In that case it is sufficient to analyse the action of relaxation on a single component. Substition of (2.105) and (2.106) in (2.104) then gives for each θ:

$$\bar{A}(\theta)e^{\imath\theta i} = (1-\omega)\tilde{A}(\theta)e^{\imath\theta i} + \frac{\omega}{2}(e^{-\imath\theta} + e^{\imath\theta})\tilde{A}(\theta)e^{\imath\theta i} \tag{2.107}$$

which gives the amplitude amplification factor of the relaxation for component θ:

$$\mu(\theta) = \left|\frac{\bar{A}(\theta)}{\tilde{A}(\theta)}\right| = |(1-\omega) + \omega\cos(\theta)| \tag{2.108}$$

This amplification factor as a function of θ for $\omega = 1$ is depicted in Figure 2.4. The figure shows that $\mu \approx 1$ for frequencies $\theta \approx 0$ and for $\theta \approx \pi$. This indicates that both these low and high frequency components are responsible for the slow convergence of the scheme. With respect to the high frequency components this can be cured very easily. For example, Figure 2.5 gives $\mu(\theta)$ for $\omega = 2/3$. In that case the bad reduction for high frequency components is replaced by a very efficient reduction. Only for $\theta \approx 0$ the amplification factor is still close to unity. Hence these smooth components will determine the asymptotic speed of convergence. In fact the Local Mode Analysis gives $\mu = 1$ for $\theta = 0$. However, an error component with $\theta = 0$ can not exist on the grid since the error in the boundary points is zero by definition. The lowest frequency that can exist on this grid has $\theta = O(h)$. Using a Taylor series expansion around zero, $\cos(\theta) = 1 - \theta^2/2 + \cdots$, Equation (2.108) gives for this lowest frequency:

$$\mu(h) = |(1-\omega) + \omega(1 - h^2/2)| = |1 - \omega h^2/2| \tag{2.109}$$

thus one finds the error reduction $1 - O(h^2)$ as was found in Section 2.7.

The effect of Gauss-Seidel relaxation on the different components of the error can be analysed in the same way. In that case the error before and after relaxation are related according to:

$$\bar{v}_i^h = (1-\omega)\tilde{v}_i^h + \frac{\omega}{2}(\bar{v}_{i-1}^h + \tilde{v}_{i+1}^h) \tag{2.110}$$

This equation states that the error after relaxation in a given point is a combination of the error before relaxation in the point itself and the error before and after relaxation in two of its neighbours.

Substition of (2.105) and (2.106) in (2.110) then gives for each θ:

$$\bar{A}(\theta)e^{\imath\theta i} = \tilde{A}(\theta)(1-\omega)e^{\imath\theta i} + \frac{\omega}{2}(\bar{A}(\theta)e^{-\imath\theta} + \tilde{A}(\theta)e^{\imath\theta})e^{\imath\theta i} \tag{2.111}$$

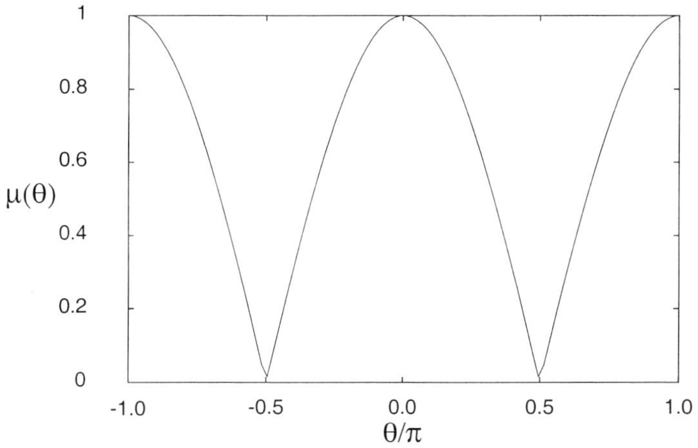

Figure 2.4: *Error amplification factor μ as a function of θ for Jacobi relaxation applied to the one dimensional model problem with $\omega = 1$.*

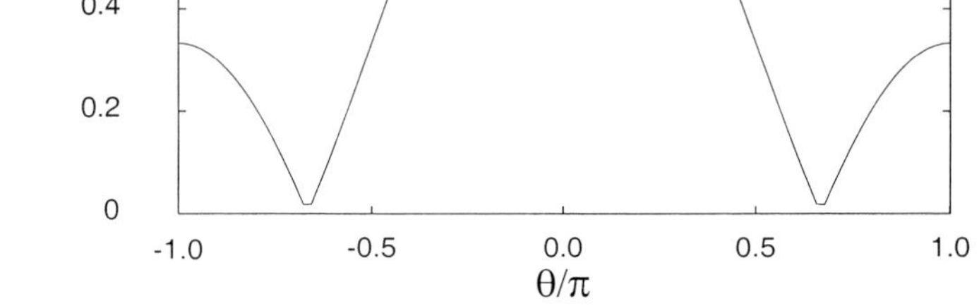

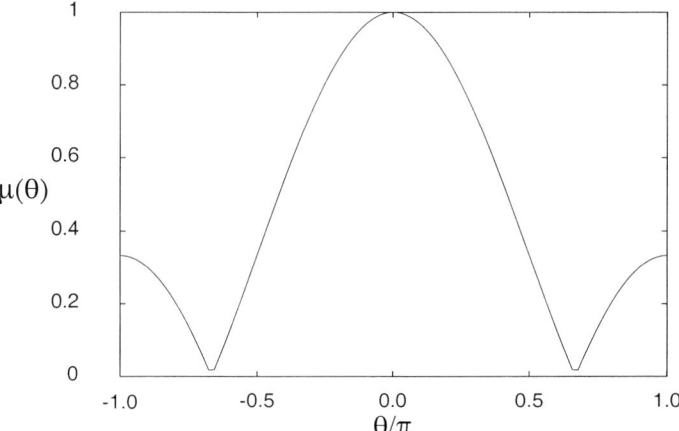

Figure 2.5: *Error amplification factor μ as a function of θ for Jacobi relaxation applied to the one dimensional model problem with $\omega = 2/3$.*

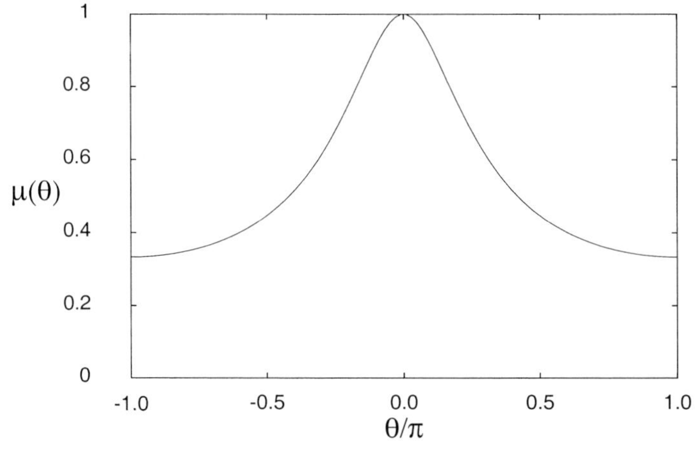

Figure 2.6: *Error amplification factor μ as a function of θ for Gauss-Seidel relaxation with lexicographic ordering applied to the one dimensional model problem with $\omega = 1$.*

which yields:

$$\frac{\bar{A}(\theta)}{\tilde{A}(\theta)} = \frac{2(1-\omega) + \omega e^{\imath\theta}}{2 - \omega e^{-\imath\theta}} \tag{2.112}$$

Subsequently, by taking the absolute value one obtains the amplitude amplification factor of the relaxation for the component θ. For $\omega = 1$ this gives:

$$\boxed{\mu(\theta) = \left|\frac{\bar{A}}{\tilde{A}}\right| = \frac{|e^{\imath\theta}|}{|2 - e^{-\imath\theta}|} = \frac{1}{|2 - \cos\theta + \imath\sin\theta|} = \frac{1}{\sqrt{5 - 4\cos\theta}}} \tag{2.113}$$

Figure 2.6 displays $\mu(\theta)$ for $(-\pi \leq \theta \leq \pi)$. This figure shows that the amplification factor is small for $\theta \approx \pi$, e.g. $\mu(\pi) = 1/3$ and close to unity for θ close to zero. Consequently, the relaxation process reduces high-frequency components $(\pi/2 \leq |\theta| \leq \pi)$ very efficiently. However, low frequency components are only reduced a little. The frequency component with $\theta = 0$ is not reduced at all, according to this figure. In reality this error component can not exist on the domain because of the boundary conditions. The smoothest component actually occurring on the grid is the one with $\theta = O(h)$ for which the asymptotic error amplification factor is $1 - O(h^2)$, which was exactly the asymptotic speed of convergence shown in the previous section.

The analysis presented above leads to the conclusion that Jacobi and Gauss-Seidel relaxation applied to the one dimensional problem are very ineffective in reducing low frequency error components. High frequency error components on the other hand can be reduced very effectively (provided some underrelaxation is used for the Jacobi scheme). The result will be that after some relaxation sweeps, the high frequency error components will have disappeared from the solution, and the remaining error will be smooth compared to the mesh size. These remaining low frequency components will only disappear very slowly and determine the asymptotic speed of convergence. Consequently, the relaxation routine can thus be characterized as a very good 'smoother' (error averager: (2.104)) but it is a bad 'solver'.

2.8.2 2d Problem

The effect of relaxation on the error for the discrete two dimensional Poisson Equation (2.57) can be analysed in the same way. If $\tilde{v}^h_{i,j} = u^h_{i,j} - \tilde{u}^h_{i,j}$ and $\bar{v}^h_{i,j} = u^h_{i,j} - \bar{u}^h_{i,j}$ are used to denote the error before and after the relaxation sweep respectively, the equivalent of Equation (2.104) for Jacobi relaxation applied to this problem is:

$$\bar{v}^h_{i,j} = (1-\omega)\tilde{v}^h_{i,j} + \frac{\omega\alpha^2}{2(1+\alpha^2)}(\tilde{v}^h_{i-1,j} + \tilde{v}^h_{i+1,j}) + \frac{\omega}{2(1+\alpha^2)}(\tilde{v}^h_{i,j-1} + \tilde{v}^h_{i,j+1}) \tag{2.114}$$

where $\alpha = h_y/h_x$. The error before and after relaxation are written as Fourier series:

$$\tilde{v}^h_{i,j} = \sum_{\underline{\theta}} \tilde{A}(\underline{\theta}) e^{\imath(\theta_1 i + \theta_2 j)} \tag{2.115}$$

and

$$\bar{v}^h_{i,j} = \sum_{\underline{\theta}} \bar{A}(\underline{\theta}) e^{\imath(\theta_1 i + \theta_2 j)} \tag{2.116}$$

where $\underline{\theta}$ is (θ_1, θ_2), and the summations are over the square $|\underline{\theta}| = \max(|\theta_1|, |\theta_2|) \leq \pi$. Using the fact that the relaxation maps each component onto itself, substitution of (2.115) and (2.116) in (2.114) yields the amplitude amplification factor for the component $\underline{\theta} = (\theta_1, \theta_2)$:

$$\frac{\bar{A}(\underline{\theta})}{\tilde{A}(\underline{\theta})} = (1-\omega) + \frac{\omega\alpha^2}{2(1+\alpha^2)}\cos(\theta_1) + \frac{\omega}{2(1+\alpha^2)}\cos(\theta_2) \tag{2.117}$$

For $\alpha = h_y/h_x = 1$ this gives:

$$\mu(\underline{\theta}) = \left|\frac{\bar{A}(\underline{\theta})}{\tilde{A}(\underline{\theta})}\right| = \left|(1-\omega) + \frac{\omega}{2}(\cos(\theta_1) + \cos(\theta_2))\right| \tag{2.118}$$

Figures 2.7 and 2.8 show μ as a function of (θ_1, θ_2) for $\omega = 1$ and $\omega = 2/3$ respectively. Figure 2.7 shows that without underrelaxation both low frequencies $\underline{\theta} \approx \underline{0}$ and some specific high frequency components with $(|\theta_1|, |\theta_2|) \approx (\pi, \pi)$ are hardly affected by the relaxation. However, Figure 2.8 shows that some underrelaxation effectively takes care of the high frequency components and as a result the low frequency components remain the ones responsible for the slow asymptotic convergence.

The effect of Gauss-Seidel relaxation on the two dimensional model problem can be analysed in exactly the same way. As can be easily verified for this case one obtains that the error before and after relaxation are related according to:

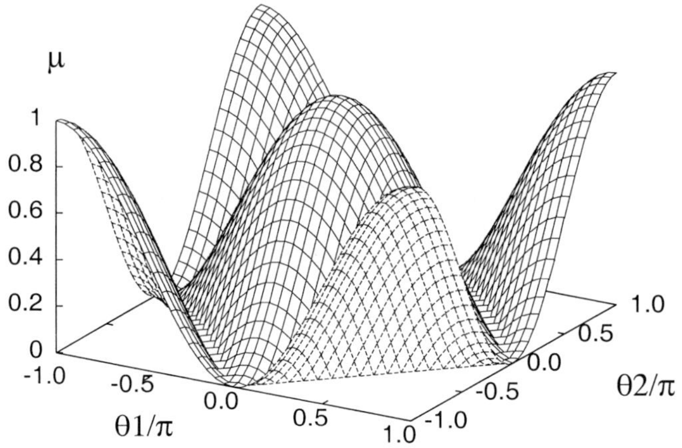

Figure 2.7: *Error amplification factor μ as a function of (θ_1, θ_2) for Jacobi relaxation applied to the two dimensional model problem, with $\omega = 1$.*

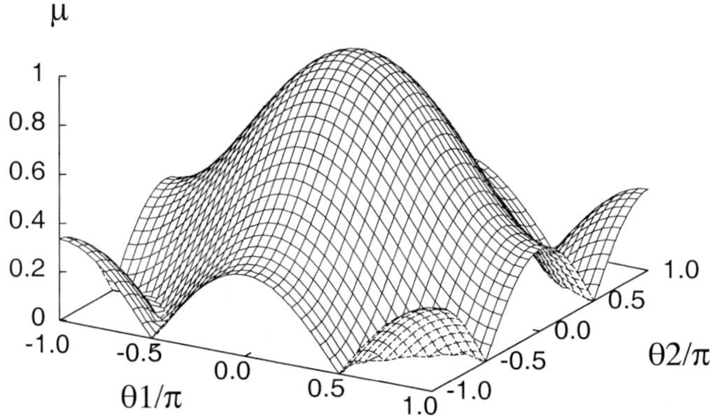

Figure 2.8: *Error amplification factor μ as a function of (θ_1, θ_2) for Jacobi relaxation applied to the two dimensional model problem, with $\omega = 2/3$.*

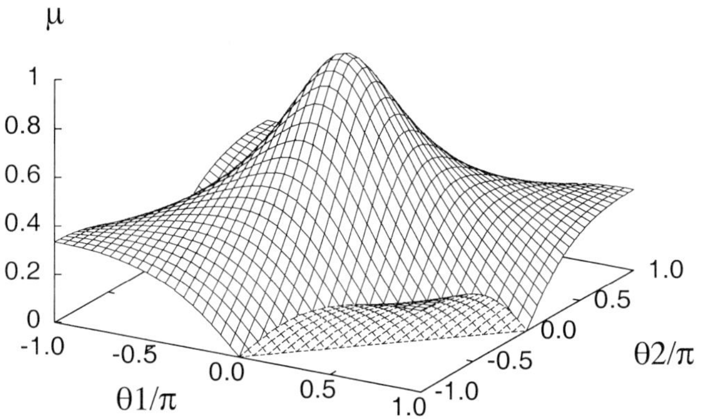

Figure 2.9: *Error amplification factor μ as a function of (θ_1, θ_2) for Gauss-Seidel relaxation with lexicographic ordering applied to the two dimensional model problem, with $\omega = 1$.*

$$\bar{v}_{i,j}^h = (1-\omega)\tilde{v}_{i,j}^h + \frac{\omega\alpha^2}{2(1+\alpha^2)}(\bar{v}_{i-1,j}^h + \tilde{v}_{i+1,j}^h) + \frac{\omega}{2(1+\alpha^2)}(\bar{v}_{i,j-1}^h + \tilde{v}_{i,j+1}^h) \quad (2.119)$$

Substitution of Equations (2.115) and (2.116) and assuming that each Fourier component is mapped onto itself then gives:

$$\frac{\bar{A}(\underline{\theta})}{\tilde{A}(\underline{\theta})} = \frac{2(1+\alpha^2)(1-\omega) + \omega(\alpha^2 e^{\imath\theta_1} + e^{\imath\theta_2})}{2(1+\alpha^2) - \omega(\alpha^2 e^{-\imath\theta_1} + e^{-\imath\theta_2})} \quad (2.120)$$

For $\alpha = h_y/h_x = 1$ and $\omega = 1$ this gives:

$$\mu(\underline{\theta}) = \left|\frac{\bar{A}(\underline{\theta})}{\tilde{A}(\underline{\theta})}\right| = \left|\frac{e^{\imath\theta_1} + e^{\imath\theta_2}}{4 - e^{-\imath\theta_1} - e^{-\imath\theta_2}}\right| \quad (2.121)$$

Figure 2.9 shows μ according to Equation (2.121) as a function of θ_1 and θ_2. From this figure it can be observed that, as for the one dimensional problem, the relaxation hardly affects the low frequency components with $\theta_1 \simeq \theta_2 \simeq 0$. However, high frequency components are very effectively reduced. For example $\mu(\pi,\pi) = 1/3$, $\mu(\pi, 0) = 0$ and $\mu(0,\pi) = 0$.

Summarizing, the Local Mode Analysis for the two dimensional problem shows that Jacobi and Gauss-Seidel relaxation are very ineffective in reducing low frequency error components. High frequency error components on the other hand are reduced very effectively (provided some underrelaxation is used for the Jacobi scheme). The result is once again that after some relaxation sweeps, the high frequency error components will have disappeared from the solution, and the remaining error will be smooth compared to the mesh size. These remaining low frequency components will only disappear very slowly and determine the asymptotic speed of convergence. Consequently, the relaxation routine can thus be characterized as a very good 'smoother' (error averager: (2.114)) but it is a bad 'solver'.

2.9 Conclusion

In this chapter the first steps towards the development of an efficient solver for a partial differential problem were explained. The first step in the numerical solution process is referred to as discretization, it replaces the continuous problem by a discrete problem that can subsequently be solved on a computer. To solve the system one can basically choose between direct and iterative methods. The general conclusion is that the complexity of the solution algorithm, defined as the amount of work needed to obtain the solution, is of $O(N^{\gamma})$ with typically $\gamma \approx 2$ for a two dimensional problem. For realistic problems with large numbers of unknowns N, and when a small discretization error is required, this leads to excessive computing times. In a way both direct methods and iterative methods are inefficient. The actual amount of information in the discrete system, i.e. the actual number of unknowns, is $O(N)$. However, the amount of work invested to retrieve this information is $O(N^{\gamma})$. Clearly in the optimal situation $\gamma = 1$ should be achieved.

For iterative methods it was shown that the cause of the inefficiency lies in the local nature of the process, i.e. the process can very efficiently reduce high frequency error components, but smooth error components are hardly affected. This was illustrated for the one and two dimensional model problems by means of a very simple tool 'Local Mode Analysis'. This particular convergence behaviour is not limited to the model problems for which it was shown here. It applies quite generally to simple iterative schemes applied to the solution of elliptical partial differential problems. The next step is to realize that it is exactly this nature of the process, the good smoothing properties, that can be exploited to obtain a solver of optimal complexity, i.e. where the problem is solved in an amount of work just proportional to the amount of information in the system, i.e. in $O(N)$ operations. This is the subject of the next chapter.

2.10 Advanced Topics

To conclude this chapter some more advanced issues are discussed. They can be omitted at first reading as they are not essential to understand the material presented in the next chapter.

2.10.1 Non Linearity

The model problems are both linear. However, in practice often non-linear problems have to be solved. How to account for the non-linearity in the numerical solution process depends on the method used. First an algorithm based on a direct method is described. Typically the discrete equation for a non linear problem can be written as:

$$L^h\langle \underline{u}^h\rangle = \underline{f}^h \tag{2.122}$$

where $L^h\langle \underline{u}^h\rangle$ is the non-linear discrete operator working on the vector with unknowns $\underline{u}^h$. In each grid point the equation reads:

$$L_i^h\langle \underline{u}^h\rangle = L_i^h\langle u_1^h, ..., u_i^h, ..., u_{N-1}^h\rangle = f_i^h \tag{2.123}$$

and $\underline{f}^h$ denotes the vector with the right hand side values.

Direct Method with Global Newton-Raphson Iteration

The non linear problem can be solved iteratively by means of a Newton-Raphson algorithm. Let $\tilde{u}^h$ denote the current solution. Writing a Taylor expansion around $\tilde{\underline{u}}^h$ one obtains that the exact solution should satisfy:

$$L^h\langle \tilde{\underline{u}}^h\rangle + \mathcal{J}[\underline{u}^h - \tilde{\underline{u}}^h] + h.o.t. = \underline{f}^h \tag{2.124}$$

where $h.o.t.$ stands for higher order terms, and $\mathcal{J}$ denotes the Jacobian matrix. The element in row k and column l of this matrix is defined as the partial derivative of the function $L_k^h\langle \underline{u}^h\rangle$ with respect to the unknown u_l^h:

$$\mathcal{J}_{k,l} = \left.\frac{\partial L_k^h\langle \underline{u}^h\rangle}{\partial u_l^h}\right|_{\underline{u}^h=\tilde{\underline{u}}^h} \tag{2.125}$$

The idea of Newton Raphson iteration is now to replace solving $\underline{u}^h$ from the non-linear problem (2.122) by solving a (hopefully better) approximation $\bar{\underline{u}}^h$ from the linear system of equations that results from (2.124) when the higher order terms are neglected, i.e. to solve $\bar{\underline{u}}^h$ from:

$$L^h\langle \tilde{\underline{u}}^h\rangle + \mathcal{J}[\bar{\underline{u}}^h - \tilde{\underline{u}}^h] = \underline{f}^h \tag{2.126}$$

This equation can be rewritten as:

$$\mathcal{J}[\bar{\underline{u}}^h - \tilde{\underline{u}}^h] = \underline{r}^h \tag{2.127}$$

where $\underline{r}^h$ is the residual vector defined as:

$$\underline{r}^h = \underline{f}^h - L^h\langle \tilde{\underline{u}}^h\rangle \tag{2.128}$$

In terms of a vector of changes $\underline{\delta}^h$ to be applied to $\tilde{\underline{u}}^h$ one can rewrite (2.127) as:

$$\bar{\underline{u}}^h = \tilde{\underline{u}}^h + \underline{\delta}^h \tag{2.129}$$

with $\underline{\delta}^h$ to be solved from:

$$\mathcal{J}\underline{\delta}^h = \underline{r}^h \tag{2.130}$$

Subsequently, using $\bar{\underline{u}}^h$ the process can be repeated to obtain an even better approximation, until a solution has been obtained for which the residuals are sufficiently small.

Summarizing, given an approximation $\underline{\tilde{u}}^h$ one iteration for the non-linear problem consists of:

- Compute the Jacobian matrix $\mathcal{J}$ as defined by Equation (2.125).
- Compute the residual vector $\underline{r}^h$ according to (2.128).
- Solve the vector of changes $\underline{\delta}^h$ from (2.130) by means of a direct solver, e.g. Gaussian Elimination with backsubstitution.
- Apply the changes according to Equation (2.129) or apply them only partly, using some underrelaxation factor ω:

$$\underline{\bar{u}}^h = \underline{\tilde{u}}^h + \omega\underline{\delta}^h \tag{2.131}$$

For a linear problem only one iteration is needed. For that case the higher order terms in Equation (2.124) are zero and this equation can be written as (2.22). This follows immediately as $\mathcal{J} = A$ and one can take $\underline{\tilde{u}}^h = \underline{0}$ such that $\underline{r}^h = \underline{f}^h$.

Summarizing, the non-linear problem is solved in an iterative procedure where at each step a linearized problem is solved by means of e.g. a direct method. Although it is simple and straightforward to apply, this iterative algorithm has several disadvantages. Firstly it has all the disadvantages of a direct method due to the second step, solving the linearized system of equations. However, for the non-linear problem this disadvantage has increased by a certain factor as now the system of equations should not be solved only once, but several times. How many times depends on the convergence of the non-linear iteration. This convergence can not be predicted and should be taken as it comes. In the vicinity of the solution Newton-Raphson is known to converge very rapidly, but to get near the solution many iterations may be needed. Also, if the non-linearity is strong, large underrelaxation may be needed because all equations are solved at once. Summarizing, solving the non-linear problem by means of a direct method incorporated in a Newton-Raphson scheme, the required cpu time is a multiple of the required cpu time to solve the linear problem. This already indicates the inefficiency of the approach.

Relaxation with Local Linearization

If a local relaxation scheme such as Jacobi or Gauss-Seidel relaxation is used, the non-linearity is much easier to account for. For simplicity the notation is restricted to a one dimensional problem. Let the equation for grid point i be written as:

$$L_i^h\langle\underline{u}^h\rangle = L_i^h\langle u_1^h, ..., u_i^h, ..., u_{N-1}^h\rangle = f_i^h \tag{2.132}$$

for $1 \leq i \leq (N-1)$. In most cases the equation to be solved is a differential equation and thus the function $L_i^h\langle\underline{u}^h\rangle$ will only contain unknowns u_j^h in the immediate vicinity of the point i. The relaxation for the non-linear problem is exactly the same as for the linear problem. It consists of one pass over the grid scanning all grid points i in a predetermined e.g. lexicographic order, at each point changing u_i^h such that Equation (2.132) is satisfied. However, as the equation is non-linear it can not be solved exactly. Instead the linearized local equation is solved. Let $\tilde{u}_i^h$ denote the current approximation in grid point i. The new approximation $\bar{u}_i^h$ is solved from:

$$\bar{u}_i^h = \tilde{u}_i^h + \omega\delta_i^h \tag{2.133}$$

with

$$\delta_i^h = \left(\frac{\partial L_i^h \langle \underline{u}^h \rangle}{\partial u_i^h}\right)^{-1}_{\underline{u}^h = \underline{\tilde{u}}^h} r_i^h \tag{2.134}$$

where r_i^h denotes the local residual. For Jacobi relaxation it is defined as:

$$r_i^h = f_i^h - L_i^h \langle \tilde{u}_1^h ..., \tilde{u}_{i-1}^h, \tilde{u}_i^h, \tilde{u}_{i+1}^h, ..., \tilde{u}_{N-1}^h \rangle \tag{2.135}$$

and for Gauss-Seidel relaxation it is the dynamic residual which for lexicographic order is given by:

$$r_i^h = f_i^h - L_i^h (\bar{u}_1^h, ..., \bar{u}_{i-1}^h, \tilde{u}_i^h, \tilde{u}_{i+1}^h, ..., \tilde{u}_{N-1}^h \rangle \tag{2.136}$$

Due to the non-linearity the local equation will not exactly be satisfied after the change is made, not even when $\omega = 1$. One could now repeat the step of solving the local linearized equation several times until the local equation is *exactly* solved. However, this would be inefficient. There is no need to solve the local equation exactly. When relaxing the next point $i+1$ the previous equation is disturbed again anyway and one gains very little from attempting to solve it exactly. It only increases the amount of work done per grid point. As one relaxation sweep doesn't solve the problem exactly anyway, per point only a crude correction is needed and it is sufficient to execute (2.134) once for each grid point. For a linear problem the algorithm automatically reduces to the relaxation as described before. After all, in that case the derivative in Equation (2.134) is simply the coefficient multiplying u_i^h in the discrete operator, e.g. $(-2/h^2)$ for the one dimensional Poisson problem, and Equation (2.134) is thus exactly the same as (2.54).

Summarizing, non-linearity can be accounted for in a very simple way in a local relaxation scheme. Instead of solving the local equation exactly, the linearized local equation is solved. There is no additional global loop needed as was the case for a direct method. All linearization that is needed is a linearization of the local equation. As a result performing a relaxation for the non-linear problem in this way is computationally hardly more expensive than for a linear problem.

2.10.2 Line Relaxation

In Section 2.6, Jacobi and Gauss-Seidel relaxation applied to the one and two dimensional model problems was described. The description was restricted to what is referred to as 'one point' relaxation, i.e. the grid points are treated one by one. Instead of solving one equation changing one unknown at the time one may consider solving the equation for several grid points changing the approximation in this group of grid points simultaneously. This leads to schemes that in general are referred to as 'Block-relaxation' schemes. In that case one relaxation sweep over the grid consists of scanning *blocks* of points. Naturally

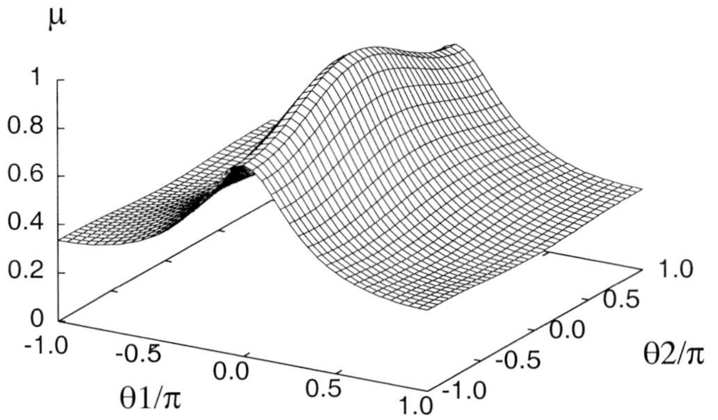

Figure 2.10: *Error amplification factor μ as a function of (θ_1, θ_2) for Gauss-Seidel relaxation with lexicographic ordering applied to the two dimensional model problem, with $\omega = 1$ and $\alpha = h_y/h_x = 0.25$.*

these blocks should not be too large, as solving a system of equations simultaneously is more expensive than solving a single equation. Again one distinguishes Jacobi relaxation and Gauss-Seidel relaxation. In the first case, the new values obtained from blocks previously relaxed are not taken into account. In the second case, the latest approximation available at all points is used when relaxing a certain block. The asymptotic speed of convergence of block relaxation schemes is generally not larger than that of a one point relaxation scheme when applied on a single grid. However, the behaviour in terms of its action on the different Fourier components of the error can be quite different. For example for some cases a one point relaxation scheme will not exhibit the characteristic smoothing behaviour explained in Section 2.8. For example, for one point Gauss-Seidel relaxation in lexicographic order applied to the two dimensional model problem it was derived that the amplification factor of a Fourier component of the error was given by Equation (2.120):

$$\mu(\theta_1, \theta_2) = \left| \frac{2(1+\alpha^2)(1-\omega) + \omega(\alpha^2 e^{\imath\theta_1} + e^{\imath\theta_2})}{2(1+\alpha^2) - \omega(\alpha^2 e^{-\imath\theta_1} + e^{-\imath\theta_2})} \right| \tag{2.137}$$

For $\omega = 1$ and $\alpha = h_y/h_x = 1$, $\mu(\theta_1, \theta_2)$ was given in Figure 2.9. Now assume that the computational domain is such that it is much wider in y direction than in x direction. In that case one would tend to choose $h_y > h_x$ and thus $\alpha > 1$. This is for example the case for the Hydrodynamic Lubrication problem discussed in Chapter 4.

Figure 2.10 shows the error amplification factor according to (2.137) for $\alpha = 4$. The graph shows, as anticipated, that low frequency components of the error ($\theta_1 \approx 0, \theta_2 \approx 0$) are hardly reduced. The graph also shows that the behaviour for high frequency components $\theta_1 \approx \pi$ or $\theta_2 \approx \pi$ is quite different now. High frequency components in the

x direction are still very efficiently reduced by one relaxation. However the reduction of components that are smooth in the x direction and oscillate in the y direction, i.e. the components with $(\theta_1 \approx 0, \theta_2 \approx \pi)$ is less impressive. This effect is even stronger for larger α. In the limit for very large α such components are hardly reduced at all. To explain this effect the reader is reminded that in a point relaxation scheme in each point a change is applied that is proportional to the residual at that point, see Equations (2.59) and (2.63). If h_y is much larger than h_x it can be seen from Equation (2.63) that changes applied to $\tilde{u}^h_{i,j-1}$ will hardly affect the dynamic residual in a point (i,j). In the same way, changes applied to a point (i,j) hardly affect the residual in points $(i,j-1)$ and $(i,j+1)$. In other words, the coupling between the discrete equations in the y direction is very weak. A smooth error must imply a smooth residual and consequently, one can only smooth the error by one point relaxation if neighbouring points appear in the discrete operator (residual) with roughly equal strength. A very simple way to restore the smoothing property of the iterative scheme is to use *line*-relaxation. By this it is meant that instead of scanning the grid point by point the grid is scanned line by line, at each line simultaneously solving the equations of that line. This line relaxation should be carried out along lines of strong coupling, i.e. in the present case one would simultaneously solve lines in x direction.

In its most simple form this relaxation can be described as follows. Let $\underline{\tilde{u}}^h$ denote the current approximation to the solution. Using Gauss-Seidel line relaxation in x direction with lexicographic order implies scanning the gridlines in order of increasing j solving simultaneously for a given j a new approximation $\tilde{u}^h_{i,j}$ for all points i of that line according to:

$$\frac{\bar{u}^h_{i-1,j} - 2\bar{u}^h_{i,j} + \bar{u}^h_{i+1,j}}{h_x^2} + \frac{\bar{u}^h_{i,j-1} - 2\bar{u}^h_{i,j} + \tilde{u}^h_{i,j+1}}{h_y^2} = f^h_{i,j} \tag{2.138}$$

for $1 \le i \le (n_x - 1)$ and j given. Notice that characteristic for Gauss-Seidel relaxation the new solution computed for line $j-1$ is already used when computing the new solution for line j.

How to solve the system of equations for a given line will be discussed at the end of this section. First the smoothing behaviour of this type of relaxation is investigated by means of Local Mode Analysis, see Section 2.8.2. Defining $\underline{\tilde{v}}^h$ as the error before relaxation and $\underline{\bar{v}}^h$ the error after relaxation one obtains from (2.138):

$$\frac{\bar{v}^h_{i-1,j} - 2\bar{v}^h_{i,j} + \bar{v}^h_{i+1,j}}{h_x^2} + \frac{\bar{v}^h_{i,j-1} - 2\bar{v}^h_{i,j} + \tilde{v}^h_{i,j+1}}{h_y^2} = 0 \tag{2.139}$$

Substitution of Equations (2.115) and (2.116) and using that each Fourier component is mapped onto itself, then gives:

$$\mu(\theta_1, \theta_2) = \left| \frac{e^{\imath\theta_2}}{2(1+\alpha^2) - 2\alpha^2\cos(\theta_1) - e^{\imath\theta_2}} \right| \tag{2.140}$$

Figure 2.11 shows $\mu(\theta_1, \theta_2)$ for $\alpha = 4.0$. The graph clearly shows that using line-relaxation along lines in x direction one obtains an iterative scheme that very efficiently reduces the

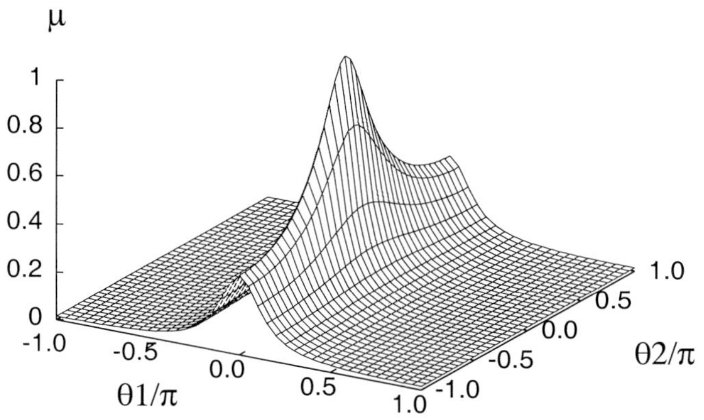

Figure 2.11: *Error amplification factor μ as a function of (θ_1, θ_2) for Gauss-Seidel line relaxation in x direction with lexicographic ordering applied to the two dimensional model problem, with $\omega = 1$ and $\alpha = h_y/h_x = 4$.*

high frequency errors in both directions. In fact, in the x direction it acts almost as an exact solver !

Finally, the question arises of how to perform the line relaxation. As it requires the solution of a system of equations for each line it may seem that such a relaxation is much more expensive than a simple point wise scheme. The system of equations for each line can be written as:

$$A^j \underline{u}_j^h = \underline{f}_j^h \tag{2.141}$$

where $\underline{u}_j^h$ is a vector with the new approximation to $\underline{u}^h$ in all points i of the line j, and $\underline{f}_j^h$ a vector containing the right hand sides in all points i of the line j. The matrix A^j is given by:

$$A^j = \frac{1}{h_y^2} \begin{pmatrix} -2(1+\alpha^2) & \alpha^2 & & & \\ \alpha^2 & -2(1+\alpha^2) & \alpha^2 & & \\ & & \cdot \quad \cdot \quad \cdot & & \\ & & \alpha^2 & -2(1+\alpha^2) & \alpha^2 \\ & & & \alpha^2 & -2(1+\alpha^2) \end{pmatrix} \tag{2.142}$$

As the number of nodes on one line is $n_x - 1$ the system to be solved is a system with a banded matrix. The bandwidth is 1 and thus by means of Gaussian Elimination with backsubstitution it can very efficiently be solved in $O(n_x)$ operations, see Section 2.4. The total work of one relaxation sweep is then to solve (2.142) for $1 \leq j \leq n_y - 1$.

Hence the total work count is $O(n_x n_y) = O(N)$, if N is the number of nodes on the grid. Consequently, one relaxation sweep still costs $O(N)$ operations.

As mentioned above, on a single grid there is little use in applying line relaxation to get a faster convergence. However, if one can design a scheme that efficiently reduces high frequency error components, i.e. smooth the error, this enables a very efficient solution of the problem as will be explained in the next chapter. Line relaxation should not be seen as a solver by itself, but as an alternative to point relaxation in all cases where the discrete equations exhibit a weak coupling in one of the directions of the grid compared to the other directions. The weak coupling may be due to a different mesh size in the x and y direction as is the case here. However, most often the weak coupling is already present in the differential equations and is just inherited by the discrete equations. This is for example the case if our two dimensional model problem (2.1) is replaced by its anisotropic version:

$$\begin{aligned} \frac{\partial^2 u}{\partial x^2} + \epsilon \frac{\partial^2 u}{\partial y^2} &= f(x,y) \quad \text{for} \quad (x,y) \in \Omega \\ u(x,y) &= g(x,y) \quad \text{for} \quad (x,y) \in \partial\Omega \end{aligned} \tag{2.143}$$

For small ϵ the coupling in y direction is much weaker than in the x direction and vice versa for large ϵ the coupling in x direction is much weaker than the coupling in the y direction. To obtain good smoothing, one should apply line relaxation along lines of strong coupling. This implies lines of constant x for small values of ϵ and lines of constant y for large values of ϵ.

Chapter 3

Multigrid

In the previous chapter it was shown that relaxation has good error smoothing properties but bad (slow) solving properties. The central point is that error components with wavelengths which are comparable to the grid size are reduced efficiently. Those with wavelengths which are large compared to the mesh size converge only slowly. This is true for many iterative solvers applied to the solution of the systems of equations resulting from the discretization of partial differential problems. In general one can say that: when for these problems an iterative process converges slowly, the error is smooth and vice versa; if the error is smooth the speed of convergence is low.

Clearly, a smooth error can in principle be accurately represented on a coarser grid, and solved there. This is the basis of the Multigrid solution method. Using coarser grids to solve the smooth error components, a grid independent convergence speed is obtained, which leads to an algorithm that solves the problem in $O(N)$ operations. In this chapter the Multigrid method is described and its performance and efficiency are illustrated using the one and two dimensional model problems introduced in the previous chapter.

3.1 General Principle

Before going into details the general principle behind a Multigrid algorithm is explained (again) in words. Consider the case that some type of relaxation is applied to solution of a partial differential problem discretized on a given grid. For simplicity it is assumed that this grid is a uniform grid and it will be referred to as the *target grid*. Assume that the relaxation process exhibits the characteristic behaviour illustrated in Chapter 2, i.e. high frequency components are reduced efficiently whereas low frequency components converge only slowly. In that case, starting from an arbitrary first approximation the convergence history will be such that initially a relatively large error reduction may be observed (e.g. measured by means of the residual) but after the first few relaxations the convergence speed reduces to a very low asymptotic speed. For the error this implies that in the first few relaxations the high frequency components have almost completely been removed from the error. The smooth components remain for which the error reduction per relaxation is very slow. As explained in Section 2.7 the amplification factor for these components per iteration is typically $1 - O(h^2)$. Thus after the first few iterations the error is dominated by components with a wavelength that is large compared to the mesh size of the grid, or in other words, the error is smooth (on the scale of the mesh size of the grid).

Naturally, if the error is smooth, one does not need a fine grid to accurately represent it. With little loss of accuracy it can be described, and even solved, on a coarser grid. This is exactly what is done in a multigrid solution algorithm. Instead of continuing the relaxation process on the target grid after the convergence speed has become slow, one switches over to a coarser grid to solve an approximation to the smooth remaining error instead. As will be shown in Section 3.2, solving the error requires solving exactly the same equation as the original problem equation, but with a different right hand side vector. Hence, also solving the error must generally be done iteratively, e.g. with the same relaxation process. However, applied to the solution of the error on a coarser grid the process is more efficient for two reasons. Firstly the speed of convergence of the iterative process is larger, because the ratio error wavelength to mesh size is smaller. In particular, an error that is smooth on the target grid is less smooth on the coarse grid. Secondly, the number of nodes on the coarse grid, and thereby the amount of work per relaxation, will be smaller.

Once an approximation to the error has been solved on the coarse grid it can be interpolated to the target grid and used as a correction to improve the approximation to the solution on this grid. Finally, a few additional iterations can be performed to eliminate high frequency errors that may have resulted from the interpolation of the correction. The sequence of pre-relaxations on the target grid, solving the coarse grid problem, interpolation of the smooth error from the coarse grid to the target grid to serve as correction, and a few post-relaxations, is referred to as a *coarse grid correction cycle.*

So far only two grids were used, i.e. a target grid on which the problem is to be solved and an auxiliary coarser grid, to solve the smooth error. If the target grid is a very fine grid, the coarser grid will still be a fine grid. (Typically in multigrid algorithms it is coarser by a factor 2 in each dimension). Assuming that the coarse grid problem is to be solved with the same iterative process it may be clear that after a few initial iterations on this grid convergence will become slow again, now at the speed $(1 - O(H^2))$ if H is the mesh size of the coarse grid. The logical next step is then to realize that now the error of the coarse grid problem has become smooth. Hence, one can introduce an even coarser grid to resolve it, and so on. This principle is applied repeatedly (and recursively) until a grid is reached on which the problem can be solved quickly with a small amount of work.

Eventually, to solve a problem on a given grid a sequence of coarser grids is employed, hence the name Multigrid. The resulting multigrid correction cycle passing from target grid through all the coarser grids and back again can be seen as a way to solve each error component on a grid at which it has a wavelength comparable to the mesh size and thus is efficiently reduced by the iterative process. This cycle will therefore give an error reduction that is determined by the efficiency with which the iterative process reduces high frequency components. As will be shown in detail in this chapter one can obtain a process with an error reduction factor that is independent of the mesh size and has a computational cost of only a small number of iterations on the target grid.

In following sections the principle of the coarse grid correction is formulated in equations. There are two different ways of implementing a Multigrid process. The simplest scheme is called the 'Correction Scheme' (CS) and is only applicable to linear equations. The 'Full Approximation Scheme' (FAS) is suited for both linear and non-linear equations. The next section introduces the Correction Scheme, whereas the Full Approximation Scheme is described in a later section.

In order to keep the notation as simple as possible, only two equidistant grids are used, the fine grid (target grid) has mesh size h in each direction, while the coarse grid (auxiliary grid) has mesh size $H = 2h$ in each direction. This type of coarsening is referred to as *standard coarsening.*

For a one dimensional problem the fine grid index is chosen as $i \in [0, n]$ and the coarse grid index as $I \in [0, n/2]$. This choice allows a simple grid implementation where the coarse grid can be chosen as a subset of the fine grid, for instance: $x_I^{H=2h} = x_{i=2I}^h$. The coarse grid thus forms the subset of fine grid points with even indices. Figure 3.1 shows a fine and coarse grid created in this way.

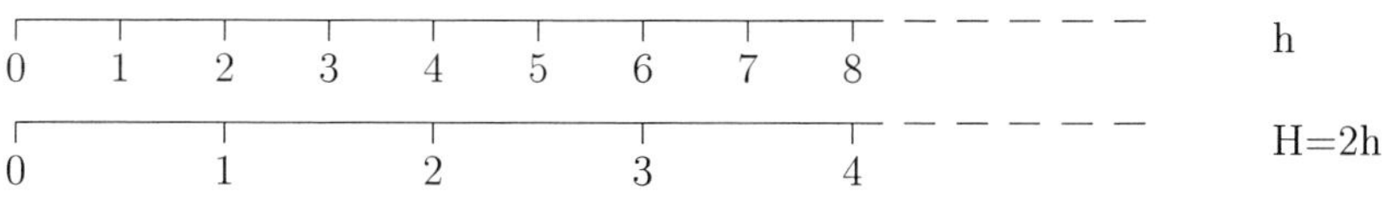

Figure 3.1: *Two one dimensional grids with mesh size h and $H = 2h$.*

For problems in higher dimensions the procedure is the same and it is applied to each direction. For a two dimensional problem this is illustrated in Figure 3.2. If the fine grid points are (x_i^h, y_j^h) with $i \in [0, n_x]$ and $j \in [0, n_y]$, the coarse grid points are $(x_I^H, y_J^H) = (x_{2i}^h, y_{2j}^h)$ with $I \in [0, n_x/2]$ and $J \in [0, n_y/2]$.

The choice of grids explained above is by far the most convenient choice for explaining multigrid algorithms and is suitable for all problems considered in this book. However, for systems of equations from fluid dynamics such as for the Stokes equations and the Navier-Stokes equations, grids based upon a discretization of the equations on the cells of the grid, rather than the points of the grid are generally preferred.

Throughout this chapter uniform grids are assumed. Also this is the most convenient choice for explaining and developing multigrid algorithms. However, it is by no means a restriction. If the problem is such that in some parts of the domain a smaller mesh size is required than in other parts, this can be realized efficiently in a multigrid algorithm using a technique of local grid refinement. This technique yields an effectively non-uniform grid by using a series of uniform grids. It is described in the advanced techniques Section 3.12.1 at the end of this chapter. For simplicity uniform grids are assumed in the following sections.

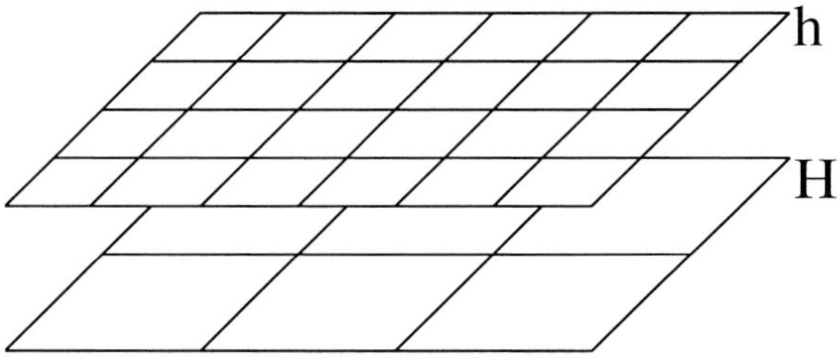

Figure 3.2: *Fine grid with mesh size h and coarse grid with mesh size $H = 2h$ for a two dimensional problem.*

3.2 Correction Scheme

Consider the discrete problem described by:

$$L^h \underline{u}^h = \underline{f}^h \tag{3.1}$$

where L^h is a linear operator such as (2.23) and (2.25). The notation of Chapter 2 is used in which $\underline{u}^h$ and $\underline{f}^h$ denote vectors on a grid with mesh size h. After a small number, say ν_1, relaxation sweeps on this equation, an approximation $\underline{\tilde{u}}^h$ to the solution vector $\underline{u}^h$ is obtained. These relaxations are referred to as pre-relaxations and serve to smoothen the error. Subsequently this smooth error will be solved on a coarser grid. The equation from which the error can be solved is obtained using the residual. Generalising the Equations (2.55) and (2.61) a residual vector $\underline{r}^h$ is defined:

$$\underline{r}^h = \underline{f}^h - L^h \underline{\tilde{u}}^h \tag{3.2}$$

Note that $\underline{r}^h = \underline{0}$ implies $\underline{\tilde{u}}^h = u^h$. After all, by definition, the exact solution satisfies:

$$\underline{f}^h = L^h \underline{u}^h \tag{3.3}$$

Substitution of (3.3) in (3.2) gives:

$$\underline{r}^h = L^h \underline{u}^h - L^h \underline{\tilde{u}}^h \tag{3.4}$$

Since it was assumed that L^h is a linear operator:

$$\underline{r}^h = L^h (\underline{u}^h - \underline{\tilde{u}}^h) \tag{3.5}$$

Note that the difference between the exact solution $\underline{u}^h$ and the approximate solution $\underline{\tilde{u}}^h$ appearing in Equation (3.5) is by definition the error in the approximation $\underline{\tilde{u}}^h$. Thus substitution of the definition of the numerical error vector $\underline{v}^h$:

$$\underline{v}^h = \underline{u}^h - \underline{\tilde{u}}^h \tag{3.6}$$

In Equation (3.5) leads to a relation between the residual and the error:

$$L^h \underline{v}^h = \underline{r}^h \tag{3.7}$$

As for a given $\underline{\tilde{u}}^h$ the residual $\underline{r}^h$ can be computed straightforwardly, the right hand side of Equation (3.7) is a known quantity and an equation has been obtained from which the error can be solved. Note that it is exactly the same equation as the original system to be solved except for a different right hand side. As was mentioned above, many iterative processes have a local nature and thus after a few relaxations the error $\underline{v}^h$ is a smooth function on the fine grid and can be approximated on the coarse grid, i.e by $\underline{v}^H$. This coarse grid error is defined by the representation of Equation (3.7) on the coarse grid:

$$L^H \underline{v}^H = I_h^H \underline{r}^h \tag{3.8}$$

with:

- L^H is a coarse grid approximation to L^h, see Section 3.4.
- I_h^H is a restriction operator from the fine to the coarse grid, see Section 3.3.1.
- $\underline{v}^H$ is a coarse grid approximation to the fine grid error $\underline{v}^h$.

Next Equation (3.8) is solved. Because this equation has exactly the same form as the original Equation (3.1) the same iterative procedure can be used. For the moment it is simply assumed that in some way the solution of $\underline{v}^H$ on the coarse grid is obtained. Subsequently it can be used to correct $\underline{\tilde{u}}^h$ according to:

$$\underline{\bar{u}}^h = \underline{\tilde{u}}^h + I_H^h \underline{v}^H \tag{3.9}$$

Where I_H^h is an interpolation operator from the coarse to the fine grid and $\underline{\bar{u}}^h$ is an improved approximation to the fine grid solution. Finally, a number of relaxation sweeps, say ν_2, is carried out to remove the high frequency errors from $\underline{\bar{u}}^h$, introduced by the interpolation $I_H^h \underline{v}^H$.

3.3 Intergrid Transfers

At two points in the coarse grid correction cycle interaction occurs between the fine and the coarse grid. To compute the right hand side of the coarse grid problem defined by Equation (3.8) the fine grid residual vector must be transferred to the coarse grid. The operator performing this task was denoted with I_h^H and is referred to as a *restriction* operator. Secondly the error solved on the coarse grid $\underline{v}^H$ is only defined in coarse grid points. To use it to correct the fine grid approximation $\underline{\tilde{u}}^h$ in Equation (3.9) requires its transfer to the fine grid. The operator performing this task was denoted with the symbol I_H^h and is referred to as an *interpolation* or *prolongation* operator. In the multigrid literature both terms are used. Here only *interpolation* will be used for the transfer from coarse to fine grid. Below some aspects of these intergrid transfers are explained in detail.

3.3.1 Restriction

The purpose of restriction is to obtain a coarse grid representation $\underline{r}^H$ given a fine grid representation $\underline{r}^h$:

$$\underline{r}^H = I_h^H \underline{r}^h \tag{3.10}$$

First consider a one dimensional problem. In that case $\underline{r}^h$ is a vector of $n+1$ points and $\underline{r}^H$ should be a vector of $n/2+1$ points, and thus I_H^h is a matrix of $n/2+1$ rows and $n+1$ columns. For generality it is assumed that $\underline{r}^h$ has a value for all grid points $i \in [0, n]$. As such the description given below applies to the restriction of an arbitrary vector from a fine grid to a coarse grid.

The simplest way to obtain a coarse grid representation given a fine grid quantity is to simply 'inject' the fine grid quantity to the coarse grid. This type of restriction is referred to as *injection* and for a one dimensional grid it is illustrated in Figure 3.3. The value in the coarse grid point is taken as the value of the coinciding fine grid point.

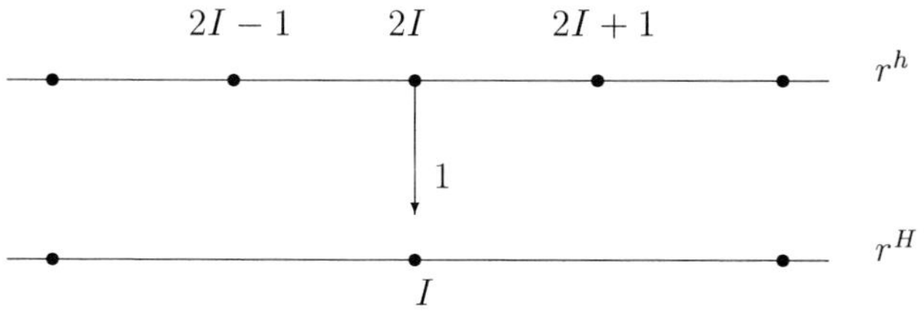

Figure 3.3: *Restriction from fine to coarse grid by injection for a one dimensional problem.*

Thus Equation (3.10) for injection reads:

$$\begin{pmatrix} r_0^H \\ r_1^H \\ \cdot \\ r_I^H \\ \cdot \\ r_{n/2-1}^H \\ r_{n/2}^H \end{pmatrix} = \begin{pmatrix} 1 & 0 & & & & & & & \\ & 0 & 1 & 0 & & & & & \\ & & & \cdot & & & & & \\ & & & & 0 \; 1 \; 0 & & & & \\ & & & & & \cdot & & & \\ & & & & & & 0 & 1 & 0 \\ & & & & & & & 0 & 1 \end{pmatrix} \begin{pmatrix} r_0^h \\ r_1^h \\ \cdot \\ r_{2I}^h \\ \cdot \\ r_{n-1}^h \\ r_n^h \end{pmatrix} \tag{3.11}$$

For the specific case of the transfer of residuals the first and last columns and rows of the system can be left out if the problem at hand has Dirichlet boundary conditions, as is the case for the one dimensional model problem.

Because the equation for each I is the same, the restriction operator can be described in a shorter way using a stencil notation:

$$I_h^H = \begin{bmatrix} 0 & 1 & 0 \end{bmatrix} \tag{3.12}$$

The stencil forms a representation of the equation for grid point I of (3.11):

$$r_I^H = 1 \times r_{2I}^h \tag{3.13}$$

A more sophisticated operator is the so-called *full weighting* operator. The value on the coarse grid is a weighted average of the value in the coinciding fine grid point and its two neighbours. This is illustrated in Figure 3.4.

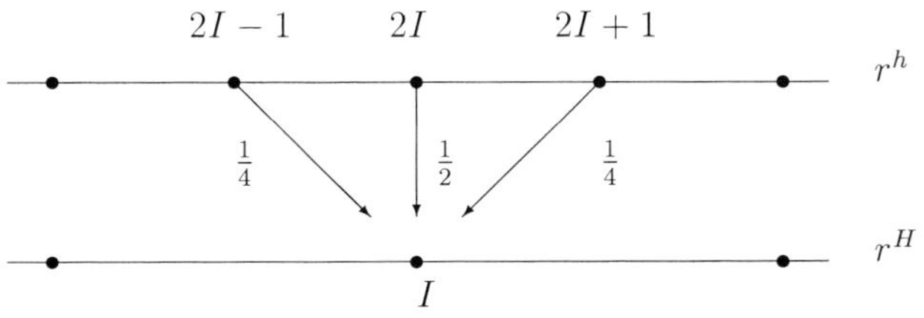

Figure 3.4: *Restriction from fine to coarse grid by full weighting for a one dimensional problem.*

The matrix equation describing the restriction of a fine grid vector $\underline{r}^h$ to the coarse grid by means of full weighting is given by:

$$\begin{pmatrix} r_0^H \\ r_1^H \\ \cdot \\ r_I^H \\ \cdot \\ r_{n/2-1}^H \\ r_{n/2}^H \end{pmatrix} = \frac{1}{4} \begin{pmatrix} 2 & 1 & & & & & & & & & \\ & 1 & 2 & 1 & & & & & & & \\ & & & & \cdot & & & & & & \\ & & & & & 1 & 2 & 1 & & & \\ & & & & & & & & \cdot & & \\ & & & & & & & & 1 & 2 & 1 \\ & & & & & & & & & 2 & 1 \end{pmatrix} \begin{pmatrix} r_0^h \\ r_1^h \\ \cdot \\ r_{2I}^h \\ \cdot \\ r_{n-1}^h \\ r_n^h \end{pmatrix} \tag{3.14}$$

Again it is noted that for the specific case of the transfer of residuals the first and last columns and rows of the system can be left out if the problem at hand has Dirichlet boundary conditions, as is the case for the one dimensional model problem.

Note that also for full weighting the equations for all interior points I are the same and as a result it can be represented by the stencil:

$$I_h^H = \frac{1}{4}\left[\begin{array}{ccc} 1 & 2 & 1 \end{array}\right] \tag{3.15}$$

which reflects the equation for the coarse grid point I of (3.14):

$$r_I^H = (1 \times r_{2I-1}^h + 2 \times r_{2I}^h + 1 \times r_{2I+1}^h)/4 \tag{3.16}$$

The Equations (3.11) and (3.14) generally need not be used in this matrix form in practice. Assuming that two arrays are available, one containing the values r_i^h for all fine grid points, and one to store r_I^H for all coarse grid points all one needs to do is to scan the coarse grid array point by point computing r_I^H according to (3.13) or (3.16). For the boundary points (3.16) can not be used as r_{2I-1}^h or r_{2I+1}^h are not defined. For these points a modified stencil should be used. If boundary values are really needed one could use injection for these points, or, as is done in (3.14) use the stencil without the contribution of the points which lie outside of the domain, r_{2I-1}^h and r_{2I+1}^h respectively for the left and right boundary. This implies that it is assumed that r^h is zero outside the domain. In any case, the main point is that one generally does not need to store the matrix, as the coefficients represented in the stencil are all that is needed.

Injection and full weighting differ in accuracy with which they represent specific aspects of the fine grid vector in the coarse grid result. For example, injection does not preserve the value of a quantity such as:

$$Ir^h = \sum_{j=1}^{n-1} hr_j^h + (r_0^h + r_n^h)h/2 \tag{3.17}$$

which represents an approximation to the integral over the function, whereas full weighting does.

A formal analysis of the restriction process can be carried out using Local Mode Analysis, see Section 3.12.3. For the present section it is sufficient to note that full weighting, because of its averaging, acts as a filter removing high frequency components. The injection operator on the other hand takes such components along to the coarser grid but, because this grid can not resolve them, they alias with low frequency components. This difference between injection and full weighting is illustrated in the Figures 3.5 and 3.6.

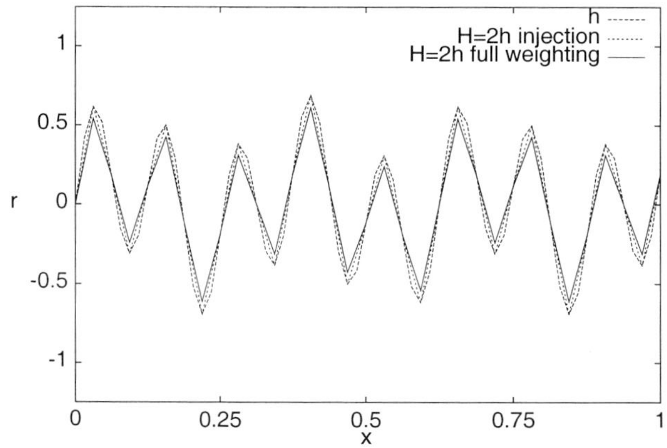

Figure 3.5: *Restriction of a fine grid vector $\underline{r}^h$ to the coarse grid with injection and full weighting for a relatively smooth function. Fine grid: $h = 1/64$, coarse grid: $H = 2h = 1/32$.*

Figure 3.5 shows the restriction of a relatively smooth function from a fine grid to a coarser grid with twice the mesh size. Because the function hardly contains frequencies that the coarse grid can not represent, both injection and full weighting represent it accurately on the coarse grid.

Figure 3.6 shows the restriction of a more oscillatory function from a fine to a coarse grid for the same choice of grids. In this case the function does contain high frequencies that the coarse grid can not represent. Figure 3.6 shows that the coarse grid representation obtained with full weighting indeed accurately approximates the behaviour of the fine grid function with respect to the components that the coarse grid can represent. However, if injection is used this is not the case. Aliasing causes high frequency components from the fine grid that can not be represented accurately on the coarse grid to appear on the coarse grid as low frequency components. Consequently the coarse grid representation obtained by injection forms a poor approximation of the smooth part of the fine grid function.

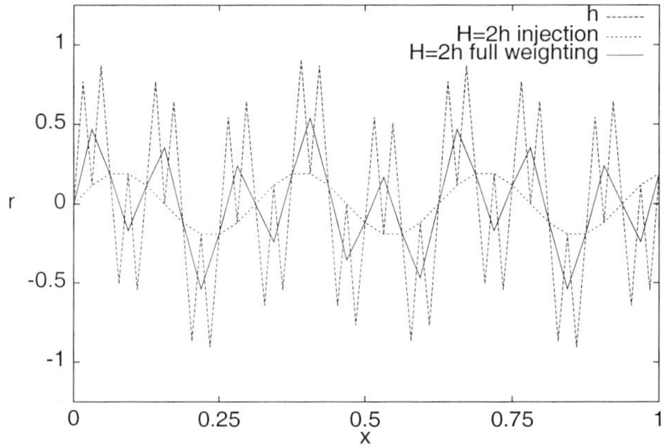

Figure 3.6: *Restriction of a fine grid vector* $\underline{r}^h$ *to the coarse grid with injection and full weighting for a function with high frequency components that can not be seen by the coarse grid. Fine grid:* $h = 1/64$, *coarse grid:* $H = 2h = 1/32$.

In the coarse grid correction process the coarse grid is introduced to solve the smooth error components. These components are associated with the smooth components of the residual. As a result the restriction should thus accurately represent all components of the fine grid residual that can also be represented on the coarse grid. If a large error is made in the representation of these components of the residuals it will lead to a poor coarse grid correction which adversely affects the performance of the cycle. Full weighting is therefore commonly preferred for residual transfers. It ensures that high frequency components still present in the residual on the fine grid after relaxation, even though they may have a small amplitude, are not transferred to the coarser grid.

Restriction operators are ranked according to their order where order refers to the magnitude of the power of h of the error resulting in the coarse grid representation. One even distinguishes a secondary order distinguishing the contribution from high frequency components on the fine grid from the contribution of low frequency components to this error. Injection has order 0 and full weighting has order 2. Higher order restrictions can also be derived. However, for the transfer of residuals they are of little use. For the case of the transfer of a function they may be needed and in that case they are generally derived from a higher order interpolation as will be shown in the following section.

So far the description was restricted to a one dimensional problem. The notation of restriction operators in terms of matrices for two dimensions and higher is cumbersome. For these problems the stencil notation is very convenient. In general for a multi-dimensional problem the stencil of a restriction can be derived from the one dimensional stencil by vector multiplication. For a two dimensional problem the stencil of injection can be written as:

$$I_h^H = \begin{bmatrix} 0 \\ 1 \\ 0 \end{bmatrix} \begin{bmatrix} 0 & 1 & 0 \end{bmatrix} = \begin{bmatrix} 0 & 0 & 0 \\ 0 & 1 & 0 \\ 0 & 0 & 0 \end{bmatrix} \tag{3.18}$$

which is short for

$$r_{I,J}^H = 1 \times r_{2I,2J}^h \tag{3.19}$$

The stencil of the full weighting operator for two dimensions is given by:

$$I_h^H = \frac{1}{4}\begin{bmatrix} 1 \\ 2 \\ 1 \end{bmatrix} \frac{1}{4}\begin{bmatrix} 1 & 2 & 1 \end{bmatrix} = \frac{1}{16}\begin{bmatrix} 1 & 2 & 1 \\ 2 & 4 & 2 \\ 1 & 2 & 1 \end{bmatrix} \tag{3.20}$$

which is short for

$$\begin{aligned} r_{I,J}^H = \; & (4 \times r_{2I,2J}^h + \\ & 2 \times (r_{2I-1,2J}^h + r_{2I+1,2J}^h + r_{2I,2J-1}^h + r_{2I,2J+1}^h) + \\ & 1 \times (r_{2I-1,2J-1}^h + r_{2I+1,2J-1}^h + r_{2I+1,2J+1}^h + r_{2I-1,2J+1}^h))/16 \end{aligned} \tag{3.21}$$

Full weighting for a two dimensional problem is illustrated in Figure 3.7.

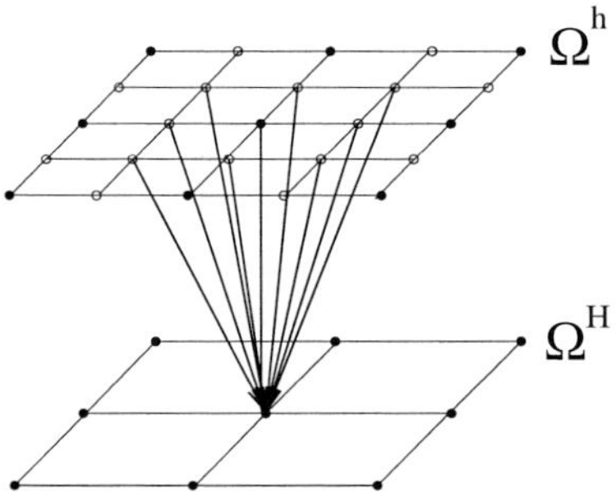

Figure 3.7: *Restriction from fine grid h to coarse grid H with full weighting for a 2 dimensional problem.*

3.3.2 Interpolation

The second inter-grid operator is the interpolation operator I_H^h. Interpolation implies that the value of a fine grid point is determined by interpolation from the values of a group of coarse grid points in its neighbourhood. In matrix form interpolation can be written as:

$$\underline{v}^h = I_H^h \underline{v}^H \tag{3.22}$$

where $\underline{v}^H$ is the coarse grid vector and $\underline{v}^h$ the fine grid vector. For simplicity of description first a one dimensional problem is assumed. In that case, with the grids as assumed here, $\underline{v}^H$ is a vector with $n/2+1$ points, $\underline{v}^h$ is a vector with $n+1$ points and I_H^h is a matrix with $n+1$ rows and $n/2+1$ columns. For this case linear interpolation is illustrated in Figure 3.8. and can be described point by point as:

$$\begin{array}{lcll} v_{2I}^h & = & v_I^H & 0 \leq I \leq n/2 \\ v_{2I+1}^h & = & (v_I^H + v_{I+1}^H)/2 & 0 \leq I < n/2 \end{array} \tag{3.23}$$

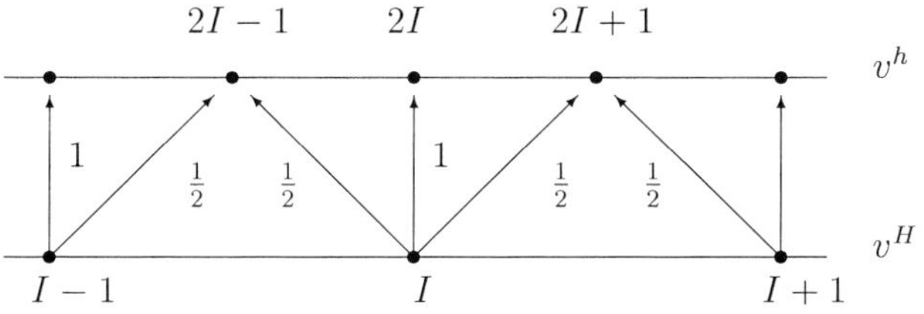

Figure 3.8: *Interpolation from coarse to fine grid by linear interpolation for a one dimensional problem.*

In matrix form the interpolation according to Equation (3.23) can be written as:

$$\begin{pmatrix} v_0^h \\ v_1^h \\ \cdot \\ v_{2I-1}^h \\ v_{2I}^h \\ v_{2I+1}^h \\ \cdot \\ v_{n-1}^h \\ v_n^h \end{pmatrix} = \frac{1}{2} \begin{pmatrix} 2 & & & & & & \\ 1 & 1 & & & & & \\ & & & & & & \\ & & 1 & 1 & & & \\ & & & 2 & & & \\ & & & 1 & 1 & & \\ & & & & & & \\ & & & & & 1 & 1 \\ & & & & & & 2 \end{pmatrix} \begin{pmatrix} v_0^H \\ v_1^H \\ \cdot \\ v_{I-1}^H \\ v_I^H \\ v_{I+1}^H \\ \cdot \\ v_{n/2-1}^H \\ v_{n/2}^H \end{pmatrix} \tag{3.24}$$

In the previous section a short notation for the restriction was obtained using a stencil notation. The same can be done for interpolation. For a restriction the equation for each interior I was the same and thus the non zero elements of a row were sufficient to represent it. However, as can be seen from the pointwise description, interpolation uses two equations. One equation that describes the injection to the fine grid points coinciding with the coarse grid points and one that describes (central) interpolation to the fine grid point in between. As a result the rows of the matrix in (3.24) form sets of 2 repeating equations. The columns on the other hand are the same for each interior I. Therefore, the short description of interpolation is obtained by taking the stencil of non zero elements of the column associated with the coarse grid point I:

$$I_H^h = \frac{1}{2}\left[\begin{array}{ccc} 1 & 2 & 1 \end{array}\right] \tag{3.25}$$

The physical meaning of the stencil becomes clear if the interpolation is viewed as a process where the coarse grid value in point I is given to the coinciding fine grid point, and some points in its neighbourhood with certain weights. The stencil contains the weights with which the coarse grid point I contributes its value to the fine grid points appearing in the stencil. For example, the stencil (3.25) is short for stating that the coarse grid point I gives its entire value to the fine grid point $i = 2I$ and half of its value to the fine grid points $2I-1$ and $2I+1$.

Applying this rule to each interior coarse grid point, gives a result identical to (3.23). After all, the coarse grid point $I+1$ will give its value to the fine grid point $2I+2$ and half of its value to the points $2I+1$ and $2I+3$. As a result the point $2I+1$ indeed collects half the sum of the values of the coarse grid points I and $I+1$ as is needed.

The fact that the interpolation can be characterized by the elements of a single column and restriction by the elements of a single row is not accidental. Comparing the Equations (3.14) and (3.24) shows that:

$$I_h^H = \frac{1}{2}(I_H^h)^T \tag{3.26}$$

In general for any interpolation, a restriction is obtained by taking its transpose multiplied with a factor $(h/H)^d$ if d is the dimension of the problem:

$$I_h^H = \left(\frac{h}{H}\right)^d (I_H^h)^T \tag{3.27}$$

Different orders of interpolation can be used and the order needed depends on the required accuracy. In the same way as restriction, interpolation is ranked according to the power of H of the error that is made. For linear interpolation the error will be $O(H^2)$ and it is referred to as second order. Also for interpolation one distinguishes primary and secondary orders referring to the behaviour for low and high frequency components, see Brandt [15]. In this section order of interpolation will simply refer to the primary order.

For the interpolation of the coarse grid error (3.9), the use of a linear interpolation is sufficient because of the smoothness of the error. In Section 3.7 the solution is also interpolated to the finer grid to serve as a first approximation. In that case a higher order interpolation may be needed. In general higher order interpolations can be derived easily using e.g. Lagrange interpolation formula. However, higher order interpolations introduce the need for extra work due to the different (non central) stencils needed close to the boundaries. For example cubic interpolation from grid H with points $I \in [0, n/2]$ to grid h with points $i \in [0, n]$ is given by:

$$\begin{array}{lcll} v^h_{2I} & = & v^H_I & 0 \le I \le n/2 \\ v^h_{2I+1} & = & (-v^H_{I-1} + 9v^H_I + 9v^H_{I+1} - v^H_{I+2})/16 & 1 \le I < n/2-1 \end{array} \quad (3.28)$$

and

$$\begin{array}{lcl} v^h_1 & = & (5v^H_0 + 15v^H_1 - 5v^H_2 + v^H_3)/16 \\ v^h_{n-1} & = & (5v^H_{n/2} + 15v^H_{n/2-1} - 5v^H_{n/2-2} + v^H_{n/2-3})/16 \end{array} \quad (3.29)$$

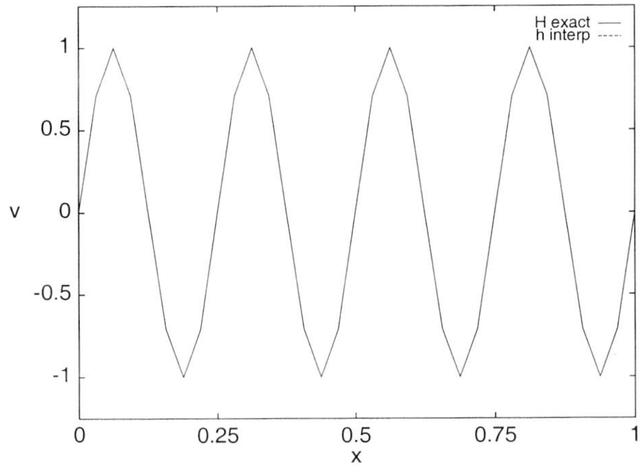

Figure 3.9: *Coarse grid vector $\underline{v}^H$ and result $\underline{v}^h$ after linear interpolation to the fine grid. Fine grid: $h = 1/64$, coarse grid: $H = 2h = 1/32$.*

The role of interpolation in a coarse grid correction cycle can be analyzed using Local Mode Analysis, see 3.12.3. However, in Section 3.2 the remark was made that post relaxations are performed to reduce a possible high frequency error introduced by the interpolation. This is illustrated below. Let $\underline{v}^h$ denote the representation of a function on the fine grid, and $\underline{v}^H$ its representation on the coarse grid. Suppose $\underline{v}^h$ is approximated by $I^h_H\underline{v}^H$, i.e. by interpolating the coarse grid representation to the fine grid. In the fine grid points that coincide with the coarse grid points the error, by definition, will be zero. As a result the error will satisfy:

$$\begin{array}{lcll} (\underline{v}^h - I^h_H\underline{v}^H)_{2I} & = & 0 & 0 \le I \le n/2 \\ (\underline{v}^h - I^h_H\underline{v}^H)_{2I+1} & = & (\gamma H)^p |v^{(p)}| & 0 \le I < n/2 \end{array} \quad (3.30)$$

where p is the order of interpolation i.e. $p = 2$ for linear interpolation, and $p = 4$ for cubic interpolation. γ is a coefficient e.g. $\gamma = 1/2$ for central interpolation. Finally, $|v^{(p)}|$ is the maximum of the pth derivative of v in the interval of interpolation.

As an example Figure 3.9 shows a smooth harmonic function on the coarse grid and the result obtained on a twice finer grid using linear interpolation. Figure 3.10 shows the error $\underline{v}^h - I^h_H\underline{v}^H$. Clearly in all points coinciding with the coarse grid points the error is zero, and in the intermediate points it is non-zero. The result is a combination of a highly

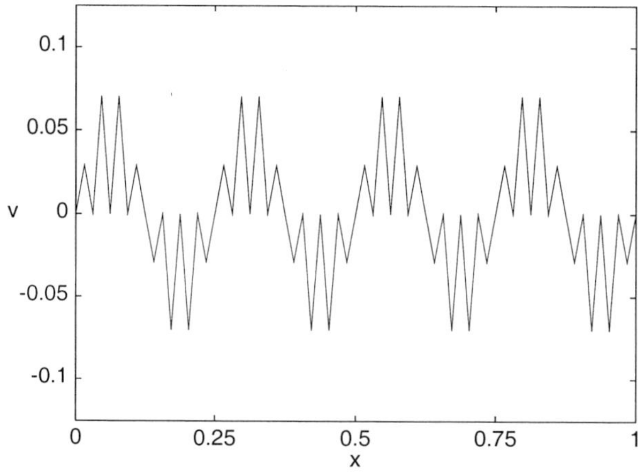

Figure 3.10: *Difference* $\underline{v}^h - I_H^h \underline{v}^H$ *using linear interpolation of the vector* $\underline{v}^H$ *from the coarse grid to the fine grid. Fine grid:* $h = 1/64$, *coarse grid:* $H = 2h = 1/32$.

oscillatory component and a smooth component (stemming from the original function v^H). It is this oscillatory component that was referred to in Section 3.2. For further details see Section 3.12.3.

In the same way as for restriction operators the stencil of interpolation operators for higher dimensions can be derived straightforwardly from the one dimensional operators using vector multiplication. For example the stencil for bi-linear interpolation in two dimensions is given by:

$$I_H^h = \frac{1}{4}\begin{bmatrix} 1 & 2 & 1 \\ 2 & 4 & 2 \\ 1 & 2 & 1 \end{bmatrix} \tag{3.31}$$

As a reminder, for interpolation the stencil contains the weights with which the coarse grid value at point I, J is contributing to the fine grid points in the neighbourhood of the fine grid point $2I, 2J$. Thus the equivalent of the stencil (3.31) is the pointwise description:

$$\begin{array}{lcll} v^h_{2I,2J} & = & v^H_{I,J} & 0 \le I \le n_x/2 \;\; 0 \le J \le n_y/2 \\ v^h_{2I+1,2J} & = & (v^H_{I,J} + v^H_{I+1,J})/2 & 0 \le I < n_x/2 \;\; 0 \le J \le n_y/2 \\ v^h_{2I,2J+1} & = & (v^H_{I,J} + v^H_{I,J+1})/2 & 0 \le I \le n_x/2 \;\; 0 \le J < n_y/2 \\ v^h_{2I+1,2J+1} & = & (v^H_{I,J} + v^H_{I,J+1} + v^H_{I+1,J} + v^H_{I+1,J+1})/4 & 0 \le I < n_x/2 \;\; 0 \le J < n_y/2 \end{array} \tag{3.32}$$

Neither the stencil nor the pointwise description yield the most efficient way to program interpolation in higher dimensions. The most efficient way is to actually perform interpolation one dimension at the time using the one dimensional interpolation stencils (which is actually the meaning of the vector multiplication of stencils in (3.31)). In that case interpolation from grid H to grid h can be done with $O(p)$ operations per fine grid point if p is the order of interpolation.

3.4 Coarse Grid Operator L^H

Now that the inter-grid operators have been discussed, the coarse grid operator L^H can be studied in more detail. The coarse grid serves to approximate and solve the low frequency error components that were slow to converge on the fine grid. Therefore, the coarse grid operator has to form a good approximation to the fine grid operator for those low frequency components.

The coarse grid operator in Equation (3.8) is formed by a restriction of Equation (3.7).

$$I_h^H(L^h \underline{v}^h) = I_h^H \underline{r}^h \tag{3.33}$$

Substitution of $\underline{v}^h = I_H^h \underline{v}^H$ yields:

$$I_h^H(L^h I_H^h \underline{v}^H) = I_h^H \underline{r}^h \tag{3.34}$$

Rearranging the brackets yields:

$$\boxed{(I_h^H L^h I_H^h)\underline{v}^H = I_h^H \underline{r}^h} \tag{3.35}$$

Thus, a possible way of defining L^H is:

$$L^H = I_h^H L^h I_H^h \tag{3.36}$$

Another way of defining the coarse grid operator is by discretising it directly on the coarse grid, as was done on the fine grid. For the one dimensional model problem using full weighting and linear interpolation there is no difference. For this case the stencil of the coarse grid operator defined by Equation (3.36) is:

$$L^H = \frac{1}{H^2}\begin{bmatrix} 1 & -2 & 1 \end{bmatrix} \tag{3.37}$$

which is indeed exactly the stencil of the discrete operator defined on grid H, i.e. it is exactly the stencil (2.24) but with H replacing h.

For the two dimensional model problem using full weighting and linear interpolation the stencil of the coarse grid operator defined by Equation (3.36) is given by:

$$L^H = \frac{1}{4H^2}\begin{bmatrix} 1 & 2 & 1 \\ 2 & -12 & 2 \\ 1 & 2 & 1 \end{bmatrix} \tag{3.38}$$

whereas the stencil of the discrete operator directly defined on the coarse grid is given by Equation (2.27):

$$L^H = \frac{1}{H^2}\begin{bmatrix} & 1 & \\ 1 & -4 & 1 \\ & 1 & \end{bmatrix} \tag{3.39}$$

By means of Taylor expansions (see Chapter 2), it can be verified easily that L^H defined by Equation (3.38) and by Equation (3.39) form an $O(H^2)$ approximation to the continuous operator and thereby also of the operator L^h. Consequently, both will give a sufficiently accurate approximation of the behaviour of L^h for smooth components.

However, as Equation (3.36) yields an operator involving 9 grid points it is computationally more expensive than the operator obtained by straightforward discretization on the coarse grid which uses only 5 grid points. Also, using straightforward discretization on the coarse grid to obtain L^H implies that the same operator is used on all grids, which is more convenient from an implementation point of view.

Only in special cases where the coarse grid operator defined by straightforward discretization on the coarse grid does not give a good approximation of the fine grid operator Equation (3.36) should be used. Generally, and in particular for the problems encountered in this book, the coarse grid operator obtained by straightforward discretization will be sufficiently accurate. Finally, the effect of different coarse grid operators on the performance of a coarse grid correction cycle can be analysed by means of Local Mode Analysis, see Section 3.12.3.

3.5 Coarse Grid Correction Cycle

The process of going from fine to coarse grids and back again can be carried out in two ways. For example one can use a self steering algorithm where the residuals are monitored and if the reduction per relaxation falls below some prescribed limit the change to a coarser grid can be made. However, a far more simple way is to use a fixed pattern of cycling between the fine and the coarse grid. Referring to a grid as a "level" for the case of two grids this leads to the following two-level cycle:

Two-Level Cycle (ν_1, ν_2) *Correction Scheme*

- ν_1 relaxations on grid h to obtain an approximation $\underline{\tilde{u}}^h$ to $\underline{u}^h$.
- Restriction of residuals from grid h to grid H.
- Solution of the grid H problem defined by (3.8):

$$L^H \underline{v}^H = I_h^H \underline{r}^h \tag{3.40}$$

- Interpolation and addition of the correction from grid H to grid h according to Equation (3.9):

$$\underline{\bar{u}}^h = \underline{\tilde{u}}^h + I_H^h \underline{v}^H \tag{3.41}$$

- ν_2 relaxations on grid h.

The cycle is characterized by two parameters ν_1 and ν_2. Typically $\nu_i = 1, 2$ as will be explained in Section 3.6. In a two-level cycle it is assumed that the coarse grid problem is solved exactly, e.g. using many relaxations. However, if the coarse grid still contains many points, the convergence speed on this grid will diminish after a few relaxation sweeps and solving the problem by relaxations only, will be very expensive. However, after a few relaxations on grid H the error on this grid will be smooth. Hence, an efficient way to solve the coarse grid problem will be to use the coarse grid correction cycle itself. This principle can be applied recursively until a grid is reached where solving by relaxation becomes cheap, i.e. a very coarse grid on which the number of nodes is small and relaxation converges rapidly.

The leads to a coarse grid correction cycle involving multiple grids. Generally the grids are referred to as levels and numbered. The coarsest grid is referred to as level 1, and the finest grid as level k. In addition the superscripts h and H are replaced by a superscript indicating the level. For standard coarsening the mesh size on level k is $h^k = 2^{-(k-1)}h^1$, where h^1 is the mesh size on grid 1. A coarse grid correction cycle for level k is given by the following recursive description:

Multi-Level Cycle $(k, \nu_1, \nu_2, \gamma)$ *Correction Scheme*

IF $k > 1$:

- perform ν_1 relaxations on:

$$L^k \underline{u}^k = \underline{f}^k \tag{3.42}$$

  yielding $\underline{\tilde{u}}^k$.

- coarsen to grid $k-1$ to define the level $k-1$ grid problem:

$$L^{k-1} \underline{u}^{k-1} = \underline{f}^{k-1} \tag{3.43}$$

  where

$$\underline{f}^{k-1} = I_k^{k-1} \underline{r}^k \tag{3.44}$$

  with

$$\underline{r}^k = \underline{f}^k - L^k \underline{\tilde{u}}^k \tag{3.45}$$

- perform γ times **Multi-Level Cycle** $(k-1, \nu_1, \nu_2, \gamma)$ yielding an approximation $\tilde{u}^{k-1}$ to the solution of the coarse grid problem defined by Equation (3.44).

- correct the grid k approximation $\underline{\tilde{u}}^k$:

$$\underline{\bar{u}}^k = \underline{\tilde{u}}^k + I_{k-1}^k \underline{\tilde{u}}^{k-1} \tag{3.46}$$

- perform ν_2 relaxations on grid k.

ELSE

- perform ν_0 relaxations on the problem $L^k \underline{u}^k = \underline{f}^k$.

Please remark that the variable on the coarse grids is no longer called v but u.

The multi-grid or multi-level coarse grid correction cycle has three main parameters: ν_1, ν_2, and γ. The ν_1 relaxations serve to smoothen the error before coarsening, the ν_2 relaxations to remove errors introduced by the interpolation of the correction. The parameter γ can be seen as a parameter determining how accurately each coarse grid problem is solved before returning to the fine grid.

To illustrate the above Figure 3.11 shows a flow diagram through the levels of a coarse grid correction cycle for the case of 4 levels and $\gamma = 1$. Relaxations on a grid are indicated by a circle with ν_i. The transfer to a coarser grid is represented by an arrow pointing down. At the coarsest grid (level 1) ν_0 relaxations are performed to solve the equation. The correction step to a finer grid is indicated by an arrow going up. Due to the shape of the flow diagram a cycle with $\gamma = 1$ is referred to as a V-cycle or $V(\nu_1, \nu_2)$ cycle.

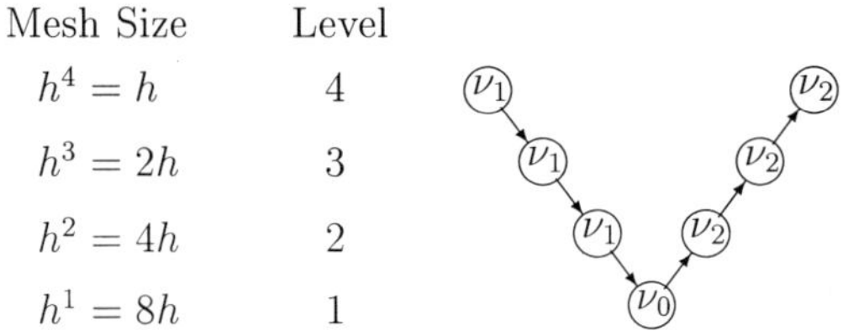

Figure 3.11: *Flow diagram of a* $V(\nu_1, \nu_2)$ *cycle for the case of* 4 *grids (levels).*

For some particular problems, for instance those depending on a global integral such as the force balance Equation (1.13), the amount of work performed in a V-cycle on coarser grids is insufficient to solve the problem accurately enough. For such type of problems cycles with $\gamma = 2$ can be used which are referred to as $W(\nu_1, \nu_2)$ cycles due to the shape of the flow diagram as is illustrated in Figure 3.12.

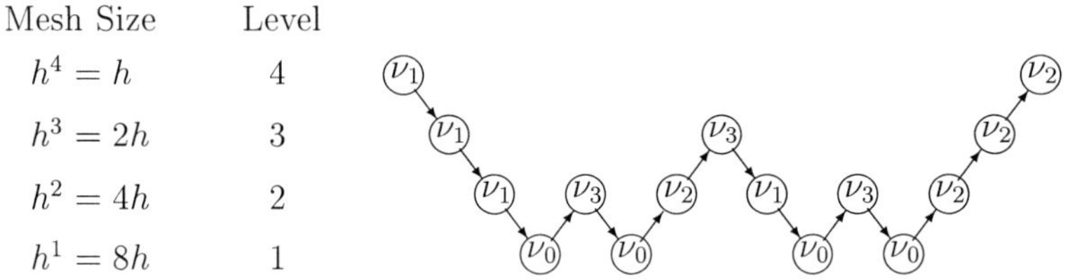

Figure 3.12: *Flow diagram of a* $W(\nu_1, \nu_2)$ *cycle for the case of* 4 *grids (levels).*

where $\nu_3 = \nu_1 + \nu_2$.

Note that a W-cycle starting on level 2 has the same shape as a V-cycle, and visits the coarsest level only once. A W-cycle starting on level 3 visits the coarsest grid twice, and one starting on level 4, four times. Generally, a W-cycle starting on level k will visit the coarsest grid 2^{k-2} times.

3.6 Cycle Performance

Having completed the description of the coarse grid correction cycle the next question is how much is gained in terms of efficiency. How much faster can the discrete problem be solved with these cycles compared to single (target) grid relaxation. To answer this question first an estimate is given for the error reduction that can be obtained with a cycle as a function of the cycle parameters, and the number of cycles needed to solve the problem up to the level of the discretization error is determined. Next the amount of work involved in a cycle is determined and the total work needed to solve the problem by means of coarse grid correction cycles.

3.6.1 Error Reduction per Cycle

An excellent estimate of the error reduction of a coarse grid correction cycle can be obtained using the results of the Local Mode Analysis for the relaxation process on a single grid as presented in Section 2.8. The estimate is based on the assumption that the coarse grid problem is solved exactly. This implies that all error components that can be represented on the coarse grid are also exactly solved in the cycle (as is indeed the case for a two-level cycle). Furthermore the effect of interpolation and restriction is neglected.

Assuming a coarse grid with $H = 2h$ in each direction this implies that one assumes that the components with angular frequency $|\underline{\theta}| \leq \pi/2$ with $|\theta| = \max(|\theta_i|)$ are solved exactly. For a one dimensional problem these are the components $0 < \theta \leq \pi/2$, and for a two dimensional problem (θ_1, θ_2) such that $0 < \theta_1 \leq \pi/2$ *and* $0 < \theta_2 \leq \pi/2$.

Under the assumptions made above, the error reduction obtained in a cycle is determined by the reduction given on the fine grid to the high frequency components, i.e. to those components that the coarse grid can not resolve. For a one dimensional problem they are given by: $\pi/2 \leq |\theta| \leq \pi$, and for a two dimensional problem by (θ_1, θ_2) such that $\pi/2 \leq \theta_1 \leq \pi$ *or* $\pi/2 \leq \theta_2 \leq \pi$.

In Section 2.8 $\mu(\underline{\theta})$ was derived as the amplification factor due to relaxation of an error component $\underline{\theta}$. To quantify the effect of relaxation for the error components that must be resolved by the fine grid the so-called *asymptotic smoothing factor* or *asymptotic smoothing rate* has been defined:

$$\bar{\mu} = \max_{\pi/2 \leq |\underline{\theta}| \leq \pi} \mu(\underline{\theta}) \tag{3.47}$$

with $|\underline{\theta}| = \max(|\theta_i|)$. This asymptotic smoothing factor is the largest amplification factor for all components to be resolved on the fine grid. In a coarse grid correction cycle on the finest grid (the target grid) in total $\nu_1 + \nu_2$ relaxations are performed. Hence the total effect of a coarse grid correction cycle on the error is given by the amplification factor μ_c defined as:

$$\mu_c = \bar{\mu}^{(\nu_1 + \nu_2)} \tag{3.48}$$

In the previous chapter it was shown that for the model problems Gauss-Seidel relaxation was very efficient in reducing high frequency components, and using some underrelaxation

Jacobi relaxation could be made very efficient too. For example for the one dimensional problem using Gauss-Seidel relaxation:

$$\mu(\theta) = \frac{1}{\sqrt{5 - 4\cos(\theta)}} \tag{3.49}$$

From this equation and from Figure 2.6 one observes that the function decreases for increasing θ. Thus the maximum value of μ over the domain $\pi/2 \leq \theta \leq \pi$ will reached on the boundary giving $\bar{\mu} = \mu(\pi/2) = 1/\sqrt{5}$. Moreover, note that $\bar{\mu}$ is a *grid independent* number

As a result a coarse grid correction cycle with $\nu_1 + \nu_2 = 3$ will give an error reduction of a factor $1/\mu_c = 5^{3/2} \simeq 11$. Hence, a single coarse grid correction cycle can reduce the error by roughly an order of magnitude, independent of the mesh size of the grid. The same is true for the two dimensional problem, and also for Jacobi relaxation with underrelaxation applied to the one and two dimensional model problems. The asymptotic smoothing factors for these cases are given in Table 3.1.

The estimates of the error reduction of a cycle as obtained above are accurate as long as ν_1 and ν_2 are not too large, i.e. $\nu_i \leq 2$. For values of $\nu_1 + \nu_2$ which are too large the predictions are too optimistic. If $\nu_1+\nu_2$ is large the effects of the intergrid transfers I_H^h and I_h^H are no longer negligible. Their effect can be analysed using the so-called Two-Level Analysis, see Section 3.12.3. Note however that when a cycle with $\nu_1 + \nu_2 = 3$ already gives an error reduction of an order of magnitude, there is no real need to use large ν_1 or ν_2.

	Jacobi		Gauss-Seidel	
	$\bar{\mu}$	ω	$\bar{\mu}$	ω
1 d Poisson	1/3	2/3	$\sqrt{5}$	1
2 d Poisson	3/5	4/5	1/2	1

Table 3.1: *Asymptotic smoothing factor $\bar{\mu}$ for Jacobi and Gauss-Seidel relaxation applied to the one and two dimensional Poisson problem. ω gives the underrelaxation factor needed to obtain this result.*

The analysis given above assumes that the coarse grid problem is solved exactly. In principle the predicted convergence rate can already be obtained if the coarse grid problem is at least solved with the same accuracy. This implies that the error reduction given on each of the coarser grids should at least be equal to the error reduction of the high frequency components on the fine grid. This requirement is already met in the V-cycle as in this cycle $\nu_1 + \nu_2$ relaxations are performed on each grid. The ν_0 relaxations on the coarsest grid should then also be large enough to yield the same error reduction. This is no problem as on a sufficiently coarse grid $\nu_0 = O(10)$ should already be sufficient. For simple problems such as the one and two dimensional model problem considered here indeed a V-cycle is sufficient to obtain the predicted performance. This will be illustrated by the results presented in Sections 3.9 and 3.10. For more complex problems, e.g. when a global constraint must be solved, W-cycles are more robust.

The estimates obtained using this very simple analysis are not just indications. In fact, the Local Mode Analysis very accurately represents what is happening in the interior of the domain, and in principle the efficiency with which the equations are solved in the interior determines the entire efficiency, see also Section 3.12.2. Hence, the estimates obtained by Local Mode Analysis should not just serve as guidelines of the error reduction that can be obtained with a multilevel cycle. Especially for simple problems they should be seen as targets. If they are not obtained for such problems most often there are still errors in the code.

To conclude this section the number of cycles M_c needed to obtain a fixed error reduction of a factor c ($c > 1$) and the number of cycles needed to solve the problem up to the level of the discretization error M_s is determined.

In Section 2.7 the number of iterations needed to obtain an error reduction of a factor c was given as:

$$M_c > \frac{\ln(c)}{\ln(1/\mu)} \tag{3.50}$$

where μ denoted the error amplification factor of the iterative process. Substitution of $\mu = \bar{\mu}^{\nu_1+\nu_2}$ in this equation gives the number of coarse grid correction cycles needed to obtain this error reduction:

$$M_c > \frac{\ln(c)}{(\nu_1 + \nu_2)\ln(1/\bar{\mu})} \tag{3.51}$$

Because $\bar{\mu}$ is grid independent M_c is independent of the mesh size and thus of the number of grid points. For $\bar{\mu} \approx 0.5$ with $\nu_1 + \nu_2 = 3$, $M_{10} \approx 1$.

The number of cycles needed to solve the problem up to the level of the discretization error follows from Equation (2.95) in the same way, i.e. by substituting $\mu = \bar{\mu}^{\nu_1+\nu_2}$:

$$M_s > \frac{p\ln(1/h)}{(\nu_1 + \nu_2)\ln(1/\bar{\mu})} \tag{3.52}$$

If n is the number of nodes in one dimension $h = 1/n$ and as $\bar{\mu}$ is independent of the mesh size one obtains:

$$M_s = O(\ln(n)) \tag{3.53}$$

and with $N \approx n^d$:

$$M_s = O(\frac{1}{d}\ln(N)) \tag{3.54}$$

3.6.2 Work

As was shown previously, to solve the problem up to the level of the discretization error requires $O(\ln(N))$ cycles. To estimate the amount of work necessary to solve the problem, the amount of work of a single cycle has to be calculated. Therefore, it is useful to introduce the concept of a Work Unit (WU). One WU is the equivalent of the amount of work for one relaxation on the finest (target) level. Assuming the fine grid to contain N

points 1 WU is usually $O(N)$ operations. One operation (relaxing the equation in a single point) generally involves a few multiplications and/or divisions to calculate the current residual, and some multiplications/divisions to change the variable in the current point. For the simpler problems around ten multiplications/divisions are sufficient per point. Having defined the Work Unit and the number of operations per point, it is now possible to define the work consumed by a cycle: W_{cycle}. This work counter W will be used below for different types of cycles, but is expressed in WU's, i.e. expressed in the work necessary for one fine grid relaxation. W_{cycle} for a problem in d dimensions is given by:

$$W_{cycle} = (\nu_1 + \nu_2)\left\{1 + \gamma\left(\frac{h}{H}\right)^d + \gamma^2\left(\frac{h}{H}\right)^{2d} + \gamma^3\left(\frac{h}{H}\right)^{3d} + \cdots\right\} \tag{3.55}$$

For $\gamma(h/H)^d < 1$

$$W_{cycle} \le \frac{(\nu_1 + \nu_2)}{1 - \gamma(h/H)^d} WU \tag{3.56}$$

This estimate neglects the number of operations effected in the intergrid routines, which for simpler problems can amount to one third of the total work. However, even with this crude assumption the analysis indicates the correct order of magnitude.

In practice $H = 2h$ and $\gamma = 1, 2$. For this case one obtains that if $WU = O(N)$ also $W_V = O(N)$ and for $d > 1$ also $W_W = O(N)$. For $d = 1$ and $\gamma = 2$ the series doesn't converge but using the fact that the number of levels will be $O(\ln(N))$ one obtains $W_W = O(N \ln(N))$.

Restricting ourselves to the cases where $\gamma(h/H)^d < 1$ and combining the work estimates of a cycle with the number of cycles needed to solve the problem to the level of the discretization error one obtains:

$$W_{cycles} = O(M_s W_{cycle}) = \frac{M_s(\nu_1 + \nu_2)}{1 - \gamma(h/H)^d} WU = O(\frac{1}{d} N \ln(N)) \tag{3.57}$$

For single grid relaxation it was shown in Chapter 2 that:

$$W_{single\ grid} = O(N^{\frac{2}{d}+1} \ln(N)) \tag{3.58}$$

Comparing these results, for large values of N, clearly a reduction in computing time of many orders of magnitude will be obtained. For example, for the one dimensional Poisson problem with $\nu_1 + \nu_2 = 3$, $W_V = 6 \times WU$. So 1 V-cycle is as expensive as 6 relaxation sweeps on the fine grid. For this case the asymptotic error reduction rate for a V-cycle is $5^{3/2} \approx 11$, thus one additional digit in precision is gained per cycle, independently of the number of grid points. The same is true for the two dimensional Poisson problem with $\nu_1 + \nu_2 = 3$, $W_V = 4WU$ and an asymptotic error reduction rate for a V-cycle with simple Gauss-Seidel relaxation of $(1/2)^3 = 1/8$. Comparing this equivalent of 6 and 4 relaxations to the numbers given in Table 2.2, shows that the gain in computing time will be substantial, already for problems with few points.

Finally, the total amount of work necessary to solve the problem to the order of the discretization error, is not yet independent of the number of points, it still contains the factor $\ln(N)$. Although this factor grows only very slowly with increasing N, a more elegant approach is possible. The factor $\ln(N)$ is caused by the discretization error which becomes smaller when h is decreased (and thus N increased). Because the error in the initial solution stays constant, the number of cycles has to increase in order to reach the level of the discretization error. The solution to this problem lies thus in obtaining a more accurate initial solution. For this a solution on a coarser grid will be used.

3.7 Full MultiGrid

The factor $\ln(N)$ in the work estimate for solving the problem with cycles stems from its occurrence in the number of cycles needed M_s, see Equation (3.52). However, this equation is based on the assumption that the initial error is $O(1)$. If the converged solution on a coarser grid H were used as the starting solution on the fine grid with $h = H/2$, the error in the initial approximation would no longer be $O(1)$ but equal to the discretization error on the coarser grid. For example for a discretization error of order p, the initial error would be $O(H^p)$. In that case the number of cycles required to reach a converged solution is:

$$\mu^M H^p \leq h^p \tag{3.59}$$

and so

$$M \geq \frac{p\ln(2)}{\ln(1/\mu)} \tag{3.60}$$

Since the error reduction rate μ of a cycle is independent of the number of points, the required number of cycles M is also independent of the number of points n. To ensure that a constant number of cycles is sufficient to reach the level of the discretization error, the solution from a coarser grid can be used as a starting solution. This principle is then applied recursively for even coarser grids, utilizing the same coarse grids that are used for convergence acceleration. In the coarse grid correction cycle the coarser grids serve to accelerate convergence on the fine (target) grid. As they are needed for this task anyway one might as well give them an additional function, i.e. to help generate an accurate first approximation on the target grid.

The resulting algorithm is the so-called Full MultiGrid (FMG) algorithm. In multigrid literature it is also referred to as *Nested Iteration*. For the case that $M = 1$, one V-cycle per level, the algorithm is depicted in Figure 3.13. The double circles represent solutions which have converged to the level of the discretization error. The interpolation of a converged solution to the next finer grid is denoted as Π_H^h, a new symbol is used because this interpolation is usually of a higher order than the interpolation of corrections I_H^h. In some cases, as in the two dimensional Poisson problem, this interpolation has to be at least third order accurate.

Note that once a coarse grid correction cycle has been built (programmed) a Full Multigrid algorithm is a very simple extension of the program. A Full Multigrid Algorithm can elegantly be described using recursion:

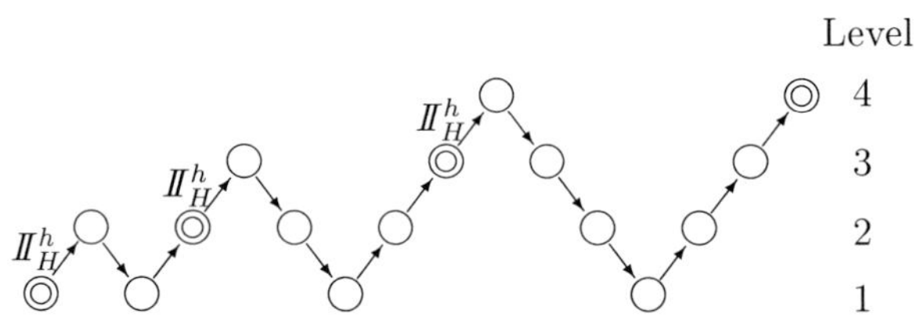

Figure 3.13: *FMG algorithm 1 V-cycle per refinement.*

Full MultiGrid $(k, M, \nu_1, \nu_2, \gamma)$

IF $k > 1$:

- Full Multigrid $(k-1, M, \nu_1, \nu_2, \gamma)$.
- Interpolate the level $k-1$ solution to level k to serve as a first approximation:

$$\underline{\tilde{u}}^k = I\!I^k_{k-1}\underline{\bar{u}}_{k-1} \tag{3.61}$$

- perform M Multi-Level cycles (k, ν_1, ν_2).

ELSE

- perform ν_0 relaxations on the problem $L^k\underline{u}^k = \underline{f}^k$.

The amount of work invested in a FMG algorithm with M cycles per level can easily be estimated:

$$W_{FMG} = M \times W_{cycle} \times \left\{1 + \left(\frac{h}{H}\right)^d + \left(\frac{h}{H}\right)^{2d} + \cdots\right\} \leq \frac{M \times W_{cycle}}{1-(h/H)^d} \tag{3.62}$$

Assuming $\gamma = 1$, $H = 2h$ and using Equation (3.56), this gives:

$$W_{FMG} \leq \frac{M(\nu_1 + \nu_2)}{(1-2^{-d})^2} WU \tag{3.63}$$

This implies that if $O(N)$ operations are needed for one relaxation sweep, the complete problem can be solved to the level of the discretization error in $O(N)$ operations. For the one dimensional Poisson problem with $M = 1$, $(\nu_1 + \nu_2) = 3$ and $d = 1$, $W_{FMG} = 12$, or the total amount of work is equal to the work of twelve relaxation sweeps on the finest grid. For the two dimensional Poisson problem with $M = 1$, $(\nu_1 + \nu_2) = 3$ and $d = 2$, $W_{FMG} = 16/3$, or the total amount of work of an FMG cycle with a single V-cycle is equivalent with the work necessary to perform five and a third relaxation sweeps on the finest grid. Even when allowing for the number operations required by the intergid transfer routines, it is obvious that a substantial gain in computing time is obtained using the FMG cycle.

3.8 Full Approximation Scheme

Having described the Full Multigrid Algorithm all necessary ingredients for a multigrid algorithm for a linear problem have been explained. However, in practice problems are often non-linear. To obtain an algorithm that is suited for non-linear problems too the Correction Scheme used in the coarse grid correction cycle must be replaced by the Full Approximation Scheme. This extension of the algorithm to non-linear operators is treated in this section.

3.8.1 Introduction

The model problem used in this chapter, the Poisson equation, is an example of a linear problem. For linear problems the differential operator working on the sum of two functions can be written as the sum of the operator working on the two functions:

$$L[u+v] = Lu + Lv \tag{3.64}$$

This property is used in the Correction Scheme (going from Equation (3.4) to (3.5)), which was described in the previous section. When the problem is non-linear this scheme can no longer be used, because the above equation is no longer valid. An example of such a non-linear problem, which resembles the Poisson equation is:

$$u\frac{\partial^2 u}{\partial x^2} + u\frac{\partial^2 u}{\partial y^2} = f(x,y) \tag{3.65}$$

A Multigrid algorithm which is applicable to both linear and non-linear operators L, is the Full Approximation Scheme. In the next section the conversion from the Correction Scheme to the Full Approximation Scheme is outlined.

3.8.2 From CS to FAS

A general way of writing a linear or non-linear differential equation is:

$$L\langle u\rangle = f \tag{3.66}$$

where the notation $L\langle u\rangle$ is used to indicate that the operator L depends on the solution u and works on u. After discretization, on a grid with mesh size h, the discrete equations can be written (as before) in vector notation as:

$$L^h\langle \underline{u}^h\rangle = \underline{f}^h \tag{3.67}$$

The solution process is started with an initial solution $\hat{\underline{u}}^h$. After some relaxation sweeps an approximate solution $\tilde{\underline{u}}^h$ is obtained. An indication of the error can be obtained from the residual defined as:

$$\underline{r}^h = \underline{f}^h - L^h\langle\tilde{\underline{u}}^h\rangle \tag{3.68}$$

It is obvious that: $\underline{r}^h = \underline{0} \Leftrightarrow \underline{v}^h = \underline{0} \Leftrightarrow \tilde{\underline{u}}^h = \underline{u}^h$.
The error vector is defined as before:

$$\underline{v}^h = \underline{u}^h - \tilde{\underline{u}}^h \tag{3.69}$$

Substitution of the definition of the error in Equation (3.68) gives:

$$\underline{r}^h = \underline{f}^h - L^h\langle\underline{u}^h - \underline{v}^h\rangle \tag{3.70}$$

When the operator L is linear the Correction Scheme can be used starting at Equation (3.8), approximating the error $\underline{v}^h$ on a coarser grid and using this coarse grid approximation to correct $\tilde{\underline{u}}^h$. In the case that L is a non-linear operator another equation for the error has to be found and another coarse grid variable has to be introduced. Since it is no longer possible to treat the error separately from the solution, the full equation will be used, hence the name Full Approximation Scheme. After a few relaxations on the level with mesh size h an approximation $\tilde{\underline{u}}^h$ is obtained. The definitions of the residual and error vector remain the same as above. Substituting Equations (3.68) and (3.69) in Equation (3.67) gives:

$$L^h\langle\tilde{\underline{u}}^h + \underline{v}^h\rangle = L^h\langle\tilde{\underline{u}}^h\rangle + \underline{r}^h \tag{3.71}$$

Equation (3.71) is the equivalent of Equation (3.8) and is used to approximate the error on the coarse grid. Now the coarse grid discrete problem is written with the coarse grid unknown $\hat{\underline{u}}^H$ as:

$$L^H\langle\hat{\underline{u}}^H\rangle = \hat{\underline{f}}^H \tag{3.72}$$

with

$$\hat{\underline{u}}^H = I_h^H(\tilde{\underline{u}}^h + \underline{v}^h) = I_h^H\tilde{\underline{u}}^h + \underline{v}^H \tag{3.73}$$

and

$$\hat{\underline{f}}^H = L^H\langle I_h^H\tilde{\underline{u}}^h\rangle + I_h^H\underline{r}^h \tag{3.74}$$

Note that: $\underline{r}^H = I_h^H\underline{r}^h = \underline{0} \Leftrightarrow \underline{v}^H = \underline{0}$. Setting the fine grid residuals (artificially) to zero in the residual transfer routine in an actual program is an extremely useful trick when debugging a multigrid program. The zero fine grid residuals should cause all coarser grids to have zero residuals, and thus to generate zero corrections. This effectively decouples the fine grid relaxation from the coarse grid correction cycles and gives the same results as performing only fine grid relaxations. Different behaviour can be easily interpreted and points readily in the direction of the incorrect subroutine.

After a number of relaxations on the coarse grid, a good approximation $\underline{\tilde{u}}^H$ to the coarse grid variable $\underline{\hat{u}}^H$ is found. The fine grid approximation $\underline{\tilde{u}}^h$ is then corrected according to:

$$\underline{\bar{u}}^h = \underline{\tilde{u}}^h + I_H^h(\underline{\tilde{u}}^H - I_h^H \underline{\tilde{u}}^h) \tag{3.75}$$

The fine grid correction in the Full Approximation Scheme (3.75) is similar to the fine grid correction of the Correction Scheme (3.9). In the CS the coarse grid error $\underline{v}^H$ appears explicitly in the equations, in the FAS the coarse grid error is written as $(\underline{\tilde{u}}^H - I_h^H \underline{\tilde{u}}^h)$. In Equation (3.75) an interpolation of the *corrections* is used because in this way the interpolation error goes to zero when $\underline{v}^H \to \underline{0}$. Naturally, the coarse grid variable $\underline{\hat{u}}^H$ is approximated using even coarser grids, employing the Equations (3.72)-(3.75) recursively. For a linear problem (such as our Poisson problem), the FAS and the CS give identical results, up to machine accuracy. The inter-grid operators for interpolation and restriction, that have been introduced in the previous section, remain valid for the Full Approximation Scheme. The same is true for the V and W-cycle described in the previous section.

As a result the only changes that have to be made going from the Correction Scheme to the Full approximation schemes are the following. In the description of the two-level cycle in Section 3.5 Equation (3.40) should be replaced by Equation (3.72) and Equation (3.41) by Equation (3.75). In the same way in the Multi-level cycle description in Section 3.5 Equation (3.44) should be replaced by Equation (3.74) and Equation (3.46) by Equation (3.75) but in this case the superscripts h and H should be replaced by k and $k-1$ respectively.

3.9 1d Results

In this section the results of the one dimensional Poisson program **MG1d** are discussed. The program is given in Appendix D, whereas the data structure is described in Appendix A. The program solves the discretised one dimensional Poisson equation:

$$\frac{\partial^2 u}{\partial x^2} = f(x) \tag{3.76}$$

on the domain $x \in [0,1]$, where the right hand side function is chosen as $f(x) = -4\pi^2 \sin(2\pi x)$ to give a simple solution $u(x) = \sin(2\pi x)$. The boundary conditions are $u(x{=}0) = u(x{=}1) = 0$. The initial solution $u_0(x) = 0$ is used. The coarsest grid contains $4+1$ points, including the boundary points, and the mesh size on this grid is 0.25. The program uses a FMG algorithm where the number of levels, the number of cycles per level and the cycle parameters can be specified. The results are compared with the results obtained when only relaxation is used on a given grid. Three different measures are used to illustrate different aspects of the performance. These measures relate to the convergence of the solution to the exact solution of the discrete problem, and to the convergence of the solution to the exact solution of the continuous differential problem.

3.9.1 Residuals

In Section 2.8 it was shown that the error reduction of the relaxation process depends on the wavelength of the error. High frequency errors which have a wavelength comparable to the mesh size h are reduced efficiently. On the other hand low frequency errors, those that have a wavelength much larger than the mesh size h, are reduced only very little. The error evolution will have a direct relation with the residual evolution, studied in this section. Two different aspects of slow convergence can be distinguished:

- First of all one observes that residuals diminish substantially during the first 'few' relaxations, but that the reduction rate decreases with increasing number of relaxation sweeps. This behaviour can be explained by the initial rapid convergence of high frequency error components followed by the slow convergence of the remaining low frequency error components.

- A second type of slowness concerns the asymptotic convergence speed which tends to decrease for increasing numbers of unknowns N. The explanation of this second slowness can also be understood using local mode analysis, see Section 2.8 and stems from the lowest frequency that can be represented on a certain grid. Due to the boundary conditions $v(x=0) = v(x=1) = 0$, the smoothest error component has an angular frequency on a grid with $N+1$ points of $\theta = \pi/N$. Thus on a finer grid, error components with a lower angular frequency can exist, and the asymptotic convergence speed on this grid will thus be lower.

These two aspects of slow convergence for one level relaxation can also be observed from the next four figures, which show the evolution of the residual norm rn^h

$$rn^h = \frac{1}{N-1} \sum_{i=1}^{N-1} |r_i^h| \tag{3.77}$$

on a grid with mesh size h as a function of the number of Gauss-Seidel relaxation sweeps s and the number of points N. The single level relaxation reduces the residual norm rn^h, but this reduction diminishes rapidly with increasing values of s and N.

To quantify the convergence speed on the levels with 64, 256 and 1024 points, a residual norm reduction by a factor of 10 is chosen. For this particular case the reduction of the residual norm from 25 to 2.5 is studied. For the grid with 64 points the residual norm of 2.5 is reached after approximately 250 relaxation sweeps. On a grid with 256 points approximately 4000 sweeps are needed, whereas on the finest grid with 1024 points around 60000 sweeps are required.

These numerical results can be compared to the theoretical predictions of Table 2.2. This table predicts that respectively 955, 15289 and 244643 relaxation sweeps are required. The numerical results give a number of required relaxation sweeps that increases by a factor of 16, indicating a dependence of $O(N^2)$ as predicted theoretically. However, the absolute number of relaxation sweeps differs for each grid by a factor of roughly four. This factor can be explained qualitatively, since the asymptotic error reduction is smaller than the measured one, which includes the rapid initial convergence. As can be seen from

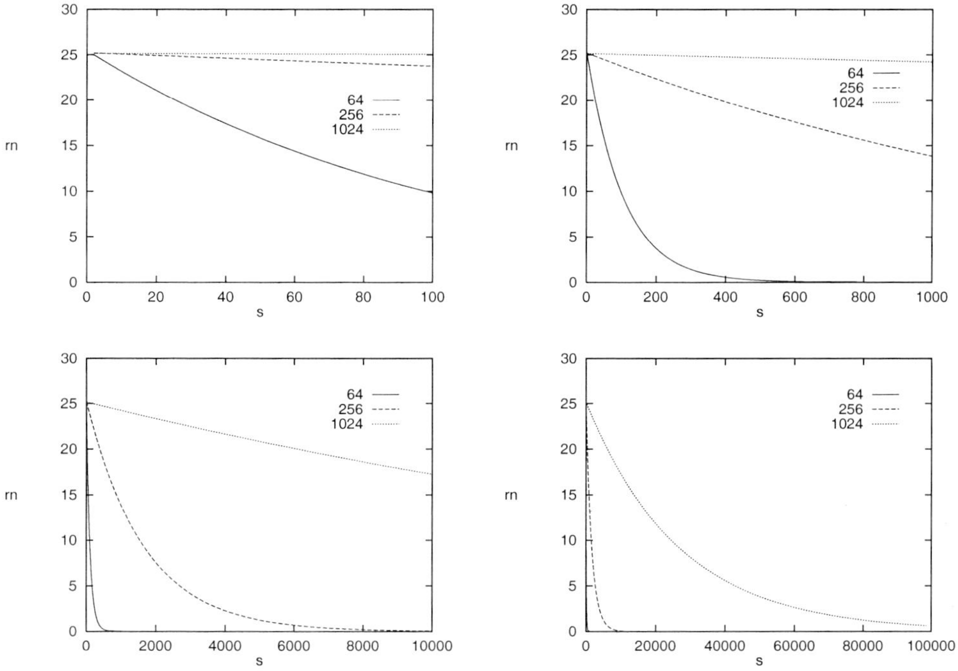

Figure 3.14: *Residual norm rn^h as a function of the number of relaxation sweeps s for* $N = 64$, $N = 256$ *and* $N = 1024$.

Figure 3.14 the residual reduction speed diminishes notably for residual norms smaller than 5.

The results of the single grid relaxation are to be compared to the Multigrid results which are given in Table 3.2 and Figure 3.15. The FMG technique has been used to generate accurate initial solutions on a certain level, by interpolation from a converged solution on the next coarser level. The V-cycles used are V(2,1) cycles, with 2 relaxation sweeps when coarsening, and a single relaxation sweep when refining.

From this table and the figure it can be concluded that the two different types of slowness have disappeared: average residual norm reduction factors of 14, 13 and 11 are found for the levels 5, 7 and 9. These factors should be compared with the theoretical result from the smoothing rate analysis, see Section 3.6, predicting a factor of 11. The level 9 result coincides with this value. On coarser grids (5 and 7), convergence is superior, due to the convergence speed on the finest grid which is not yet negligible. On even finer grids the value deteriorates slightly, (9 is obtained on level 11), due to the limited machine precision. In order to perform a detailed analysis of this limit behaviour, V-cycles should be used, starting from a random initial solution. Figure 3.15 shows graphically that the residual norm reduction factor is roughly constant on the three different grids, and that it is roughly constant as a function of the number of cycles.

V-cycle	level 5	level 7	level 9
	$N = 64$	$N = 256$	$N = 1024$
1	$2.19 \cdot 10^{-2}$	$1.76 \cdot 10^{-3}$	$1.34 \cdot 10^{-4}$
2	$9.92 \cdot 10^{-4}$	$1.52 \cdot 10^{-4}$	$1.57 \cdot 10^{-5}$
3	$7.02 \cdot 10^{-5}$	$1.21 \cdot 10^{-5}$	$1.30 \cdot 10^{-6}$
4	$4.24 \cdot 10^{-6}$	$8.63 \cdot 10^{-7}$	$1.08 \cdot 10^{-7}$
5	$2.96 \cdot 10^{-7}$	$5.56 \cdot 10^{-8}$	$8.18 \cdot 10^{-9}$
6	$2.09 \cdot 10^{-8}$	$3.69 \cdot 10^{-9}$	$6.04 \cdot 10^{-10}$
7	$1.55 \cdot 10^{-9}$	$2.74 \cdot 10^{-10}$	$6.93 \cdot 10^{-11}$
8	$1.17 \cdot 10^{-10}$	$2.08 \cdot 10^{-11}$	–
9	$9.43 \cdot 10^{-12}$	–	–
10	$7.63 \cdot 10^{-13}$	–	–

Table 3.2: *Residual norm as a function of three different levels and of the number of V-cycles, using an FMG algorithm.*

Comparing Figures 3.15 and 3.14 it can be observed that the gain is indeed very important, the more so for very fine grids: for 1024 points, an FMG algorithm using 1 V-cycle (cost: 12 WU) is much more efficient than 100000 fine grid relaxation sweeps (cost: 100000 WU). This implies a gain in convergence speed of more than 10000!

3.9.2 Errors

For this model problem the analytical solution is known; $u(x) = \sin(2\pi x)$, and therefore it is possible to study the error reduction generated by one level relaxation and by V-cycles. Therefore, the error norm is introduced :

$$en^h = \frac{1}{N-1} \sum_{i=1}^{N-1} |\tilde{u}_i^h - u(x_i^h)| \tag{3.78}$$

This error norm consists of two different components:

the numerical error

$$\frac{1}{N-1} \sum_{i=1}^{N-1} |v_i^h| = \frac{1}{N-1} \sum_{i=1}^{N-1} |u_i^h - \tilde{u}_i^h| \tag{3.79}$$

and the discretization error

$$\frac{1}{N-1} \sum_{i=1}^{N-1} |e_i^h| = \frac{1}{N-1} \sum_{i=1}^{N-1} |u(x_i^h) - u_i^h| \tag{3.80}$$

Please note that the numerical error is related to the residual, and can therefore be reduced to zero, whereas the discretization error depends on the mesh size h and is thus fixed on a certain grid. It only disappears in the limit case of very fine grids: $h \to 0$.

The next four figures represent the evolution of the error norm en^h as a function of the number of single level relaxations s and of the number of points N.

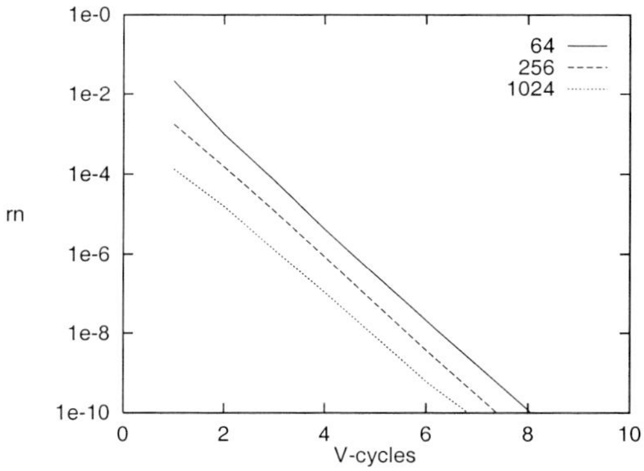

Figure 3.15: *Residual norm as a function of the number of V-cycles for $N = 64$, $N = 256$ and $N = 1024$, using FMG.*

The evolution of the error norm is similar to the evolution of the residual norm studied in the previous section. The reduction of the error norm reduces with increasing values of s and N, however, the error norm does not go to zero, as was the case for the residual norm. It tends asymptotically to the discretization error, as can be clearly observed from the next table!

s	level 5	level 7	level 9
	$N = 64$	$N = 256$	$N = 1024$
10^1	$5.87 \cdot 10^{-1}$	$6.35 \cdot 10^{-1}$	$6.37 \cdot 10^{-1}$
10^2	$2.45 \cdot 10^{-1}$	$6.01 \cdot 10^{-1}$	$6.35 \cdot 10^{-1}$
10^3	$3.28 \cdot 10^{-3}$	$3.51 \cdot 10^{-1}$	$6.14 \cdot 10^{-1}$
10^4	$5.19 \cdot 10^{-4}$	$2.20 \cdot 10^{-3}$	$4.37 \cdot 10^{-1}$
10^5	$5.19 \cdot 10^{-4}$	$3.21 \cdot 10^{-5}$	$1.48 \cdot 10^{-2}$
10^6	$5.19 \cdot 10^{-4}$	$3.21 \cdot 10^{-5}$	$2.00 \cdot 10^{-6}$
10^7	$5.19 \cdot 10^{-4}$	$3.21 \cdot 10^{-5}$	$2.00 \cdot 10^{-6}$

Table 3.3: *Error norm as a function of three different levels and of the number of relaxations.*

The level 7 discretization error is sixteen times smaller than the level 5 discretization error, and the same is true for level 9 compared to level 7. As the discretization is second order accurate and there are four times as many points, thus the meshsize h is divided by 4, this ratio confirms the theoretical accuracy of the discretization ($O(h^2)$). Note that $O(10^4)$ relaxation sweeps are needed for level 5 compared with $O(10^5)$ and $O(10^6)$ relaxation sweeps on levels 7 and 9.

Once the discretization error level attained, there is no point continuing the relaxation process on a certain level. To reduce the total error even further, the solution process has to be continued on a finer grid. Apart from this small difference, the error norm

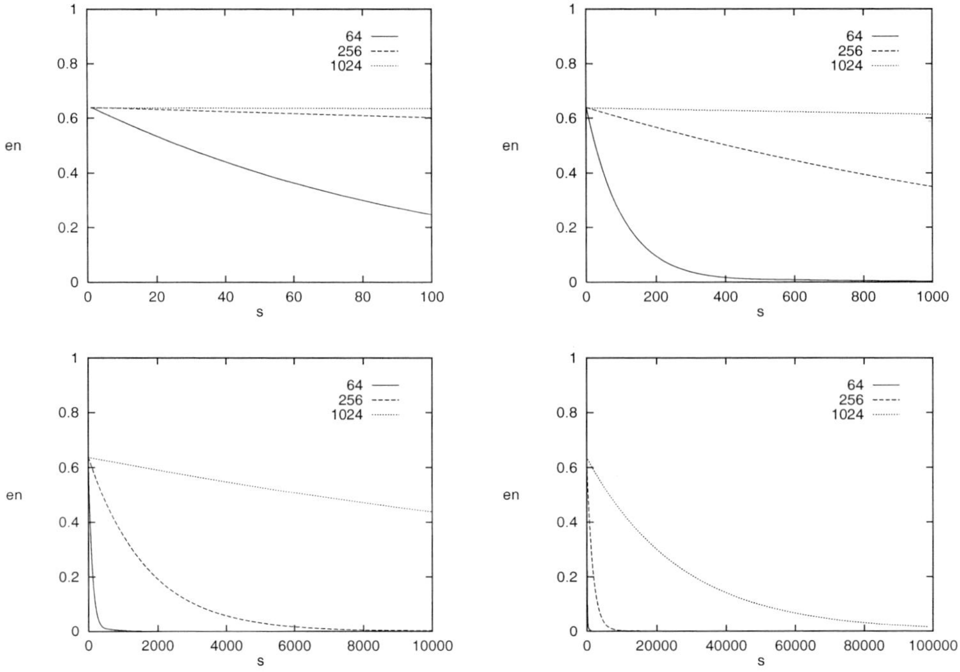

Figure 3.16: *Error norm* en^h *as a function of the number of relaxation sweeps* s *for* $N = 64$, $N = 256$ *and* $N = 1024$.

behaves in a similar way as the residual norm, and it displays the same aspects of slow convergence: with respect to large numbers of relaxation sweeps s, and with respect to large numbers of points N.

In order to improve the convergence speed, the application of the Multigrid techniques gives the same result as previously seen for the residuals. In the next table the error norm for the FMG algorithm is given as a function of the number of V-cycles, similar to Table 3.2 for the residual norm. From this table it can be concluded that the FMG algorithm solves the error to the level of the discretization error within one V-cycle, independently of the level. Additional V-cycles do hardly improve the precision of the result, and are therefore unnecessary. Comparing these results with the single level relaxation of Table 3.3, one appreciates the gain in convergence speed achieved. Remember that for the one dimensional problem, a $V(2,1)$ FMG-cycle costs as little as 12 fine grid relaxations, see Section 3.7. The reduction in computing time thus varies from a factor of 10^3 on level 5 to a factor of 10^5 on level 9.

3.9.3 Approximate Errors

In a real world problem it is impossible to know the error norm (3.78) as it involves knowing the solution $u(x)$, and thus makes the calculation of u^h irrelevant. In general

V-cycle	level 5	level 7	level 9
	$N = 64$	$N = 256$	$N = 1024$
1	$5.835 \cdot 10^{-4}$	$3.623 \cdot 10^{-5}$	$2.259 \cdot 10^{-6}$
2	$5.210 \cdot 10^{-4}$	$3.219 \cdot 10^{-5}$	$2.006 \cdot 10^{-6}$
3	$5.193 \cdot 10^{-4}$	$3.208 \cdot 10^{-5}$	$1.999 \cdot 10^{-6}$
4	$5.193 \cdot 10^{-4}$	$3.208 \cdot 10^{-5}$	$1.999 \cdot 10^{-6}$

Table 3.4: *Error norm as a function of three different levels and of the number of V-cycles, using an FMG algorithm.*

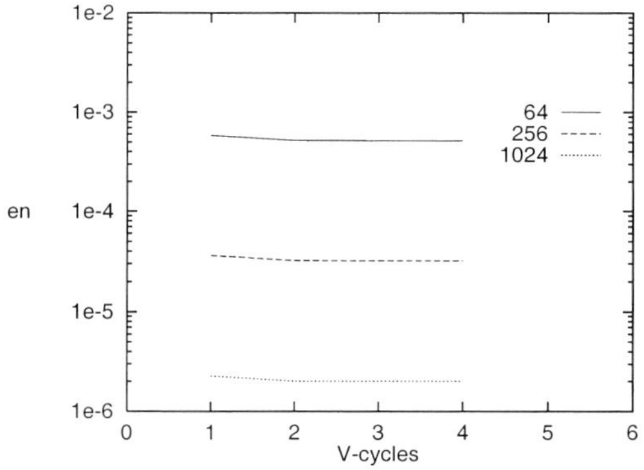

Figure 3.17: *Error norm as a function of the number of V-cycles for* $N = 64$, $N = 256$ *and* $N = 1024$.

this solution $u(x)$ is unknown, and the calculation of u^h is of interest. However, is it really necessary to know the error norm en^h exactly, or is a good approximation sufficient? As will be shown in a moment, the next question is closely related: is it really necessary to know the solution $u(x)$ exactly, or is a good approximation to $u(x)$ sufficient? The usefulness of en^h is to establish the average error in the solution, in order to see if the solution obtained has the desired accuracy. Thus only a correct order of magnitude of the error norm is required.

As a consequence, a good approximation to en^h is sufficient, and it is possible to construct one using a good approximation to $u(x)$. The next question is just how good this approximation should be. The answer is that the approximation to $u(x)$ should be much better than $\tilde{u}_i^h$. That means that the error in the approximation to u^h should be much smaller than the error in $\tilde{u}_i^h$. Thus, when evaluating the error norm of the approximate solution $\underline{\tilde{u}}^H$ on a particular grid with mesh size H, it is possible to use the approximate solution $\underline{\tilde{u}}^h$ to approximate $u(x)$, where $H = 2h$. Using the FMG technique these solutions can be compared naturally, storing them at the moment before interpolation to the next finer grid (double circles in Figure 3.13). Assuming that the grid with mesh size H consists of $N/2$ intervals, an approximate error norm aen^H is introduced:

$$en^H \simeq aen^H = \frac{1}{N/2-1}\sum_{I=1}^{N/2-1} |\tilde{u}_I^H - (I_h^H \tilde{u}_.^h)_I| \tag{3.81}$$

Where the summation is taken over all $N/2-1$ coarse grid points. Note that $u(x_I^H)$ is approximated by $(I_h^H \tilde{u}_.^h)_I$, a restriction of the fine grid solution around the point $i = 2I$. Whether an injection or a full weighted average are used is not very important. In the following analysis an injection is assumed to simplify the notation.

The approximate error norm aen^H consists of four different components: the discretization error of grids h (3.82) and H (3.83), and the numerical error on these grids (3.84) and (3.85).

$$\frac{1}{N/2-1}\sum_{I=1}^{N/2-1} |e_{2I}^h| = \frac{1}{N/2-1}\sum_{I=1}^{N/2-1} |u(x_{2I}^h) - u_{2I}^h| \tag{3.82}$$

$$\frac{1}{N/2-1}\sum_{I=1}^{N/2-1} |e_I^H| = \frac{1}{N/2-1}\sum_{I=1}^{N/2-1} |u(x_I^H) - u_I^H| \tag{3.83}$$

$$\frac{1}{N/2-1}\sum_{I=1}^{N/2-1} |v_{2I}^h| = \frac{1}{N/2-1}\sum_{I=1}^{N/2-1} |u_{2I}^h - \tilde{u}_{2I}^h| \tag{3.84}$$

$$\frac{1}{N/2-1}\sum_{I=1}^{N/2-1} |v_I^H| = \frac{1}{N/2-1}\sum_{I=1}^{N/2-1} |u_I^H - \tilde{u}_I^H| \tag{3.85}$$

In principle the discretization error is unknown, only its evolution can be predicted between a set of grids and the next finer set. In the case of the second order discretization of our model problem the discretization error will reduce by a factor of four. The numerical error is also unknown, however, one additional V-cycle will reduce this error substantially, while it will not affect the discretization error. Let us now study the evolution of the approximate error norm in our model problem.

V-cycle	level 5	level 7	level 9
	$N = 64$	$N = 256$	$N = 1024$
1	$2.168 \cdot 10^{-4}$	$1.356 \cdot 10^{-5}$	$8.466 \cdot 10^{-7}$
2	$1.938 \cdot 10^{-4}$	$1.205 \cdot 10^{-5}$	$7.520 \cdot 10^{-7}$
3	$1.932 \cdot 10^{-4}$	$1.201 \cdot 10^{-5}$	$7.494 \cdot 10^{-7}$
4	$1.932 \cdot 10^{-4}$	$1.201 \cdot 10^{-5}$	$7.494 \cdot 10^{-7}$

Table 3.5: *Approximate error norm between finest and one but finest grid, as a function of three different levels and of the number of V-cycles, using an FMG algorithm.*

From the small difference of the values of aen^H obtained with one and two or more V-cycles one can conclude that the numerical errors in 1 V-cycle aen^H are already small compared to the discretization error. By comparing the results of level 6, 8 and 10, one can conclude that the (dominating) discretization error is indeed $O(H^2)$ as the errors are reduced by a factor of 16 from level 6 to 8 and from level 8 to 10.

Finally, one can use the aen^H to determine if a solution is sufficiently accurate, i.e. if the numerical errors are small compared to the discretization error. As such aen^H can be used as an accurate and advanced stop criterion.

In conclusion it can be stated that apart from supplying a cheap initial approximation on the fine grid, the FMG algorithm provides an approximation to the numerical error and the discretization error, at small additional cost. The only disadvantage of aen^H is that it does not allow a direct approximation of the error on the finest grid, since the finest grid solution is used as a reference solution to the one but finest grid. The level 8 results of Table 3.5 are used to approximate the level 7 results of Table 3.4, and so on. The error on the finest grid can only be approximated using extrapolation.

3.10 2d Results

In this section the results of the two dimensional Poisson program **MG2d** are discussed. The program is given in Appendix E, whereas the data structure is described in Appendix A. The program solves the discretised two dimensional Poisson equation:

$$\frac{\partial^2 u}{\partial x^2} + \frac{\partial^2 u}{\partial y^2} = f(x, y) \tag{3.86}$$

on the domain $x \in [0,1], y \in [0,1]$, where the right hand side function is chosen as $f(x,y) = -8\pi^2 \sin(2\pi x)\sin(2\pi y)$ to give a simple solution $u(x,y) = \sin(2\pi x)\sin(2\pi y)$. The boundary conditions are $u(x=0,y) = u(x=1,y)\ u(x,y=0) = u(x,y=1) = 0$. The initial solution $u_0(x,y) = 0$ is used. The coarsest grid contains $(4+1) \times (4+1)$ points, including the boundary points. The mesh size on this grid is identical in x and y direction: $h_x = h_y = 0.25$.

The program uses a *FMG* algorithm where the number levels, the number of cycles per level and the cycle parameters can be specified. The results are compared with the results obtained when only relaxation is used on a given grid. Three different measures are used to illustrate different aspects of the performance. These measures relate to the convergence of the solution to the exact solution of the discrete problem, and to the convergence of the solution to the exact solution of the continuous differential problem.

3.10.1 Residuals

The same two aspects of slow convergence for one level relaxation, as were observed in the previous section for the one dimensional problem, are again encountered. The next four figures, which show the slowness of the evolution of the residual norm:

$$rn^h = \frac{1}{(n-1)^2} \sum_{i=1}^{n-1} \sum_{j=1}^{n-1} |r_{i,j}^h| \tag{3.87}$$

on a grid with mesh size $h = h_x = h_y$, as a function of the number of relaxation sweeps s and the total number of points N. The single level relaxation reduces the residual norm rn^h, but this reduction diminishes with increasing values of s and N.

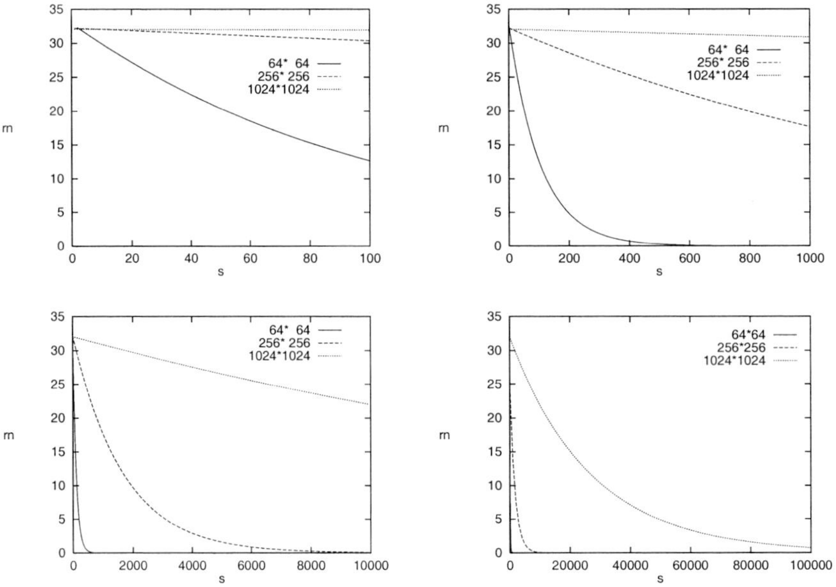

Figure 3.18: *Residual norm* rn^h *as a function of the number of relaxation sweeps* s *for* $N = 64^2$, $N = 256^2$ *and* $N = 1024^2$.

Once again a residual norm reduction of a factor of 10 is analysed, from 32 to 3. For the grid with 64×64 intervals, some 250 relaxations are needed. For the grid with 256×256 intervals, some 4000 relaxations are needed. Finally, for the grid with 1024×1024 intervals, some 60000 relaxations are needed. These numerical results can be compared with the theoretical predictions of Table 2.2. Very similar results to the one dimensional case are obtained: the numerically obtained results are roughly a factor of 4 smaller than the theoretically predicted values, due to the rapid initial convergence. However, it should be remarked that due to the second dimension the number of relaxation sweeps increases as $O(N)$, where $N = n^2$. In the one dimensional problem the number of sweeps increased as $O(N^2)$. The one and two dimensional results give the same results in terms of the number of points in one dimension n: for both problems the number of required relaxation sweeps increases with n^2.

The results of the single grid relaxation are to be compared to the results of the Multigrid program **MG2d** which are given in Figure 3.19 and Table 3.6. This program uses the FMG technique to generate accurate initial solutions on a certain level, by interpolation from a converged solution on the next coarser level. The V-cycles used are V(2,1) cycles, with 2 relaxation sweeps when coarsening, and a single relaxation sweep when refining.

From Table 3.6 it can be concluded that the FMG algorithm also removes these two types of slowness for the two dimensional problem. An additional V-cycle reduces the residual norm by a factor of approximately 10, independent of the level. A 1 V-cycle FMG algorithm on level 9 is much more efficient than 100000 relaxation sweeps on the finest grid, however, it consumes less than 6 WU, that is it requires less cpu time than 6 fine grid relaxations.

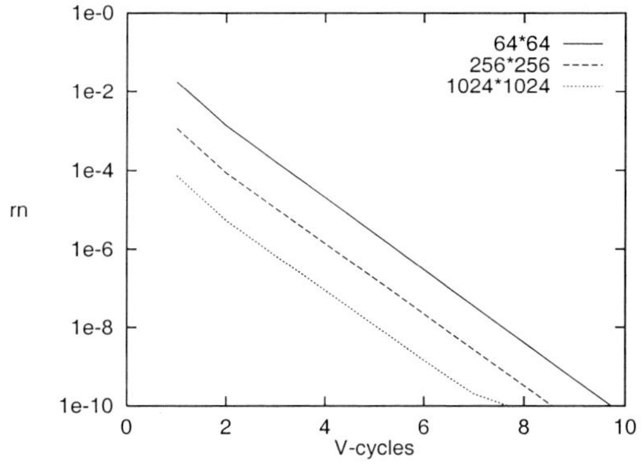

Figure 3.19: *Residual norm as a function of the number of V-cycles for* $N = 64 \times 64$, $N = 256 \times 256$ *and* $N = 1024 \times 1024$, *using FMG.*

level	N	1 V-cycle	2 V-cycles	3 V-cycles
2	8^2	$9.15 \cdot 10^{-1}$	$6.69 \cdot 10^{-2}$	$5.36 \cdot 10^{-3}$
3	16^2	$2.49 \cdot 10^{-1}$	$1.73 \cdot 10^{-2}$	$1.78 \cdot 10^{-3}$
4	32^2	$6.84 \cdot 10^{-2}$	$5.19 \cdot 10^{-3}$	$5.99 \cdot 10^{-4}$
5	64^2	$1.79 \cdot 10^{-2}$	$1.36 \cdot 10^{-3}$	$1.66 \cdot 10^{-4}$
6	128^2	$4.61 \cdot 10^{-3}$	$3.46 \cdot 10^{-4}$	$4.30 \cdot 10^{-5}$
7	256^2	$1.17 \cdot 10^{-3}$	$8.64 \cdot 10^{-5}$	$1.08 \cdot 10^{-5}$
8	512^2	$2.94 \cdot 10^{-4}$	$2.15 \cdot 10^{-5}$	$2.70 \cdot 10^{-6}$
9	1024^2	$7.38 \cdot 10^{-5}$	$5.36 \cdot 10^{-6}$	$6.74 \cdot 10^{-7}$
10	2048^2	$1.85 \cdot 10^{-5}$	$1.34 \cdot 10^{-6}$	$1.68 \cdot 10^{-7}$

Table 3.6: *Residual norm, as a function of the level and of the number of V-cycles, using an FMG algorithm.*

3.10.2 Errors

For this two dimensional model problem it is possible to calculate the error norm for each solution. Similarly to the residual norm it is defined as:

$$en^h = \frac{1}{(n-1)^2} \sum_{i=1}^{n-1} \sum_{j=1}^{n-1} |\tilde{u}_{i,j}^h - u(x_i^h, y_j^h)| \tag{3.88}$$

From the figure and the table the same conclusion can be draw as for the one dimensional problem: the numerical error in the 1 V-cycle results is already small (50% at most) compared to the discretization error. Additional cycles reduce only the residual, but do not result in a significant reduction of the total error.

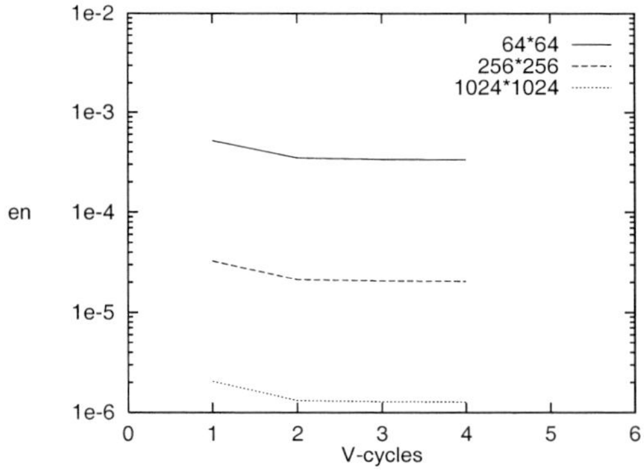

Figure 3.20: *Error norm as a function of the number of V-cycles for* $N = 64 \times 64$, $N = 256 \times 256$ *and* $N = 1024 \times 1024$.

level	N	1 V-cycle	2 V-cycles	3 V-cycles
2	8^2	$3.47 \cdot 10^{-2}$	$2.58 \cdot 10^{-2}$	$2.52 \cdot 10^{-2}$
3	16^2	$8.28 \cdot 10^{-3}$	$5.98 \cdot 10^{-3}$	$5.83 \cdot 10^{-3}$
4	32^2	$2.06 \cdot 10^{-3}$	$1.43 \cdot 10^{-3}$	$1.39 \cdot 10^{-3}$
5	64^2	$5.20 \cdot 10^{-4}$	$3.49 \cdot 10^{-4}$	$3.37 \cdot 10^{-4}$
6	128^2	$1.31 \cdot 10^{-4}$	$8.59 \cdot 10^{-5}$	$8.30 \cdot 10^{-5}$
7	256^2	$3.28 \cdot 10^{-5}$	$2.13 \cdot 10^{-5}$	$2.06 \cdot 10^{-5}$
8	512^2	$8.20 \cdot 10^{-6}$	$5.31 \cdot 10^{-6}$	$5.12 \cdot 10^{-6}$
9	1024^2	$2.05 \cdot 10^{-6}$	$1.32 \cdot 10^{-6}$	$1.28 \cdot 10^{-6}$
10	2048^2	$5.13 \cdot 10^{-7}$	$3.31 \cdot 10^{-7}$	$3.19 \cdot 10^{-7}$

Table 3.7: *Error norm, as a function of the level and of the number of V-cycles, using an FMG algorithm.*

3.10.3 Approximate Errors

In the general case it is impossible to compute the error norm, and therefore an approximated error norm is again introduced. It is a straightforward extension of the one dimensional norm (3.81) and the coarse grid with meshsize H is assumed to consist of $n/2 \times n/2$ intervals:

$$aen^H = \frac{1}{(n/2-1)^2} \sum_{I=1}^{n/2-1} \sum_{J=1}^{n/2-1} |\tilde{u}^H_{I,J} - \tilde{u}^h_{2I,2J}| \tag{3.89}$$

The approximate error norm is computed in the following table using a FMG algorithm with 1, 2 or 3 V-cycles as a function of the level.

From this table the same conclusion can be drawn as for the one dimensional case. One V-cycle is sufficient to attain solutions which have a numerical error small compared to the discretization error. In other words, the total error is close to the discretization error

level	N	1 V-cycle	2 V-cycles	3 V-cycles
2	8^2	$2.58 \cdot 10^{-2}$	$1.95 \cdot 10^{-2}$	$1.91 \cdot 10^{-3}$
3	16^2	$6.09 \cdot 10^{-3}$	$4.48 \cdot 10^{-3}$	$4.38 \cdot 10^{-3}$
4	32^2	$1.52 \cdot 10^{-3}$	$1.07 \cdot 10^{-3}$	$1.04 \cdot 10^{-3}$
5	64^2	$3.87 \cdot 10^{-4}$	$2.61 \cdot 10^{-4}$	$2.53 \cdot 10^{-4}$
6	128^2	$9.79 \cdot 10^{-5}$	$6.44 \cdot 10^{-5}$	$6.22 \cdot 10^{-5}$
7	256^2	$2.46 \cdot 10^{-5}$	$1.60 \cdot 10^{-5}$	$1.54 \cdot 10^{-5}$
8	512^2	$6.15 \cdot 10^{-6}$	$3.98 \cdot 10^{-6}$	$3.84 \cdot 10^{-6}$
9	1024^2	$1.54 \cdot 10^{-6}$	$9.93 \cdot 10^{-7}$	$9.59 \cdot 10^{-7}$

Table 3.8: *Approximate error norm, as a function of the level and of the number of V-cycles, using an FMG algorithm.*

even after one V-cycle. Furthermore, it can be concluded that the accuracy of the discrete equation is indeed $O(h^2)$. In this case it is even possible to quantitatively approximate the error, using the fact that the discretization error on the finer grid will be 1/4 of that of the coarser grid, and will be in the same 'direction'. Thus the approximate error norm should be 3/4 of the error norm itself. This is indeed the case when comparing the level 9 error norm: $2.05 \cdot 10^{-6}$ to the level 10 approximate error norm $1.54 \cdot 10^{-6}$. The same is true for the results using more V-cycles.

3.11 Conclusion

In this chapter the Multigrid techniques have been introduced. It was shown that using cycles that involve coarser and coarser grids, a grid independent convergence speed can be obtained. Furthermore, the theoretical predictions of slow convergence of the previous chapter were reproduced numerically for the one and two dimensional model problem. The fast convergence predicted using the Multigrid techniques was also demonstrated numerically in one and two dimensions. For fine grids the obtained reduction in computing time can attain many orders of magnitude. Next, using Full MultiGrid (FMG) techniques it was shown how the solutions on coarser grids can be used as cheap starting solutions, and how they can contribute to the analysis of the numerical error as well as the discretization error.

Finally, it is important to stress that the increase in convergence speed, and thus the reduction in computing time, results from the solution of the smooth *error* on a coarser grid. The *solution* itself, however, can be very non-smooth, and in that case it can not be accurately represented on the coarser grids. The Multigrid process will not be affected, and will continue to reduce the errors and thus to converge rapidly. It is therefore essential to think of the coarser grids, in terms of error correction grids.

3.12 Advanced Techniques

3.12.1 Local Grid Refinements

Occasionally, a problem requires a grid which is locally much finer than the grid required in the rest of the domain. Using a fine grid everywhere would only augment the overall computing time and memory requirements, without increasing the overall precision. In such cases the finer grids can be restricted to smaller and smaller sub domains, whereas the coarser grids cover the entire calculational domain. The resulting grid strongly resembles a non equidistant grid as is shown by Figure 3.21, while retaining the advantages of an equidistant grid. An example of such a series of finer grids covering smaller and smaller domains is given in Figure 3.22. The FAS requires only a small extension of Equation (3.74), in which the fine grid residuals r_i^h are only calculated there where the fine grid exists, and the coarse grid right hand side uses zero residuals, there where a finer grid does not exist. Likewise, the coarse grid correction (3.75) is only carried out in the part where a finer grid exists. In terms of the program this can be simply implemented by defining an i_{start} and an i_{end} which cover smaller and smaller parts of the domain. The extension to higher dimensions is straightforward. Furthermore, the process of local refinements can be automatic, when measuring the difference between the solution $\underline{u}^h$ and $\underline{u}^H$. There where the difference is large, an even finer grid is required. For more information the reader is referred to Bai and Brandt [4].

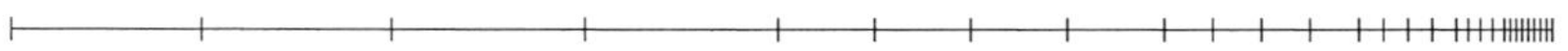

Figure 3.21: *Grid refinement towards the right boundary.*

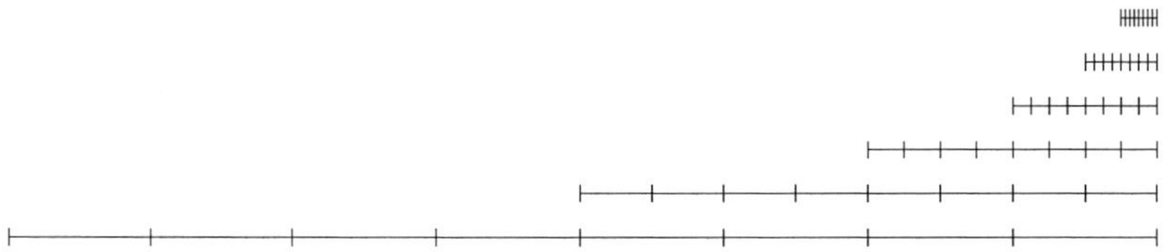

Figure 3.22: *Implementation using different grids of constant mesh size covering smaller and smaller parts of the domain.*

3.12.2 Neumann Boundary

The boundary conditions that have been studied in this chapter, and will be studied in the next chapters, are all Dirichlet type conditions. This means that the solution u is prescribed on the boundary. Using the discrete one dimensional Poisson equation as an example the equations read:

$$\frac{u_{i-1}^h - 2u_i^h + u_{i+1}^h}{h^2} = f_i^h \qquad i \in [1, n-1] \tag{3.90}$$

with the boundary conditions $u_0^h = u_0$ and $u_n^h = u_n$.

Another type of boundary condition, called Neumann condition, prescribes the derivative at the boundary. In the case of the one dimensional problem the Neumann condition imposes for instance in $x = 0$ a value of the derivative n_0:

$$\partial u / \partial x|_{x=0} = n_0 \tag{3.91}$$

Apart from having to discretize the equation in the interior points, it is now necessary to discretize the boundary condition as well:

$$\frac{u_{i-1}^h - 2u_i^h + u_{i+1}^h}{h^2} = f_i^h \qquad i \in [0, n-1] \tag{3.92}$$

and

$$\frac{u_1^h - u_{-1}^h}{2h} = n_0 \tag{3.93}$$

Please note that the number of equations and unknowns has increased by two, compared to the equation with the Dirichlet conditions. It is also possible to discretize the Neumann condition in $x = 0$ using the points $i = 0$, $i = 1$ and $i = 2$. Once again a second order accurate discretization is obtained, but now only a single unknown is added compared to the Dirichlet condition. The implementation is very similar to the implementation of the second order central discretization described below. When discretising the boundary equations, it is important to ensure that the discretization error of the boundary equation is the same as that of the equation in the interior. In the above cases both are $O(h^2)$.

In order to obtain a Multigrid solver with the same efficiency as the one employing Dirichlet conditions, two criteria have to be satisfied:

- the condition has to be consistent on the various grids,
- the condition should not introduce locally large (non-smooth) errors.

For this particular case both conditions can be satisfied when the point u_{-1} is not actually implemented. This non-implementation avoids transfer problems in the various transfer operators. Instead, the Neumann condition is used when relaxing the interior equation in the point $i = 0$, thereby effectively modifying the equation in this points. In effect $u_{-1} = u_1 - 2hn_0$ is used when relaxing the interior Equation (3.92) in the point $i = 0$.

This simple modification allows the program to obtain the same efficiency, as for the Dirichlet conditions. Obviously, the Neumann boundary condition can be implemented in a similar way in the point x_n. This approach can straightforwardly be extended to higher dimensions.

3.12.3 Two Level Analysis

In Section 3.6 an estimate of the error reduction that can be obtained per cycle was given based on the assumption that it is determined by the reduction of the high frequency components on the fine grid. The influence of the restriction, the choice of the coarse grid operator, and the interpolation were neglected. For small ν_1 and ν_2 the estimates thus obtained are very accurate as was shown by the results presented in Section 3.9 and 3.10. However, as one can easily verify using the programs MG1d or MG2d they are too optimistic for larger values of ν_1 and ν_2. In that case the predicted reduction of high frequency components on the fine grid becomes so large that the quality of the coarse grid correction determines the convergence rate of the cycle. In order to obtain an accurate estimate for these cases a more advanced analysis is required, referred to as a *two level analysis*. Below, this analysis will be briefly presented. For more details the reader is referred to e.g. Brandt [15], Hemker [59] and Trottenberg [98], and the general multigrid literature. In the description given here a coarse grid with mesh size $H = 2h$ and a one dimensional problem are assumed for simplicity.

To analyze the effect of relaxation on the error it was sufficient to look at its effect on a single Fourier component of the error of the form:

$$v_j^h = Ae^{\imath\theta j} \tag{3.94}$$

with $-\pi \leq \theta \leq \pi$. However, to analyze the effect of a cycle including the coarse grid, this is no longer sufficient. The coarse grid can only uniquely represent half of the frequencies, i.e. those with $0 < |\theta| \leq \pi/2$. The higher freqencies on the fine grid alias with the lower frequencies. For example the fine grid component $\theta = \pi$ appears on the coarse grid as a component with $\theta = 0$. To be precise, each pair of components θ and θ' with $|\theta - \theta'| = 0$ or π on the coarse grid coincide with each other. Such a pair of *harmonics* θ, θ' is coupled to each other by the two-level processes. Therefore, the analysis taking into account the details of the coarse grid correction is carried out looking at a pair of components $\theta, \theta + \pi$, where $0 < |\theta| \leq \pi/2$. Let such a pair be represented by a vector:

$$\underline{v}_\theta^h = \begin{bmatrix} A_\theta \\ A_{\theta+\pi} \end{bmatrix} \tag{3.95}$$

A two-level cycle was defined as follows: make ν_1 relaxations on grid h. Then transfer the residual problem to grid H and solved it exactly, then interpolate the grid H solution to grid h as a correction to the former grid h solution, then make ν_2 relaxations. A pair of the form (3.95) will be mapped onto itself with amplitudes after the cycle that form a linear combination of the amplitudes before relaxation. It can now be shown that the error before a cycle and the error after a cycle are related according to:

$$\bar{\underline{v}}_\theta^h = M(\theta)\tilde{\underline{v}}_\theta^h \tag{3.96}$$

where $M(\theta)$ is a 2×2 matrix referred to as the two-level amplification matrix given by:

$$M(\theta) = \tilde{\underline{R}}^{\nu_1}[I - \tilde{\underline{I}}_H^h(\tilde{\underline{L}}^H)^{-1}\tilde{\underline{I}}_h^H\tilde{\underline{L}}^h]\tilde{\underline{R}}^{\nu_2} \tag{3.97}$$

In this equation I is the 2×2 unit matrix. Furthermore, the symbol $\underline{\tilde{R}}$ stands for a 2×2 matrix representing the action of relaxation upon a pair of components (3.95). Assuming that each error components is mapped on itself:

$$\underline{\tilde{R}} = \begin{bmatrix} \mu(\theta) & 0 \\ 0 & \mu(\theta+\pi) \end{bmatrix} \tag{3.98}$$

with $\mu(\theta)$ standing for the error amplification function of relaxation.

The symbol $\underline{\tilde{L}}^h$ stands for the response of the fine grid operator to a pair of components (3.95) and is given by a 2×2 matrix:

$$\underline{\tilde{L}}^h = \begin{bmatrix} \tilde{L}^h(\theta) & 0 \\ 0 & \tilde{L}^h(\theta+\pi) \end{bmatrix} \tag{3.99}$$

For example for the one dimensional model problem:

$$\tilde{L}^h(\theta) = (2 - 2cos(\theta))/h^2 \tag{3.100}$$

The symbol $\tilde{L}^H$ stands for the response of the coarse grid operator to the component θ. This is not a vector but a single number as the two components on the coarse grid alias into a single component. For the one dimensional model problem choosing straightforward discretization to obtain the coarse grid operator one finds:

$$\tilde{L}^H(\theta) = (2 - 2cos(\theta))/H^2 \tag{3.101}$$

The symbol $\underline{\tilde{I}}_h^H$ is a vector representing the action of the restriction upon the pair (3.95). As restriction aliases the two components on the coarse grid, hence it is 1×2 vector.

$$\underline{\tilde{I}}_h^H = \begin{bmatrix} \tilde{I}_h^H(\theta), & \tilde{I}_h^H(\theta+\pi) \end{bmatrix} \tag{3.102}$$

For full weighting:

$$\underline{\tilde{I}}_h^H = \begin{bmatrix} (1+cos(\theta))/2, & (1-cos(\theta))/2 \end{bmatrix} \tag{3.103}$$

Finally the symbol $\underline{\tilde{I}}_H^h$ is a vector representing the action of the interpolation. Given a component θ on the coarse grid the result is two components θ and $\theta + \pi$ on the fine grid. Therefore $\underline{\tilde{I}}_H^h$ is a 2×1 vector.

$$\underline{\tilde{I}}_H^h = \begin{bmatrix} (1+cos(\theta))/2 \\ (1-cos(\theta))/2 \end{bmatrix} \tag{3.104}$$

The error reduction per cycle is now given by the largest eigenvalue of the matrix $M(\theta)$ for $0 < |\theta| \leq \pi/2$.

For small values of $\nu_1 + \nu_2$ this eigenvalue will be determined by the relaxation. For larger values by the other operators. For example for $\nu_2 = 2$, $\nu_1 = 1$ the largest eigenvalue is $\rho = 0.088$, whereas the prediction given in Section 3.6 was $\bar{\mu}^3 = (1/\sqrt{5})^3 = 0.089$. Clearly the prediction neglecting the intergrid transfers and the effect of the coarse grid operator is quite accurate. However, using $\nu_2 = 4$ and $\nu_1 = 2$ this is no longer true. The analysis of Section 3.6 predicts an error multiplication facter per cycle of: $\bar{\mu}^6 = 0.008$

whereas the two level analysis yields $\rho = 0.017$. This shows that now the effects of the intergrid transfers and the coarse grid correction have become the factors determining the performance of the cycle.

The analysis for higher dimensions can be carried out along the same lines. For example for a two dimensional problem there are 4 components aliasing on the coarse grid, i.e. (θ_1, θ_2), $(\theta_1, \theta_2 + \pi)$, $(\theta_1 + \pi, \theta_2)$, and $(\theta_1 + \pi, \theta_2 + \pi)$. As a result $\underline{M}$, $\underline{\tilde{R}}$ and $\underline{\tilde{L}}^h$ will be 4×4 matrices, $\underline{\tilde{I}}_h^H$ will be a 1×4 vector, and $\underline{\tilde{I}}_H^h$ a 4×1 vector. $\tilde{L}^H$ remains a single number. For further details the reader is referred to [15], [59] and [98].

3.12.4 Problematic Components

The multigrid techniques described here are based on the fact that an iterative process reduces high frequency components efficiently whereas low frequency components are hardly affected. As a result a coarse grid can be used to obtain an efficient solver. This is the standard scheme. The multigrid principle is much more general. As can be seen from the description of the relaxation processes in Chapter 2, they all have in common that the residual is used to compute corrections to the solution. Assuming for simplicity a linear problem. In Section 3.2 it was shown that the residual satisfies:

$$L^h \underline{v}^h = \underline{r}^h \tag{3.105}$$

Consequently, if there are components of the error for which $|\underline{v}^h| = O(1)$ whereas $L^h \underline{v}^h$ is almost zero this implies that the residual does not reflect the true magnitude of the error component in the solution. Hence, any process using the residual as a means to update the solution will only give small corrections and for this particular component converge very slowly. For operators resulting from the discretization of elliptic partial differential equations generally this applies to the smoothest components. For non-elliptic problems, or bad discretizations there may be other components for which this holds. In that case these components should be identified and dealt with separately. Well known problems in this respect are Navier Stokes equations for high Reynolds numbers, see Brandt [21], or the Helmholtz standing wave equation, see [22]. In that case the standard techniques as described here are not sufficient to obtain a fast solver and auxiliary measures need to be taken.

The general rule of the Multigrid development is therefore not to just solve smooth errors on coarse grids. This is only one aspect of the general principle to identify which are the components that are slow to converge and design a measure to deal with them.

Chapter 4

Hydrodynamic Lubrication

In this chapter the efficient solution of the problem of isoviscous Hydrodynamic Lubrication (HL) in a journal bearing is treated as a Poisson problem with coefficients (film thickness cubed) that vary over the domain, but are known in advance. This will allow us to solve the problem of a journal bearing using only small extensions of the two dimensional Poisson program.

We will see that the solution obtained, consists of negative and positive pressures. In order to obtain a physically meaningful solution, extra conditions will be added to this equation. This creates a complementarity problem, where the Reynolds equation is valid in a certain part of the domain (Ω_1) and the cavitation 'equation' is valid in the remaining part of the domain (Ω_2). The division into the two sub domains is a priori unknown and has to be established while solving the equation.

4.1 Equations

As a model problem for hydrodynamic lubrication a journal bearing is taken. Figure 4.1 shows the so-called *full* journal bearing (as opposed to a partial journal bearing where the bearing only surrounds part of the journal). As mentioned in Chapter 1 it was the experimental work of Beauchamp Tower demonstrating the pressure generation in this type of bearing which led to the classical publication of Reynolds [92]. In his paper Reynolds not only derives the basic equation now bearing his name but he also gives solutions among

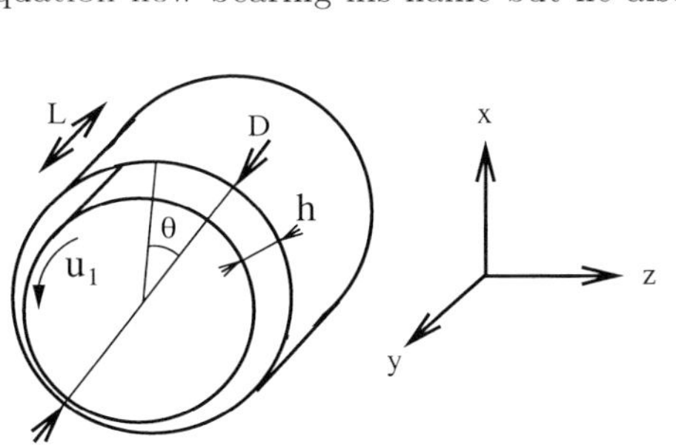

Figure 4.1: *Finite length journal bearing.*

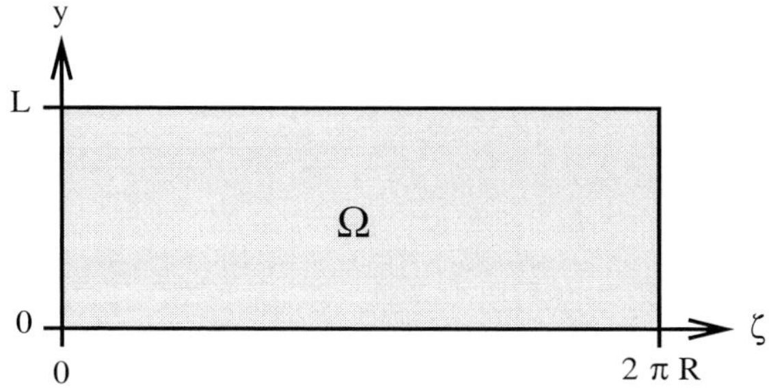

Figure 4.2: *Developed journal bearing.*

which approximate solutions of the pressure profile in a journal bearing. A subsequent major contribution to the subject was presented by Sommerfeld [96]. The first numerical solutions were presented about half a century later by Cameron and Wood [27] and by Sassenfeld and Walther [95]. For a historical review of the various contributions the reader is referred to Cameron [28] and to Dowson [37].

In this chapter the numerical solution of the pressure field in a full journal bearing of finite width is revisited. The essential steps are described to extend the Multigrid algorithm for the 2 dimensional Poisson problem presented in the previous chapter into an efficient Multigrid solver for the journal bearing problem. The resulting program is given in Appendix F. It can serve as the basis of a solver for a variety of hydrodynamic lubrication problems. In addition, compared to the two dimensional Poisson problem, it forms a next step towards an efficient Multigrid solver for the EHL problem.

The geometry of the journal bearing is outlined in Figure 4.1. The radii of the journal and the bearing differ by the clearance c which is (very) small compared to the average radius $R = D/2$. The bearing has a length L, and lubricant is supplied at the top of the bearing ($\theta = 0$) through an axial groove at ambient pressure $p = 0$. The gap height between the journal and the bearing is called h.

Assuming the radius of curvature is very large compared to the film thickness ($c \ll R$), the curvature can be neglected and the bearing can be developed onto a flat surface as in Figure 4.2, where ζ is a local circumferential coordinate.

The general Reynolds equation (1.1) for this developed bearing reads:

$$\frac{\partial}{\partial \zeta}(\frac{\rho h^3}{12\eta}\frac{\partial p}{\partial \zeta}) + \frac{\partial}{\partial y}(\frac{\rho h^3}{12\eta}\frac{\partial p}{\partial y}) - \frac{\partial(u_m \rho h)}{\partial \zeta} - \frac{\partial(\rho h)}{\partial t} = 0 \tag{4.1}$$

with the boundary conditions $p(\zeta,0) = p(\zeta,L) = 0$ and $p(0,y) = p(2\pi R,y) = 0$. The first two terms are Poiseuille flow terms, whereas the third term is the Couette term. Assuming stationary hydrodynamic conditions, with constant viscosity η, constant density ρ and constant entrainment velocity u_m, the Reynolds equation simplifies to:

$$\frac{\partial}{\partial \zeta}(h^3\frac{\partial p}{\partial \zeta}) + \frac{\partial}{\partial y}(h^3\frac{\partial p}{\partial y}) - 12\eta u_m \frac{\partial h}{\partial \zeta} = 0 \tag{4.2}$$

For $c \ll R$ the gap height $h(\theta)$ can be approximated by:

$$h(\theta) = c + e\cos(\theta) \tag{4.3}$$

where e is the eccentricity, i.e. the distance between the centre of the journal and the centre of the bearing. Note that the minimum film thickness will be given by:

$$h_m = c - e \tag{4.4}$$

Finally force balance applies, which leads to the following conditions:

$$\begin{aligned} \mathrm{w}_n &= \int_0^{2\pi R}\int_0^L p(\zeta, y)\cos(\zeta/R)\,d\zeta\,dy \\ \mathrm{w}_t &= \int_0^{2\pi R}\int_0^L p(\zeta, y)\sin(\zeta/R)\,d\zeta\,dy \end{aligned} \tag{4.5}$$

The total load generated by the bearing is subsequently defined as:

$$\mathrm{w} = \sqrt{\mathrm{w}_n^2 + \mathrm{w}_t^2} \tag{4.6}$$

and it acts at an attitude angle:

$$\Psi = \arctan(\mathrm{w}_t/\mathrm{w}_n) \tag{4.7}$$

The problem can be solved in two different ways:

- For a given eccentricity determine the load capacity of the bearing w. This is the most straightforward case and therefore it is the approach chosen in this chapter. At first sight prescribing the eccentricity may seem odd. However, this is the equivalent of requiring a given minimum film thickness if the clearance c is known.

- For the prescribed load components w_n and w_t (or w and Ψ) determine the eccentricity e. When the load is prescribed, the problem becomes an inverse problem, in which the current eccentricity has to be adjusted, to ensure that the generated pressure field balances the applied load. How to extend an algorithm solving the pressure (and load) for a given eccentricity to a solver for this inverse problem is explained briefly in Section 4.10.3.

From the Equations (4.2)-(4.7) it follows that the minimum film thickness h_m, the load capacity w, and the attitude angle Ψ will be a function of 6 central parameters, i.e. e, c, η, R, L, u_m. However, the number of independent parameters is only 2 as immediately follows from the introduction of the following dimensionless quantities:

$$\begin{aligned} H &= h/c \\ P &= pc^2/(12\eta u_m R) \\ \theta &= \zeta/R \\ Y &= y/L \end{aligned}$$

Substitution of these dimensionless quantities in Equation (4.2) gives the following dimensionless Reynolds equation:

$$\frac{\partial}{\partial\theta}(H^3\frac{\partial P}{\partial\theta}) + \frac{R^2}{L^2}\frac{\partial}{\partial Y}(H^3\frac{\partial P}{\partial Y}) - \frac{\partial H}{\partial\theta} = 0 \tag{4.8}$$

with $\theta \in [0, 2\pi]$, $Y \in [0, 1]$, and the boundary conditions $P(0, Y) = P(2\pi, Y) = 0$ and $P(\theta, 0) = P(\theta, 1) = 0$.

The dimensionless film thickness H is defined by:

$$H(\theta) = 1 + \epsilon\cos(\theta) \tag{4.9}$$

with $\epsilon = e/c$.
Consequently, the dimensionless minimum film thickness will be:

$$H_m = 1 - \epsilon \tag{4.10}$$

Finally the dimensionless force balance conditions are:

$$W_n = \mathrm{w}_n\left(\frac{c^2}{12\eta L u_m R^2}\right) = \int_0^{2\pi}\int_0^1 P(\theta, Y)\cos(\theta)\,d\theta\,dY$$
$$W_t = \mathrm{w}_t\left(\frac{c^2}{12\eta L u_m R^2}\right) = \int_0^{2\pi}\int_0^1 P(\theta, Y)\sin(\theta)\,d\theta\,dY \tag{4.11}$$

The expression for the attitude angle is now $\Psi = \arctan(W_t/W_n)$ and the dimensionless load capacity of the bearing is given by:

$$W = \sqrt{W_n^2 + W_t^2} = \frac{1}{6}\frac{\mathrm{w}}{L\eta u_s}\frac{c^2}{R^2} = \frac{S}{6} \tag{4.12}$$

where S is the so-called Sommerfeld number defined as:

$$S = \frac{\mathrm{w}}{L\eta u_s}\frac{c^2}{R^2} \tag{4.13}$$

In European literature a modified version of this number is often used denoted by S_0. It is obtained by substituting $u_s = \omega D/2$:

$$S = \frac{2\mathrm{w}}{\eta L D\omega}\frac{c^2}{R^2} = 2S_0 \tag{4.14}$$

where ω is the *effective* angular journal frequency defined as:

$$\omega = (\omega_j - \omega_\mathrm{w}) + (\omega_s - \omega_\mathrm{w}) \tag{4.15}$$

with ω_j and ω_s the angular frequencies of the journal and the bearing respectively and ω_w is the frequency at which the load rotates. For the present case where only the journal rotates $\omega = \omega_j$.

The non-dimensional Equations (4.8)-(4.12) contain only two parameters, L/R and ϵ. Thus the dimensionless load W (or the Sommerfeld number) and the attitude angle Ψ are a function of ϵ, and L/R only. This is a significant reduction in control parameters, and as a result design graphs can easily be constructed as will be shown later.

The task at hand is thus to solve P from Equations (4.8) with (4.9) for a given ϵ and L/R. As for a given ϵ, H is known, the first derivative term in Equation (4.8) and the coefficients H^3 are known. If H were a constant the equation would reduce to the 2 dimensional Poisson problem considered in the previous chapters. In its present form it can be characterized as a 2 dimensional Poisson problem with varying coefficients. Note that for a given film thickness the $\partial H/\partial\theta$ term can either be computed analytically or be given in a discrete manner, and moved to the right hand side. However, for generality this option is not chosen.

One may therefore expect that the Multigrid solution algorithm developed for the 2 dimensional Poisson problem can be extended, yielding an efficient Multigrid solver for the pressure in a hydrodynamically lubricated journal bearing. However, such extensions should not be taken lightly and carried out step by step. The reader is reminded of the general rule of Multigrid development stated in the previous chapter. In this chapter the steps to be taken for the hydrodynamic lubrication problem are detailed. First the discrete equations are derived. Next the question of how to relax the equation is addressed. Finally coarse grid correction and Full Multigrid are treated.

4.2 Discrete Equations

The dimensionless Reynolds equation (4.8) can be written as:

$$\frac{\partial}{\partial\theta}(\xi\frac{\partial P}{\partial\theta}) + \frac{1}{k^2}\frac{\partial}{\partial Y}(\xi\frac{\partial P}{\partial Y}) - \frac{\partial H}{\partial\theta} = 0 \tag{4.16}$$

for $\theta \in [0, 2\pi]$, $Y \in [0, 1]$ and with the boundary conditions $P(0, Y) = P(2\pi, Y) = P(\theta, 0) = P(\theta, 1) = 0$. The coefficient $k = L/R$ is a constant for a bearing of given dimensions and $\xi = H^3$ with H defined by (4.9). Note that ξ varies over the domain, but does not depend on the pressure.

The domain is discretised using a rectangular grid of constant mesh size h_θ and h_y, hence the grid points are (θ_i, Y_j) with $\theta_i = ih_\theta$, $Y = jh_y$ for $0 \le i \le n_\theta$ and $0 \le j \le n_y$. The discrete approximation to the solution will be denoted by P^h. In the boundary points $P^h_{i,j} = 0$ is prescribed. In each interior point of the grid a discrete approximation of Equation (4.16) is derived using finite differences.

First consider the Poiseuille term for the θ direction. Discretising the outer differential $\partial/\partial\theta$ using a short central second order accurate discretization:

$$\left(\frac{\partial}{\partial\theta}(\xi\frac{\partial P}{\partial\theta})\right)_{i,j} \doteq \frac{(\xi\frac{\partial P}{\partial\theta})_{i+1/2,j} - (\xi\frac{\partial P}{\partial\theta})_{i-1/2,j}}{h_\theta} \tag{4.17}$$

where the two dots over the equality sign symbol indicates that it is an $O(h_\theta^2)$ approximation and the subscripts $i-1/2$ and $i+1/2$ refer to the *intermediate* locations $\theta = \theta_i + h_\theta/2$.

and $\theta = \theta_i - h_\theta/2$. In a second step the inner derivatives are developed, again using a short central discretization:

$$\left(\xi\frac{\partial P}{\partial \theta}\right)_{i+1/2,j} \doteqdot \xi^h_{i+1/2,j}\frac{(P^h_{i+1,j} - P^h_{i,j})}{h_\theta} \tag{4.18}$$

and

$$\left(\xi\frac{\partial P}{\partial \theta}\right)_{i-1/2,j} \doteqdot \xi^h_{i-1/2,j}\frac{(P^h_{i,j} - P^h_{i-1,j})}{h_\theta} \tag{4.19}$$

A combination of these two steps gives:

$$\left(\frac{\partial}{\partial \theta}(\xi\frac{\partial P}{\partial \theta})\right)_{i,j} \doteqdot \frac{\xi^h_{i+1/2,j}P^h_{i+1,j} - (\xi^h_{i+1/2,j} + \xi^h_{i-1/2,j})P^h_{i,j} + \xi^h_{i-1/2,j}P^h_{i-1,j}}{h^2_\theta} \tag{4.20}$$

where $\xi^h_{i\pm1/2,j}$ stands for ξ at the intermediate location $(\theta_i \pm h_\theta/2, Y_j)$.

A similar discretization can be used for the Poiseuille term in Y direction giving:

$$\left(\frac{\partial}{\partial Y}(\xi\frac{\partial P}{\partial Y})\right)_{i,j} \doteqdot \frac{\xi^h_{i,j+1/2}P^h_{i,j+1} - (\xi^h_{i,j+1/2} + \xi^h_{i,j-1/2})P^h_{i,j} + \xi^h_{i,j-1/2}P^h_{i,j-1}}{h^2_y} \tag{4.21}$$

where $\xi^h_{i,j\pm1/2}$ stands for ξ at the intermediate location $(\theta_i, Y_j \pm h_y/2)$. In the present case $\xi = H^3$ and H is known analytically so $\xi^h_{i\pm1/2,j}$ and $\xi^h_{i,j\pm1/2}$ can be computed exactly. Anticipating the EHL problem, which is to be considered later it will be assumed that H and P are only given/obtained at the grid points. Thus the discrete film thickness equation will be:

$$H^h_{i,j} = 1 + \epsilon\cos(\theta_i) \tag{4.22}$$

for all grid points (θ_i, Y_j). In that case an approximation is required for the values of ξ in the intermediate points. Examples of approximations are simple averaging such as:

$$\begin{aligned} \xi^h_{i\pm1/2,j} &\doteqdot (\xi^h_{i\pm1,j} + \xi^h_{i,j})/2 \\ \xi^h_{i,j\pm1/2} &\doteqdot (\xi^h_{i,j\pm1} + \xi^h_{i,j})/2 \end{aligned} \tag{4.23}$$

and

$$\begin{aligned} \xi^h_{i\pm1/2,j} = (H^h_{i\pm1/2})^3 &\doteqdot \left[(H^h_{i\pm1,j} + H^h_{i,j})/2\right]^3 \\ \xi^h_{i,j\pm1/2} = (H^h_{i\pm1/2})^3 &\doteqdot \left[(H^h_{i,j\pm1} + H^h_{i,j})/2\right]^3 \end{aligned} \tag{4.24}$$

Alternatively, a harmonic averaging can be used:

$$\frac{1}{\xi^h_{i\pm1/2,j}} \doteq \frac{1}{2}\left[\frac{1}{\xi^h_{i\pm1,j}} + \frac{1}{\xi^h_{i,j}}\right]$$
$$\frac{1}{\xi^h_{i,j\pm1/2}} \doteq \frac{1}{2}\left[\frac{1}{\xi^h_{i,j\pm1}} + \frac{1}{\xi^h_{i,j}}\right] \tag{4.25}$$

When a single grid is used it does not make much difference which approximation is used. However, in a Multigrid algorithm the operator is used on coarser grids too, but to solve a correction for the finer grids. As explained in Chapter 3 the coarse grid operator should be such that it gives a good approximation to the fine grid operator for smooth components. If the coefficients vary smoothly over the grid simple averaging is generally good enough. However, in the case of strongly varying coefficients or discontinuities the harmonic averaging may give better results.

Finally a discrete approximation to the wedge term is derived. A first order discrete approximation is given by:

$$\frac{\partial H}{\partial \theta}|_{i,j} \doteq \frac{H^h_{i,j} - H^h_{i-1,j}}{h_\theta} \tag{4.26}$$

A short second order central discretization is:

$$\frac{\partial H}{\partial \theta}|_{i,j} \doteq \frac{H^h_{i+1/2,j} - H^h_{i-1/2,j}}{h_\theta} \tag{4.27}$$

and the long second order central discretization is:

$$\frac{\partial H}{\partial \theta}|_{i,j} \doteq \frac{H^h_{i+1,j} - H^h_{i-1,j}}{2h_\theta} \tag{4.28}$$

The second order upstream approximation discretization given by Equation (2.11) yields:

$$\frac{\partial H}{\partial \theta}|_{i,j} \doteq \frac{3H^h_{i,j} - 4H^h_{i-1,j} + H^h_{i-2,j}}{2h_\theta} \tag{4.29}$$

Because the discretization of the Poiseuille terms is of second order accuracy also the discrete wedge term should be second order accurate. As H does not depend on P the long second order central discretization is the most obvious choice. Thus discretizing Equation (4.16) yields:

$$\begin{aligned}&\frac{\xi^h_{i+1/2,j}P^h_{i+1,j} - (\xi^h_{i+1/2,j} + \xi^h_{i-1/2,j})P^h_{i,j} + \xi^h_{i-1/2,j}P^h_{i-1,j}}{h^2_\theta}\\ +&\frac{1}{k^2}\frac{\xi^h_{i,j+1/2}P^h_{i,j+1} - (\xi^h_{i,j+1/2} + \xi^h_{i,j-1/2})P^h_{i,j} + \xi^h_{i,j-1/2}P^h_{i,j-1}}{h^2_y}\\ &\qquad\qquad -\frac{H^h_{i+1,j} - H^h_{i-1,j}}{2h_\theta} = {}_Pf^h_{i,j}\end{aligned} \tag{4.30}$$

Where the symbol $_Pf^h$ is used in the right hand side only for generality. On a single grid it will be zero by definition, and when Multigrid is used it will still be zero on the current finest grid. However, on coarser grids it will be defined by the Full Approximation Scheme.

At this point it is noted that in the present case with H^h only defined at the grid points using the second order short central discretization for the wedge term has no real advantage over the long central discretization. It uses $H^h_{i\pm1/2,j}$ and these have to be approximated using values of H^h in grid points. In fact, using:

$$H^h_{i\pm1/2} \doteqdot \left[H^h_i + H^h_{i\pm1}\right]/2 \tag{4.31}$$

it yields the long central discretization as in Equation (4.30).

To complete this section the discrete expressions for the load capacity are presented:

$$W^h_n = h_\theta h_y \sum_i \sum_j P^h_{i,j} \sin(ih_\theta) \tag{4.32}$$

$$W^h_t = h_\theta h_y \sum_i \sum_j P^h_{i,j} \cos(ih_\theta) \tag{4.33}$$

giving the discrete load capacity:

$$W^h = \sqrt{W^h_n + W^h_t} = \frac{1}{3} S^h_0 \tag{4.34}$$

and attitude angle:

$$\Psi^h = \arctan(W^h_t / W^h_n) \tag{4.35}$$

4.3 Relaxation

The objective is to solve P^h from Equation (4.30) with H^h given by Equation (4.22) for all interior grid points (θ_i, Y_j), with $P^h = 0$ given on the boundary. For $H^h = constant$ this would exactly be the two dimensional model problem considered in the previous chapter.

In Chapter 2 it was explained that relaxation is a local process. The effect of changes at a given grid point decays rapidly with increasing distance. As a result relaxation formed a quick local error averager, but a bad solver. Because of the varying coefficients the Local Mode analysis can not be used for the present problem. However, its prediction for the model problem will also apply to the hydrodynamic problem if H (and thereby H^3) varies smoothly over the grid, where smoothly refers to smooth on the scale of the mesh size of the grid. In other words, on a sufficiently fine grid the coefficients in the vicinity of a point (i, j) will roughly be constant as a result of which the local mode analysis at least locally will be justified. As a result, with decreasing mesh size the behaviour of Gauss-Seidel or Jacobi relaxation applied to the solution of this Hydrodynamic problem should be like the behaviour of these relaxations applied to the 2 dimensional Poisson problem.

For Gauss-Seidel and Jacobi relaxation, given a current (or initial) approximation $\tilde{P}^h$ on a grid with mesh size h, a new approximation $\bar{P}^h$ in the point (i,j) is calculated according to:

$$\bar{P}^h_{i,j} = \tilde{P}^h_{i,j} + \delta^h_{i,j} \tag{4.36}$$

where $\delta^h_{i,j}$ is given by:

$$\delta^h_{i,j} = \frac{-r^h_{i,j}}{(\xi^h_{i+1/2,j} + \xi^h_{i-1/2,j})/h^2_\theta + (\xi^h_{i,j+1/2} + \xi^h_{i,j-1/2})/h^2_y} \tag{4.37}$$

and where the residual $r^h_{i,j}$ in the case of Jacobi relaxation is given by:

$$\begin{aligned} r^h_{i,j} = {}_P f^h_{i,j} &- \frac{\xi^h_{i+1/2,j}\tilde{P}^h_{i+1,j} - (\xi^h_{i+1/2,j} + \xi^h_{i-1/2,j})\tilde{P}^h_{i,j} + \xi^h_{i-1/2,j}\tilde{P}^h_{i-1,j}}{h^2_\theta} \\ &- \frac{1}{k^2}\frac{\xi^h_{i,j+1/2}\tilde{P}^h_{i,j+1} - (\xi^h_{i,j+1/2} + \xi^h_{i,j-1/2})\tilde{P}^h_{i,j} + \xi^h_{i,j-1/2}\tilde{P}^h_{i,j-1}}{h^2_y} \\ &+ \frac{H^h_{i+1,j} - H^h_{i-1,j}}{2h_\theta} \end{aligned} \tag{4.38}$$

and the residual $r^h_{i,j}$ for Gauss-Seidel relaxation using lexicographic ordering is given by:

$$\begin{aligned} r^h_{i,j} = {}_P f^h_{i,j} &- \frac{\xi^h_{i+1/2,j}\tilde{P}^h_{i+1,j} - (\xi^h_{i+1/2,j} + \xi^h_{i-1/2,j})\tilde{P}^h_{i,j} + \xi^h_{i-1/2,j}\bar{P}^h_{i-1,j}}{h^2_\theta} \\ &- \frac{1}{k^2}\frac{\xi^h_{i,j+1/2}\tilde{P}^h_{i,j+1} - (\xi^h_{i,j+1/2} + \xi^h_{i,j-1/2})\tilde{P}^h_{i,j} + \xi^h_{i,j-1/2}\bar{P}^h_{i,j-1}}{h^2_y} \\ &+ \frac{H^h_{i+1,j} - H^h_{i-1,j}}{2h_\theta} \end{aligned} \tag{4.39}$$

Figure 4.3 shows the convergence speed of Gauss-Seidel relaxation (4.39) applied to this problem on grids (levels) with a different mesh size covering the domain from $0 \leq \theta \leq 2\pi$ and from $0 \leq Y \leq 1$. The coarsest grid for which results are shown contains $(24+1) \times (4+1)$ points. The results marked with a 2 are obtained on a grid with $(48+1) \times (8+1)$ points and the results marked with a 3 using a grid with $(96+1) \times (16+1)$ points, etc. Please note that this choice of the number of nodes in θ and Y direction for the domain considered results in a mesh size in both directions that is approximately equal: on the coarsest grid $h_\theta = \pi/12 = 0.2618\ldots$ and $h_y = 0.25$.

From Figure 4.3 it can be concluded that the speed of convergence is reduced by a factor of four, each time one halves the mesh size. This behaviour suggests an asymptotic error amplification of $1 - O(h^2)$ per iteration and is identical to the slow convergence of the Poisson equation discussed in Sections 3.9.1 and 3.10.1. As H varies smoothly on the grid, the slow convergence can again be attributed to the bad reduction of low frequency components. Consequently, it can be expected that measures similar to the ones used in the previous chapter, will be able to increase the convergence speed on fine grids.

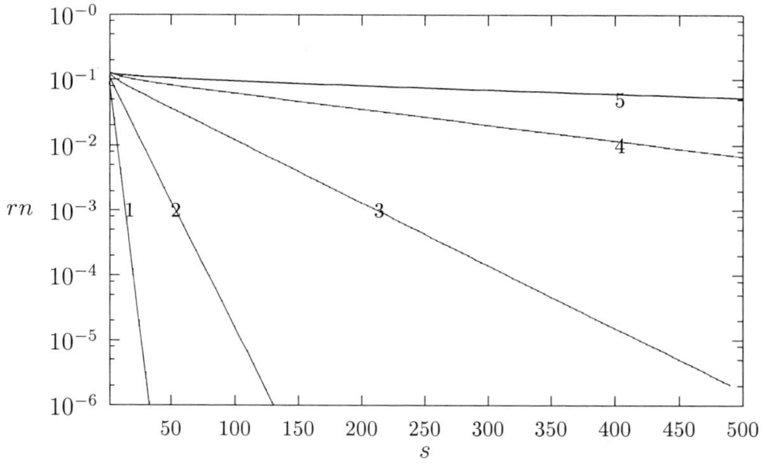

Figure 4.3: *Residual norm as a function of the number of relaxation sweeps* s *on a single grid with* $(2^{l-1}24+1) \times (2^{l-1}4+1)$ *points, for* $l = 1, 2, 3, 4, 5$.

4.4 Cavitation and Complementarity

Before explaining the details of the Multigrid techniques, a closer look is taken at the solution of the problem. Figure 4.4 shows the computed pressure distribution for $\epsilon = 0.2$ and $k = L/R = 1$ as a function of θ and Y on a grid with $(96+1) \times (16+1)$ points. The figure shows that the obtained pressure distribution is antisymmetric, i.e. the pressures are larger than zero for $\theta < \pi$ and smaller than zero for $\theta > \pi$. This is not surprising as the film shape H is symmetric, and thus the coefficients H^3 are symmetric too. However, the $\partial H/\partial\theta$ term in Equation (4.8) which causes the pressure generation is antisymmetric. With the boundary condition $P^h = 0$ at $\theta = 0$ and $\theta = 2\pi$ this leads to the antisymmetric pressure profile.

Figure 4.5 shows the pressure distribution as a function of θ at the line $Y = 1/2$. In this figure also results obtained for other values of ϵ are shown. In that case the behaviour is the same, except that for a larger eccentricity the generated pressure values are larger. The solutions of the type as shown in the Figures 4.4 and 4.5 where first obtained by Sommerfeld for the infinitely wide bearing and are since referred to as the *full Sommerfeld* solutions.

The reader is reminded that physically speaking the pressure distribution obtained is the pressure relative to the pressure of the inlet groove and journal sides (the boundary conditions). When this surrounding pressure is equal to the ambient pressure, the negative pressures generated in the bearing can imply a real pressure that is smaller than the oil vapour pressure. In this case a particular physical phenomenon will occur, the oil will start to evaporate, and the pressure can not descend below this vapour pressure. This process is called cavitation. In general, the vapour pressure is relatively small compared

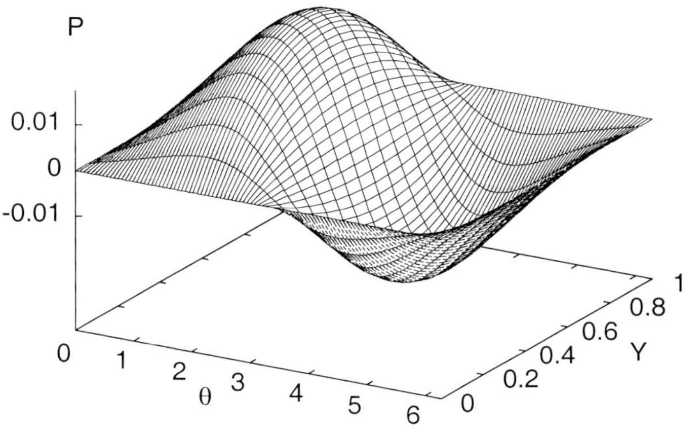

Figure 4.4: *Computed dimensionless pressure* P^h *as a function of* θ *and* Y *for* $\epsilon = 0.2$ *and* $k = L/R = 1$.

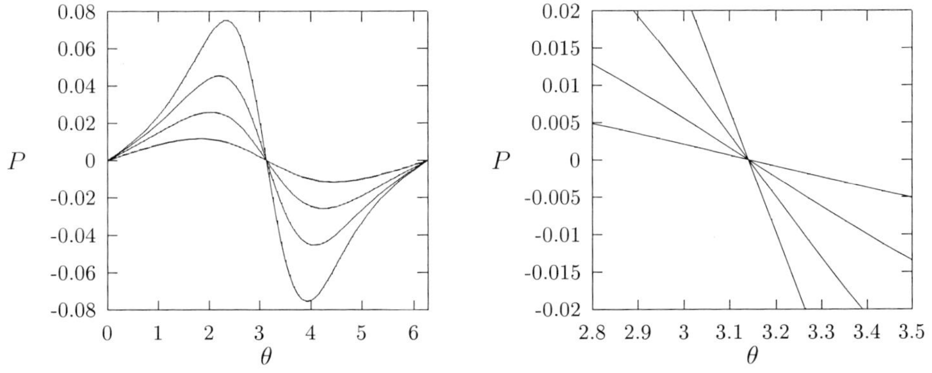

Figure 4.5: *Computed dimensionless pressure* P^h *at the centreline* $Y = 1/2$ *as a function of* θ *for* $\epsilon = 0.1, 0.2, 0.3$ *and* 0.4 *and* $k = L/R = 1$.

to the pressures generated in the journal bearing, and it is taken equal to the ambient pressure. For the case considered here cavitation will start to occur just past $\theta = \pi$. Due to the widening of the film the pressure drops and once it reaches the vapour pressure it will remain at this value as long as the gap diverges. Also, in the cavitated region there is no longer a continuous lubricant film. Because the amount of lubricant is not sufficient to fully fill the gap the film ruptures changing into a mixture of entrained air and oil.

To obtain a physically more realistic estimate of the pressure profile in the gap and the load capacity of the bearing for the present problem there are two alternatives. The simplest one leads to what is referred to as the *half Sommerfeld* solution. In this case one only accepts the positive pressures and assumes a zero pressure for $\theta \geq \pi$. Thus taking a solution as shown in the Figures 4.4 and 4.5 the load capacity of the bearing is computed using only the positive pressure values. Naturally the result will only be an approximation but it can be argued that the error in a quantity such as the load capacity will be relatively small. In reality the onset of cavitation will not be at $\theta = \pi$ but a little later, i.e. at $\theta = \pi + \alpha_c$.

The second alternative involves extension of the model with a cavitation condition. This condition changes the problem into a complementarity problem and is also referred to as the complementarity condition. The need for such a condition was already outlined by Reynolds [92]. The complementarity condition which is added to the Reynolds equation then reads $P \geq 0$, leading to:

$$\begin{array}{lll} \Omega_1 & \dfrac{\partial}{\partial \theta}(H^3\dfrac{\partial P}{\partial \theta}) + \dfrac{1}{k^2}\dfrac{\partial}{\partial Y}(H^3\dfrac{\partial P}{\partial Y}) - \dfrac{\partial H}{\partial \theta} = 0 \quad \& & P > 0 \\ \Omega_2 & & P = 0 \end{array} \tag{4.40}$$

where the domain is split into two sub-domains Ω_1 and Ω_2, and the boundary between the two is 'floating' i.e. not known in advance. Therefore the cavitation condition should be implemented in the numerical solver in such a way that the cavitated zone emerges 'automatically'. The simplest way to do so, on a computer, is to set the pressure to zero whenever a negative (approximate) pressure is computed. This implies that at each such point the discrete Reynolds equation is replaced by the equation $P^h_{i,j} = 0$. The reader is reminded that indeed the Reynolds equation is no longer valid in the cavitated region. As a result, whenever residuals are computed, they should only be computed for points with non-zero pressure and by definition the residuals are zero in the cavitated region.

The Figures 4.6 and 4.7 show the dimensionless pressure obtained taking into account the cavitation condition for the cases shown in the Figures 4.4 and 4.5. Clearly the solution resembles the positive pressure half obtained in these latter figures. Figure 4.7 also shows a detail of the centre line pressure distribution around the cavitation point. Two observations can be made: first of all the pressurized zone indeed extends a little beyond $\theta = \pi$, that is into the divergent zone, and secondly that the pressure gradient goes to zero near the cavitation point. This last observation shows that indeed both P and $\partial P/\partial n$, in this case $\partial P/\partial \theta$, go to zero simultaneously. This is a physical requirement since the pressure has to be continuous over the cavitation boundary, because of mass flux continuity.

The one level convergence speed of the relaxation is not affected by the complementarity condition and remains identical to the speed of the non-cavitating problem given in Figure 4.3. However, the complementarity condition requires special attention in the design of the coarse grid correction cycle. In the following sections the necessary steps to be taken to obtain an efficient coarse grid correction cycle are explained. First the simple case is considered where negative pressures are allowed, i.e. where the full Sommerfeld solution is computed. Next the modifications needed when the cavitation condition is incorporated are discussed.

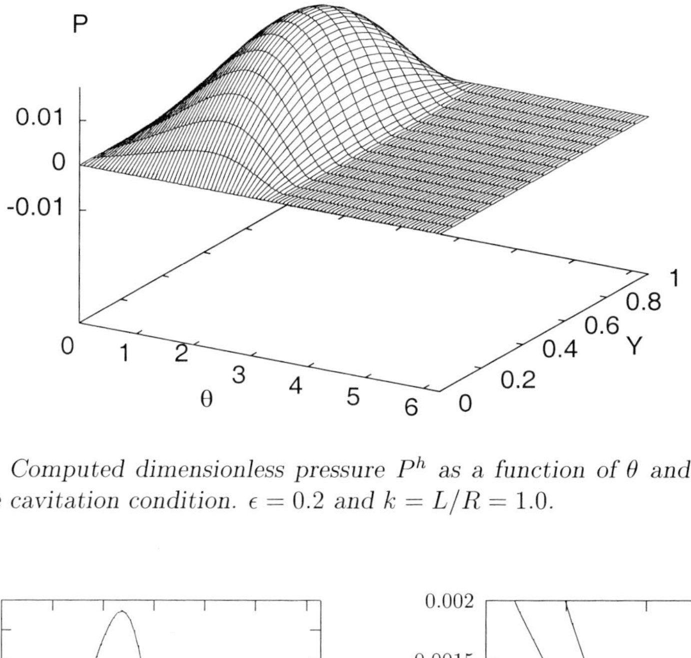

Figure 4.6: *Computed dimensionless pressure P^h as a function of θ and Y taking into account the cavitation condition.* $\epsilon = 0.2$ *and* $k = L/R = 1.0$.

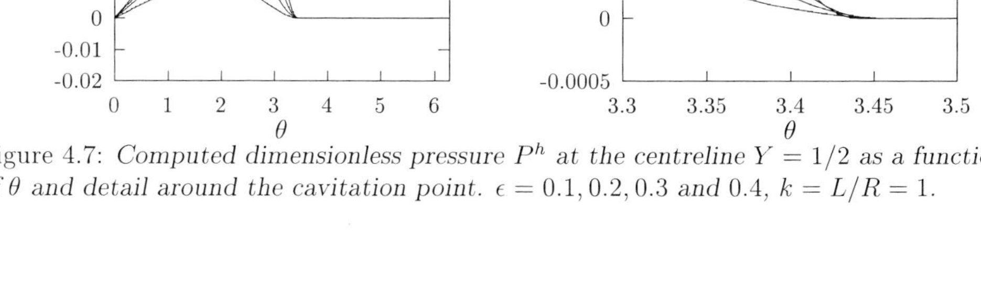

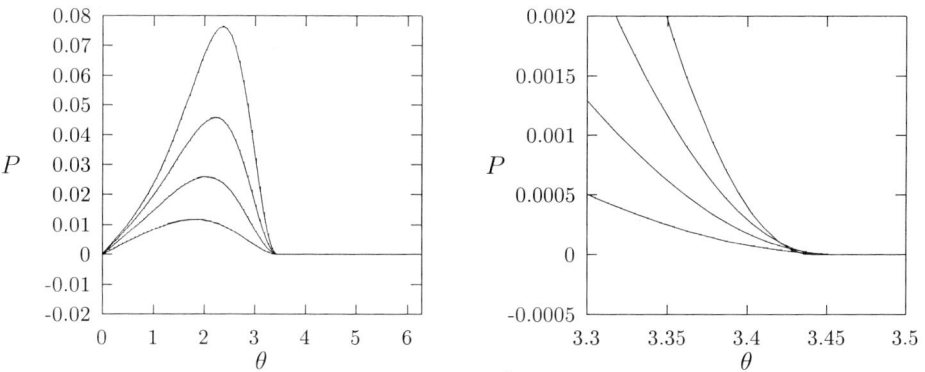

Figure 4.7: *Computed dimensionless pressure P^h at the centreline $Y = 1/2$ as a function of θ and detail around the cavitation point.* $\epsilon = 0.1, 0.2, 0.3$ *and* 0.4, $k = L/R = 1$.

4.5 Coarse Grid Correction

As relaxation gives very slow convergence on fine grids (Figure 4.3), we will implement Multigrid techniques to remove this slowness. The first step towards a Multigrid cycle is to construct the coarse grid correction. In the next paragraph the problem is analysed without the cavitation condition. After good convergence is obtained, the cavitation problem is studied in the consequent section.

4.5.1 Sommerfeld Problem

In this paragraph the problem without the complementarity condition is considered. For generality the coarse grid correction is described using the Full Approximation Scheme, see Section 3.8.2, although strictly FAS is only needed if cavitation is taken into account as the complementarity condition causes the problem to become non-linear. For simplicity only two grids are assumed. A fine grid with mesh size h_θ and h_y and a coarse grid with H_θ and H_Y. The fine grid discretised dimensionless Reynolds equation (4.30) reads:

$$\frac{\xi^h_{i+1/2,j}P^h_{i+1,j} - (\xi^h_{i+1/2,j} + \xi^h_{i-1/2,j})P^h_{i,j} + \xi^h_{i-1/2,j}P^h_{i-1,j}}{h_\theta^2} + \frac{1}{k^2}\frac{\xi^h_{i,j+1/2}P^h_{i,j+1} - (\xi^h_{i,j+1/2} + \xi^h_{i,j-1/2})P^h_{i,j} + \xi^h_{i,j-1/2}P^h_{i,j-1}}{h_y^2} - \frac{H^h_{i+1,j} - H^h_{i-1,j}}{2h_\theta} = {}_Pf^h_{i,j} \tag{4.41}$$

Assuming that after a few relaxations an approximate solution $\tilde{P}^h_{i,j}$ is obtained. The fine grid residual is then defined for all interior points (θ_i, Y_j) as:

$$r^h_{i,j} = {}_Pf^h_{i,j} - \frac{\xi^h_{i+1/2,j}\tilde{P}^h_{i+1,j} - (\xi^h_{i+1/2,j} + \xi^h_{i-1/2,j})\tilde{P}^h_{i,j} + \xi^h_{i-1/2,j}\tilde{P}^h_{i-1,j}}{h_\theta^2} - \frac{1}{k^2}\frac{\xi^h_{i,j+1/2}\tilde{P}^h_{i,j+1} - (\xi^h_{i,j+1/2} + \xi^h_{i,j-1/2})\tilde{P}^h_{i,j} + \xi^h_{i,j-1/2}\tilde{P}^h_{i,j-1}}{h_y^2} + \frac{H^h_{i+1,j} - H^h_{i-1,j}}{2h_\theta} \tag{4.42}$$

The FAS coarse grid Equation (3.72) for each interior point (θ_I, Y_J) reads:

$$\frac{\xi^H_{I+1/2,J}P^H_{I+1,J} - (\xi^H_{I+1/2,J} + \xi^H_{I-1/2,j})P^H_{I,J} + \xi^H_{I-1/2,J}P^H_{I-1,J}}{H_\theta^2} + \frac{1}{k^2}\frac{\xi^H_{I,J+1/2}P^H_{I,J+1} - (\xi^H_{I,J+1/2} + \xi^H_{I,J-1/2})P^H_{I,J} + \xi^H_{I,J-1/2}P^H_{I,J-1}}{H_Y^2} - \frac{H^H_{I+1,J} - H^H_{I-1,J}}{2H_\theta} = {}_P\hat{f}^H_{I,J} \tag{4.43}$$

where the initial value of P^H is defined by $\tilde{P}^H_{I,J} = [I_h^H \tilde{P}^h_{.,.}]_{I,J}$. Note that H_θ and H_Y denote the coarse grid mesh size, $H_\theta = 2h_\theta$ and $H_Y = 2h_y$. The right hand side ${}_Pf^h_{i,j} = 0$ on the finest grid. On coarser grids the right hand side is ${}_P\hat{f}^H_{I,J}$ and defined using (3.74) as:

$$
\begin{aligned}
{}_P\hat{f}^H_{I,J} = [I_h^H r^h_{.,.}]_{I,J} &+ \frac{\xi^H_{I+1/2,J}\tilde{P}^H_{I+1,J} - (\xi^H_{I+1/2,J} + \xi^H_{I-1/2,J})\tilde{P}^H_{I,J} + \xi^H_{I-1/2,J}\tilde{P}^H_{I-1,J}}{H_\theta^2} \\
&+ \frac{1}{k^2}\frac{\xi^H_{I,J+1/2}\tilde{P}^H_{I,J+1} - (\xi^H_{I,J+1/2} + \xi^H_{I,J-1/2})\tilde{P}^H_{I,J} + \xi^H_{I,J-1/2}\tilde{P}^H_{I,J-1}}{H_Y^2} \\
&\qquad - \frac{H^H_{I+1,J} - H^H_{I-1,J}}{2H_\theta}
\end{aligned}
\tag{4.44}
$$

After an approximation $\bar{P}^H$ to P^H has been obtained from Equation (4.43), the fine grid approximation $\tilde{P}^h$ is corrected according to Equation (3.75) which gives:

$$
\underline{\bar{P}}^h = \underline{\tilde{P}}^h + I_H^h(\underline{\bar{P}}^H - I_h^H \underline{\tilde{P}}^h) \tag{4.45}
$$

For the restriction of the solution and the residuals, the full weighting operator can be used in (4.44) and (4.45) as introduced in the previous chapter.

The next table studies the convergence as a function of the number of V(2,1) cycles and as a function of the level, using the initial approximation $P^h = 0$. All results in this section have been obtained for $\epsilon = 0.2$, $k = L/R = 1$ and 24×4 points on the coarsest grid. As can be seen from Table 4.1, the residual reduction per V-cycle is $O(10)$ and is not affected by either the number of preceding V-cycles, nor by the number of points on a level. On the whole, the convergence of the hydrodynamic lubrication solver is as fast as the convergence of the Poisson (2d) program. This is exactly what one could expect. On a sufficiently fine grid the coefficient ξ will vary smoothly over the grid and the problem for $k = L/R = 1$ locally approximates the Poisson 2d problem. For that problem the worst amplification factor for high frequency components was $\bar{\mu} = 0.5$ which predicts an error reduction of a factor of 8 per V(2,1) cycle. At this point it is noted that this convergence speed is valid as long as the coefficient $h_\theta^2/(k^2h_y^2) = O(1)$, see Section 4.10.1.

Having shown good convergence in terms of the reduction of residuals next convergence of the solution with decreasing mesh size can be investigated. For this purpose the load capacity of the bearing is taken. In accordance with the Sommerfeld approximation only the positive pressures are taken. This implies that Equation (4.32) is used with $\max(0, P^h_{i,j})$ replacing $P^h_{i,j}$. The value of W^h thus obtained can be compared on different grids. As such it allows us to estimate the discretization and the numerical error in the solution. Table 4.2 gives the computed load capacity as a function of the gridlevel and the number of $V(2,1)$ cycles that is used.

The table shows that after three to five cycles the solution has converged to an error that is small compared to the discretization error. This can be seen from the fact that the difference between the solution on a grid after some cycles compared to the value on the same grid after many cycles is small compared to the difference between the solution on that grid after many cycles and the one on the next finer grid after many cycles. With respect to the discretization error the table shows that the difference between the

V-cycle	level 5	level 6	level 7
	384×64	768×128	1536×256
1	$8.99 \cdot 10^{-3}$	$8.96 \cdot 10^{-2}$	$8.98 \cdot 10^{-3}$
2	$7.88 \cdot 10^{-4}$	$7.93 \cdot 10^{-4}$	$7.98 \cdot 10^{-4}$
3	$7.41 \cdot 10^{-5}$	$7.89 \cdot 10^{-5}$	$8.10 \cdot 10^{-5}$
4	$6.82 \cdot 10^{-6}$	$8.07 \cdot 10^{-6}$	$8.75 \cdot 10^{-6}$
5	$6.41 \cdot 10^{-7}$	$8.21 \cdot 10^{-7}$	$9.29 \cdot 10^{-7}$
6	$6.10 \cdot 10^{-8}$	$8.23 \cdot 10^{-8}$	$9.75 \cdot 10^{-8}$
7	$5.76 \cdot 10^{-9}$	$7.95 \cdot 10^{-9}$	$1.01 \cdot 10^{-8}$
8	$5.31 \cdot 10^{-10}$	$7.61 \cdot 10^{-10}$	$1.03 \cdot 10^{-9}$

Table 4.1: *Residual norm on three different levels as a function of the number of V(2,1) cycles.* $\epsilon = 0.2$, $k = L/R = 1$.

V-cycle	level 5	level 6	level 7
	384×64	768×128	1536×256
1	$2.39707 \cdot 10^{-2}$	$2.39728 \cdot 10^{-2}$	$2.39733 \cdot 10^{-2}$
2	$2.55129 \cdot 10^{-2}$	$2.55185 \cdot 10^{-2}$	$2.55199 \cdot 10^{-2}$
3	$2.56097 \cdot 10^{-2}$	$2.56157 \cdot 10^{-2}$	$2.56172 \cdot 10^{-2}$
4	$2.56156 \cdot 10^{-2}$	$2.56216 \cdot 10^{-2}$	$2.56232 \cdot 10^{-2}$
5	$2.56159 \cdot 10^{-2}$	$2.56220 \cdot 10^{-2}$	$2.56235 \cdot 10^{-2}$
6	$2.56159 \cdot 10^{-2}$	$2.56220 \cdot 10^{-2}$	$2.56235 \cdot 10^{-2}$
20	$2.56159 \cdot 10^{-2}$	$2.56220 \cdot 10^{-2}$	$2.56235 \cdot 10^{-2}$

Table 4.2: *Sommerfeld approximation to the load capacity* W^h *as a function of the number of V(2,1) cycles on different levels.* $\epsilon = 0.2$, $k = L/R = 1$.

computed load capacity on level 6 and 7 is four times smaller than the difference between the computed load capacity on level 5 and 6. This is in accordance with the second order accuracy of the discretization that is used. The subject of accuracy will be addressed in more detail in Section 4.7.

The fact that with decreasing mesh size an increasing number of cycles is required to attain this precision is caused by the bad initial approximation ($P = 0$), and implies that using an FMG algorithm one would reduce the amount of work needed to attain this precision.

4.5.2 Cavitation

When the cavitation condition limits the pressure to positive or zero values, the pressures obtained during relaxation have to be submitted to this constraint. This means that after the calculation of a new pressure value in the point (i, j), the pressure is immediately set to zero if it is negative. The coarse grid correction cycle outlined in the previous section can still be used. However, because of the complementarity character of the cavitation problem, special attention should be given to the restriction of residuals and to the interpolation of the corrections. Brandt and Cryer [14] have investigated the effects of different modifications on the asymptotic convergence of a cycle for a free boundary problem. The measures presented below are taken from this work.

The first main point is that in the residual transfer and in the transfer of the solution mixing of information from the pressure and cavitated zone should be avoided as much as possible. This is for example already achieved by using injection everywhere. It is easy to see that the full weighting operator applied to the pressures can in some cases significantly displace the cavitation boundary into the cavitated zone. Such displacements are highly undesirable as they can lead to a situation where the asymptotic convergence of the coarse grid correction cycle stalls. As a result of the displacement the coarse grid incorrectly treats some points as pressurized and eventually it results into correction of cavitated points into pressurized points on the fine grid. The post relaxations carried out in the cycle then return these points into cavitated points again. This process may repeat itself and cause the situation where the residuals are small everywhere except for a narrow band around the cavition boundary where they remain fixed at a certain level and do not converge because of this switching back and forth between cavitated and non-cavitated. The main idea in the entire treatment of the cavitation condition in the intergrid transfers is that the (exact location of) the cavitation boundary should be determined by the relaxation on the fine grid, and that coarser grids should not move it.

The second point is that the interpolation of the pressure correction should be adapted. Classical linear interpolation can be used provided the following modifications are implemented:

- The correction should be applied only to points where the pressure is non-zero.
- After the entire field has been corrected one should perform a sweep over the grid checking the value of the pressures in each point and changing negative pressures to zero to ensure that the solution after correction but before relaxation satisfies the cavitation condition.

V-cycle	level 5	level 6	level 7
	384×64	768×128	1536×256
1	$1.17 \cdot 10^{-2}$	$1.72 \cdot 10^{-2}$	$2.58 \cdot 10^{-2}$
2	$7.13 \cdot 10^{-3}$	$1.92 \cdot 10^{-2}$	$4.53 \cdot 10^{-2}$
3	$2.83 \cdot 10^{-3}$	$1.26 \cdot 10^{-2}$	$3.81 \cdot 10^{-2}$
4	$7.66 \cdot 10^{-4}$	$5.81 \cdot 10^{-3}$	$2.75 \cdot 10^{-2}$
5	$1.03 \cdot 10^{-4}$	$2.79 \cdot 10^{-3}$	$2.04 \cdot 10^{-2}$
6	$1.11 \cdot 10^{-5}$	$7.23 \cdot 10^{-4}$	$1.35 \cdot 10^{-2}$
7	$2.23 \cdot 10^{-6}$	$1.48 \cdot 10^{-4}$	$8.14 \cdot 10^{-3}$
8	$8.15 \cdot 10^{-6}$	$1.76 \cdot 10^{-5}$	$3.93 \cdot 10^{-3}$
9	$4.11 \cdot 10^{-7}$	$6.82 \cdot 10^{-6}$	$1.46 \cdot 10^{-3}$
10	$2.08 \cdot 10^{-7}$	$2.99 \cdot 10^{-6}$	$5.70 \cdot 10^{-4}$

Table 4.3: *Residual on three different levels as a function of the number of $V(2,1)$ cycles for the problem with cavitation condition.* $\epsilon = 0.2$, $k = L/R = 1$.

Table 4.3 shows the evolution of the residual as a function of the gridlevel and the number of $V(2,1)$ cycles for the same cases considered in the previous paragraph. The observed reduction of the residual norm is far from satisfactory: not only is the residual reduction small and irregular compared to the factor of 10 that was obtained in Table 4.1, but this reduction diminishes on finer grids. For the level 5 results the effective error amplification per fine grid relaxation is 0.74 and for the level 7 results it is 0.84. Comparing this to the asymptotic smoothing factor $\bar{\mu} = 0.5$ this is indeed rather disappointing. Since the previous section showed that the residuals in the pressure zone were effectively reduced, and since the residuals in the cavitation zone are zero by definition, this behaviour can only stem from the vicinity of the cavitation boundary.

The fact that the residual reduction is perturbed by the cavitation condition suggests that the origin of the trouble lies in the change of nature of the problem across the cavitation boundary (the complementarity problem). When studying the local residuals, it is indeed observed that the residuals in the neighbourhood of the cavitation boundary are not reduced efficiently. It is therefore possible to envisage a local solution, one that only applies to the neighbourhood of the cavitation boundary. A possible, and by far the simplest, solution is to add a number of local relaxation sweeps, in a pre-relaxation phase. In the **HL2d** program three pre-relaxations over a few points before and after the cavitational boundary are used. These pre-relaxations are carried out in θ-direction, and then skipping to the next Y-line.

To find the optimum domain over which the pre-relaxations are to be carried out, a simple test can be used. Performing the three pre-relaxations over the entire domain, the pre-relaxation routine becomes effectively identical to the relaxation routine (*debugging*). Thus the convergence is effectively one of a V(8,4) cycle! The obtained convergence should fulfill the speed requirements, and especially be grid independent. Then the pre-relaxation domain can be reduced to a small portion of the full domain, in such a way that say 90% of the performance is obtained at 1% of the cost (work/number of points involved). Table 4.4 shows the convergence history of the residuals for the case when 3 local pre-relaxations are applied.

V-cycle	level 5	level 6	level 7
	384×64	768×128	1536×256
1	$9.67 \cdot 10^{-3}$	$1.60 \cdot 10^{-2}$	$2.74 \cdot 10^{-2}$
2	$3.30 \cdot 10^{-3}$	$1.06 \cdot 10^{-2}$	$2.62 \cdot 10^{-2}$
3	$5.32 \cdot 10^{-4}$	$5.39 \cdot 10^{-3}$	$1.96 \cdot 10^{-2}$
4	$4.22 \cdot 10^{-5}$	$1.72 \cdot 10^{-3}$	$1.24 \cdot 10^{-2}$
5	$4.06 \cdot 10^{-6}$	$2.97 \cdot 10^{-4}$	$7.38 \cdot 10^{-3}$
6	$4.26 \cdot 10^{-7}$	$2.22 \cdot 10^{-5}$	$3.70 \cdot 10^{-3}$
7	$5.07 \cdot 10^{-8}$	$2.40 \cdot 10^{-6}$	$1.36 \cdot 10^{-3}$
8	$7.84 \cdot 10^{-9}$	$3.16 \cdot 10^{-7}$	$3.40 \cdot 10^{-4}$
9	$1.38 \cdot 10^{-9}$	$5.52 \cdot 10^{-8}$	$3.24 \cdot 10^{-5}$
10	$2.67 \cdot 10^{-10}$	$1.20 \cdot 10^{-8}$	$3.05 \cdot 10^{-6}$

Table 4.4: *Residual norm on three different levels as a function of the number of $V(2,1)$ cycles, for the problem with cavitation when three local pre-relaxations are used.* $\epsilon = 0.2$, $k = L/R = 1$.

The reduction of the residual norm has improved considerably in comparison with Table 4.3 and the additional cost is negligible as the cavitation boundary by definition is a one dimensional phenomenon. However, during the first few cycles, the convergence speed still tends to be rather low. A second negative point is that the number of cycles required until the convergence improves, increases with the number of grids used. It seems that only once a relatively accurate description of the location of the cavitation boundary is obtained, the solver can realise its full potential.

Now that the issue of the convergence speed has been addressed, it is time to direct our attention again towards the solution that has been obtained. Table 4.5 shows the load capacity of the bearing as a function of the gridlevel and the number of cycles. The table shows the same tendencies as observed in Table 4.2. About 3 to 5 cycles are needed for the solution to reach the level of the discretization error and this increasing number of cycles needed can be attributed to the poor quality of the first approximation. Also here the difference between the level 5 and 6 results is 4 times larger than the difference between the level 6 and 7 results which confirms the second order of the discretization. Finally, note that the computed load capacity differs very little from the values obtained when cavitation is not taken into account and negative pressures are disregarded. The differences are of the order of 2 %.

4.6 Full MultiGrid

In the previous section it was observed that the number of cycles needed to obtain a solution with an error comparable to the discretization error increased with decreasing mesh size. This was attributed to the fact that the same starting solution $P^h = 0$ was used

V-cycle	level 5	level 6	level 7
	384×64	768×128	1536×256
1	$2.55207 \cdot 10^{-2}$	$2.55182 \cdot 10^{-2}$	$2.55154 \cdot 10^{-2}$
2	$2.60724 \cdot 10^{-2}$	$2.60738 \cdot 10^{-2}$	$2.60628 \cdot 10^{-2}$
3	$2.61049 \cdot 10^{-2}$	$2.61107 \cdot 10^{-2}$	$2.61020 \cdot 10^{-2}$
4	$2.61028 \cdot 10^{-2}$	$2.61108 \cdot 10^{-2}$	$2.61081 \cdot 10^{-2}$
5	$2.61028 \cdot 10^{-2}$	$2.61094 \cdot 10^{-2}$	$2.61145 \cdot 10^{-2}$
6	$2.61029 \cdot 10^{-2}$	$2.61086 \cdot 10^{-2}$	$2.61125 \cdot 10^{-2}$
20	$2.61029 \cdot 10^{-2}$	$2.61088 \cdot 10^{-2}$	$2.61103 \cdot 10^{-2}$

Table 4.5: *Load capacity W^h as a function of the gridlevel and the number of $V(2,1)$ cycles.* $\epsilon = 0.2$, $k = L/R = 1$.

on each grid. On a finer grid the discretization error is smaller. If the starting solution is the same also the initial error is the same. Consequently, to obtain an error smaller than the discretization error on a finer grid a larger error reduction is needed, and thus more cycles. To be precise, as was explained in Chapter 3, the number of cycles needed is roughly $O(\ln(N))$ if N is the total number of nodes on the grid. The reader is reminded that the computational cost of each cycle is the equivalent of a constant times the work of a single relaxation on the finest grid. As each relaxation requires $O(N)$ operations the total work invested to obtain a converged solution by means of cycles only will be $O(N\ln(N))$.

As was shown in Chapter 3 the logarithmic factor can be removed from the work count if a better first approximation is used. This is exactly what is done in a Full MultiGrid (FMG) algorithm, see Section 3.7. For the present problem there is another reason why an accurate first approximation is needed. It was shown that for the solver to converge rapidly a certain precision in the location of the cavitation boundary is needed. In a FMG algorithm the starting solution on a certain grid is obtained by interpolating the converged solution from the next coarser grid. Subsequently, coarse grid correction cycles are used to reduce the error. This coarser grid solution is obtained in exactly the same way. Therefore, in a FMG algorithm automatically a sequence of solutions to the problem is created on finer and finer meshes. This leads to another advantage of the FMG algorithm. The sequence of solutions facilitates an a-posteriori verification whether the numerical error in the final solution is indeed smaller than the discretization error. These subjects are addressed in the present section.

Table 4.6 shows results obtained for the journal bearing problem with $\epsilon = 0.2$, $k = L/R = 1$ as before, but using a FMG algorithm. The table gives the residual on the grid as a function of the number of $V(2,1)$ cycles. From the table it can be seen that using a FMG process with a single $V(2,1)$ cycle per level the residual norm for the final solution on level 7 is already smaller than the residual norm for the solution that was obtained applying 9 cycles starting with $P^h = 0$ as an initial solution, see Table 4.4. As the total work invested to obtain the FMG solution is only 4/3 times the work of a single cycle, starting off with the converged solution of a coarser grid really pays off. The same precision is reached in an amount of work that is about 8 times smaller.

To determine if the solution obtained using a FMG algorithm with a specific number of cycles has converged to the level of the discretization error the approximate error norm presented in Chapter 3 can be used, see (3.89). This approximate error was defined as the difference between the (converged) solutions on two consecutive grids and its value as a function of the gridlevel and the number of coarse grid correction cycles performed per level is presented in Table 4.7. One can conclude that the one V(2,1) cycle FMG process gives completely converged solutions, on all levels shown. That is, on all levels the numerical error is small compared to the discretization error on that grid. Note that on each finer grid the approximate error norm is smaller by a factor 4. This is in accordance with the second order discretization that is used as will be explained in more detail in Section 4.7.

Having demonstrated convergence in terms of the approximate error, the load capacity of the bearing is considered in the next study. Table 4.8 gives the computed load capacity W^h for the solutions presented in Table 4.6 and 4.7. If a FMG algorithm with a single $V(2,1)$ cycle per level already yields a converged solution this table should show that the changes in the computed load capacity for larger numbers of cycles are small compared to the differences between the load capacity computed on the different grid. This is indeed the case.

V-cycle	level 5	level 6	level 7	level 8
	384×64	768×128	1536×256	3072×512
1	$3.17 \cdot 10^{-6}$	$8.90 \cdot 10^{-7}$	$8.06 \cdot 10^{-7}$	$3.14 \cdot 10^{-7}$
2	$5.38 \cdot 10^{-7}$	$1.35 \cdot 10^{-7}$	$2.40 \cdot 10^{-7}$	$1.34 \cdot 10^{-7}$
3	$1.05 \cdot 10^{-7}$	$2.05 \cdot 10^{-8}$	$8.52 \cdot 10^{-8}$	$2.34 \cdot 10^{-8}$
4	$2.17 \cdot 10^{-8}$	$5.00 \cdot 10^{-9}$	$3.29 \cdot 10^{-8}$	$1.65 \cdot 10^{-8}$

Table 4.6: *Residual norm on the grid as a function of the number of $V(2,1)$ cycles for the FMG algorithm applied to the problem with cavitation.* $\epsilon = 0.2$, $k = L/R = 1$.

V-cycle	level 4	level 5	level 6	level 7
	192×32	384×64	768×128	1536×256
1	$1.15 \cdot 10^{-6}$	$2.67 \cdot 10^{-7}$	$7.76 \cdot 10^{-8}$	$1.96 \cdot 10^{-8}$
2	$1.04 \cdot 10^{-6}$	$2.39 \cdot 10^{-7}$	$6.12 \cdot 10^{-8}$	$1.57 \cdot 10^{-8}$
3	$1.05 \cdot 10^{-6}$	$2.30 \cdot 10^{-7}$	$6.22 \cdot 10^{-8}$	$1.41 \cdot 10^{-8}$
4	$1.04 \cdot 10^{-6}$	$2.31 \cdot 10^{-7}$	$6.14 \cdot 10^{-8}$	$1.46 \cdot 10^{-8}$

Table 4.7: *Approximate error norm on different levels as a function of the number of $V(2,1)$ cycles for the FMG algorithm applied to the problem with cavitation.* $\epsilon = 0.2$, $k = L/R = 1$.

The FMG algorithm completes what is obviously an efficient solver for the journal bearing problem. Using a single $V(2,1)$ cycle per gridlevel a solution is obtained with an error that is small compared to the discretization error. Apart from a few remarks little

V-cycle	level 5	level 6	level 7	level 8
	384 × 64	768 × 128	1536 × 256	3072 × 512
1	$2.61028 \cdot 10^{-2}$	$2.61087 \cdot 10^{-2}$	$2.61103 \cdot 10^{-2}$	$2.61102 \cdot 10^{-2}$
2	$2.61028 \cdot 10^{-2}$	$2.61088 \cdot 10^{-2}$	$2.61103 \cdot 10^{-2}$	$2.61107 \cdot 10^{-2}$
3	$2.61029 \cdot 10^{-2}$	$2.61088 \cdot 10^{-2}$	$2.61103 \cdot 10^{-2}$	$2.61107 \cdot 10^{-2}$
4	$2.61029 \cdot 10^{-2}$	$2.61088 \cdot 10^{-2}$	$2.61103 \cdot 10^{-2}$	$2.61107 \cdot 10^{-2}$

Table 4.8: *Computed dimensionless load capacity W^h on different grids as a function of the number of $V(2,1)$ cycles for the FMG algorithm applied to the problem with cavitation.* $\epsilon = 0.2$, $k = L/R = 1$.

has been said regarding the magnitude of this discretization error itself. This is the topic of the next section.

4.7 Accuracy

Having developed an FMG algorithm for the journal bearing problem including cavitation, the precision of the generated solutions will now be analysed by comparing the results with those obtained when the second order central discretization for the $\partial H/\partial\theta$ term, Equation (4.28), is replaced by the first order upstream (backward) discretization given by Equation (4.26). The variables used for this analysis are the maximum pressure and the approximate error norm. The maximum pressure P^h_{max} is defined as:

$$P^h_{max} = \max_{i\in[0,n_\theta]} \max_{j\in[0,n_y]} |P^h_{i,j}| \tag{4.46}$$

To really obtain a significant maximum pressure the conditions are changed to $\epsilon = 0.9$ with $k = L/R = 1$. Table 4.9 and 4.10 give the maximum pressure as a function of the gridlevel using an FMG algorithm with one, two, and three $V(2,1)$ cycles per level for the first and second order discretization.

A first conclusion to be drawn from these tables is that the problem is indeed solved up to the level of the discretization error using FMG with a single $V(2,1)$ cycle per level. Adding one or two additional cycles does not significantly increase the accuracy of the solution, when comparing it to the fine grid solutions. A second conclusion is that the difference between the values of P^h_{max} on two consecutive grids decreases for finer and finer grids. For a s order discretization one should have:

$$P^h_{max} = P_{max} + O(h^s) \tag{4.47}$$

As a result the difference between the values computed at two consecutive levels with mesh sizes H and $h = H/2$ should satisfy:

$$|P^H_{max} - P^h_{max}| = O(H^s) - O(h^s) = O((2^s - 1)h^s) \tag{4.48}$$

This implies that using the first order scheme the difference between the values obtained on two consecutive grids should decrease with a factor 2 each time the mesh size decreases, and with a factor 4 using a second order scheme. This is indeed what the tables

first order				
level	$n_\theta \times n_y$	1 V-cycle	2 V-cycles	3 V-cycles
1	24×4	5.813423	5.815648	5.815675
2	48×8	5.221731	5.206182	5.205764
3	96×16	4.723120	4.693977	4.693969
4	192×32	4.370199	4.365616	4.365446
5	384×64	4.220399	4.217919	4.217739
6	768×128	4.141380	4.140629	4.140353
7	1536×256	4.102187	4.101872	4.101638
8	3072×512	4.082588	4.082461	4.082296
∞	$\infty \times \infty$	4.062989	4.063050	4.062954

Table 4.9: P^h_{max} *as a function of the grid level computed with a FMG algorithm with 1,2, and 3* $V(2,1)$ *cycles per level. First order upstream discretization of the* $\partial H/\partial\theta$ *term.* $\epsilon = 0.9$, $k = L/R = 1$.

show. Knowing this asymptotic convergence behaviour with decreasing mesh size one can approximate the discretization error on the fine grid. For the first order scheme the discretization error on the fine grid should be equal to the difference between the fine grid and the coarse grid solution. For the second order scheme the fine grid discretization error should be equal to the difference between the fine grid and coarse grid divided by three. As a result using the level 8 and level 7 solutions given in the Tables 4.9 and 4.10 one can extrapolate to obtain an approximation of:

$$P_{max} = \lim_{h \to 0} P^h_{max} \tag{4.49}$$

Note that this extrapolation allows an important increase in precision in case of the first order scheme. Also note that in order to obtain an accurate extrapolation for the second order scheme, the additional $V(2,1)$ cycles are indeed necessary.

Next the Tables 4.11 and 4.12 show the approximate error norm aen^H, see Section 3.10.3, for the different levels and FMG solutions obtained with 1,2, and 3 $V(2,1)$ cycles per level. These tables confirm once more that even after a single $V(2,1)$ cycle the discretization error is the dominant term, and hence, that the approximate error norm indeed follows the order of the discretization errors, decreasing by a factor of two for the first order scheme and by a factor of four for the second order scheme each time the mesh size is halved. Furthermore, as the approximate error is defined as the difference between a fine grid and a coarse grid solution it by definition forms an approximation to the discretization error. Please recall that the error norm on level 2 approximates the error in the solution P^h on level 1, etc. So aen^k on level k approximates the error on level $k-1$. Thus the approximate error norm serves as an indication of the average error in the solution.

The level 8 first order results indicate an average error of 10^{-3}, whereas the error occurring in the maximum pressure is 10^{-2}. The difference of a factor larger than 10 can be explained because the maximum pressure is around 4, whereas the mean pressure is

second order				
level	$n_\theta \times n_y$	1 V-cycle	2 V-cycles	3 V-cycles
1	24×4	3.762386	3.764264	3.764288
2	48×8	3.929965	3.935462	3.935783
3	96×16	4.021637	4.021282	4.021273
4	192×32	4.057139	4.057243	4.057276
5	384×64	4.060327	4.060185	4.060241
6	768×128	4.062710	4.062758	4.062769
7	1536×256	4.062957	4.062927	4.062944
8	3072×512	4.063011	4.063019	4.063023
∞	$\infty \times \infty$	4.063029	4.063050	4.063049

Table 4.10: P^h_{max} *as a function of the grid level computed with a FMG algorithm with 1,2, and 3* $V(2,1)$ *cycles per level. Second order central discretization of the* $\partial H/\partial\theta$ *term.* $\epsilon = 0.9$, $k = L/R = 1$.

level	$n_\theta \times n_y$	1 V-cycle	2 V-cycles	3 V-cycles
1	24×4	$9.59 \cdot 10^{-2}$	$9.68 \cdot 10^{-2}$	$9.68 \cdot 10^{-2}$
2	48×8	$5.15 \cdot 10^{-2}$	$5.23 \cdot 10^{-2}$	$5.23 \cdot 10^{-2}$
3	96×16	$2.45 \cdot 10^{-2}$	$2.31 \cdot 10^{-2}$	$2.31 \cdot 10^{-2}$
4	192×32	$1.12 \cdot 10^{-2}$	$1.12 \cdot 10^{-2}$	$1.12 \cdot 10^{-2}$
5	384×64	$5.52 \cdot 10^{-3}$	$5.43 \cdot 10^{-3}$	$5.45 \cdot 10^{-3}$
6	768×128	$2.67 \cdot 10^{-3}$	$2.69 \cdot 10^{-3}$	$2.68 \cdot 10^{-3}$
7	1536×256	$1.32 \cdot 10^{-3}$	$1.34 \cdot 10^{-3}$	$1.33 \cdot 10^{-3}$

Table 4.11: aen^H *using first order discretization of the* $\partial H/\partial\theta$ *term, FMG with 1,2 and 3* $V(2,1)$ *cycles per level.* $\epsilon = 0.9$, $k = L/R = 1$.

around 0.3. Similarly, the mean error of 10^{-6} indicated by the *aen* on level 8, predicts a maximum error of 10^{-5} in P^h_{max} for the second order discretization.

4.8 Other L/R Ratios: Bearing Design

To demonstrate performance and accuracy so far the results were restricted to a few single cases with $k = L/R = 1$ and $n_\theta = 24$ and $n_y = 4$ on the coarsest grid. This implies that on all grids the mesh size in θ and y direction are approximately equal which ensures good smoothing behaviour of the relaxation on each grid and thereby good cycle convergence. The developed solver can be used for other L/R ratios too but for good convergence behaviour it is then necessary to choose the number of nodes on the coarsest grid such that $(R/L)(h_\theta/h_y) \approx 1$. Hence, for a different L/R one should change the number of nodes on the coarsest grid in θ direction. The reason for this need is explained in detail in Section 4.10.1. For a few values of L/R Table 4.13 gives the coarsest grid satisfying this

level	$n_\theta \times n_y$	1 V-cycle	2 V-cycles	3 V-cycles
1	24×4	$1.83 \cdot 10^{-2}$	$1.88 \cdot 10^{-2}$	$1.89 \cdot 10^{-2}$
2	48×8	$6.25 \cdot 10^{-3}$	$5.88 \cdot 10^{-3}$	$5.86 \cdot 10^{-3}$
3	96×16	$1.07 \cdot 10^{-3}$	$1.03 \cdot 10^{-3}$	$1.02 \cdot 10^{-3}$
4	192×32	$3.62 \cdot 10^{-4}$	$3.22 \cdot 10^{-4}$	$3.24 \cdot 10^{-4}$
5	384×64	$4.89 \cdot 10^{-5}$	$5.99 \cdot 10^{-5}$	$5.53 \cdot 10^{-5}$
6	768×128	$2.38 \cdot 10^{-5}$	$1.79 \cdot 10^{-5}$	$1.85 \cdot 10^{-5}$
7	1536×256	$4.89 \cdot 10^{-6}$	$4.68 \cdot 10^{-6}$	$3.22 \cdot 10^{-6}$

Table 4.12: *aen^H using second order discretization of the $\partial H/\partial\theta$ term, FMG with 1, 2 and 3 $V(2,1)$ cycles.* $\epsilon = 0.9$, $k = L/R = 1$.

criterion. Table 4.14 subsequently gives the residual on grid level 6 as a function of the number of $V(2,1)$ cycles in a FMG algorithm with this choice of the coarsest grid. The results apply to the case $\epsilon = 0.2$ and three values of L/R. The results show very similar convergence behaviour as is demonstrated by the Table 4.14.

L/R	$1/4$	$1/2$	1	2	4
$n_\theta \times n_y$	96×4	48×4	24×4	12×4	6×4

Table 4.13: *Number of points on the coarsest grid as a function of L/R.*

	$L/R = 1/4$	$L/R = 1$	$L/R = 4$
	3072×128	768×128	192×128
1	$1.16 \cdot 10^{-7}$	$8.90 \cdot 10^{-7}$	$8.68 \cdot 10^{-6}$
2	$2.39 \cdot 10^{-8}$	$1.35 \cdot 10^{-7}$	$2.43 \cdot 10^{-6}$
3	$6.79 \cdot 10^{-9}$	$2.05 \cdot 10^{-8}$	$5.10 \cdot 10^{-7}$
4	$1.97 \cdot 10^{-9}$	$5.00 \cdot 10^{-9}$	$1.31 \cdot 10^{-7}$
5	$5.68 \cdot 10^{-10}$	$1.16 \cdot 10^{-9}$	$3.27 \cdot 10^{-8}$
6	$1.64 \cdot 10^{-10}$	$2.80 \cdot 10^{-10}$	$8.34 \cdot 10^{-9}$
7	$4.73 \cdot 10^{-11}$	$6.78 \cdot 10^{-11}$	$2.13 \cdot 10^{-9}$

Table 4.14: *Level 6 residual norm for $\epsilon = 0.2$ as a function of three different L/R ratios and of the number of $V(2,1)$ cycles, using an FMG algorithm.*

Thus having developed a program that allows us to compute the pressure distribution in a journal bearing for a given ϵ and L/R, one can use it to obtain *design* information. For example, Table 4.15 and 4.16 give the computed value of the Sommerfeld number $S_0^h = 3W^h$ and the attitude angle Ψ^h as a function of ϵ and L/D. Note that now L/D is used instead of L/R. From a mathematical point of view indeed the parameter L/R appears in the equations. However, it is engineering practice to use the parameter L/D, as it directly relates to the size of the bearing. The results presented in the Tables 4.15

and 4.16 were generated using 6 levels and a FMG algorithm and 3 $V(2,1)$ cycles per level. For each value of L/D the coarsest grid was chosen as indicated in Table 4.13.

	L/D					
ϵ	1/8	1/4	1/2	1	2	4
0.1	$2.496 \cdot 10^{-3}$	$9.810 \cdot 10^{-3}$	$3.679 \cdot 10^{-2}$	$1.196 \cdot 10^{-1}$	$2.849 \cdot 10^{-1}$	$4.550 \cdot 10^{-1}$
0.2	$5.354 \cdot 10^{-3}$	$2.100 \cdot 10^{-2}$	$7.833 \cdot 10^{-2}$	$2.519 \cdot 10^{-1}$	$5.863 \cdot 10^{-1}$	$9.110 \cdot 10^{-1}$
0.3	$9.060 \cdot 10^{-3}$	$3.539 \cdot 10^{-2}$	$1.305 \cdot 10^{-1}$	$4.093 \cdot 10^{-1}$	$9.183 \cdot 10^{-1}$	$1.382 \cdot 10^{0}$
0.4	$1.443 \cdot 10^{-2}$	$5.601 \cdot 10^{-2}$	$2.026 \cdot 10^{-1}$	$6.111 \cdot 10^{-1}$	$1.303 \cdot 10^{0}$	$1.891 \cdot 10^{0}$
0.5	$2.311 \cdot 10^{-2}$	$8.884 \cdot 10^{-2}$	$3.125 \cdot 10^{-1}$	$8.909 \cdot 10^{-1}$	$1.780 \cdot 10^{0}$	$2.481 \cdot 10^{0}$
0.6	$3.897 \cdot 10^{-2}$	$1.476 \cdot 10^{-1}$	$4.984 \cdot 10^{-1}$	$1.316 \cdot 10^{0}$	$2.427 \cdot 10^{0}$	$3.237 \cdot 10^{0}$
0.7	$7.322 \cdot 10^{-2}$	$2.709 \cdot 10^{-1}$	$8.600 \cdot 10^{-1}$	$2.045 \cdot 10^{0}$	$3.426 \cdot 10^{0}$	$4.355 \cdot 10^{0}$
0.8	$1.706 \cdot 10^{-1}$	$6.045 \cdot 10^{-1}$	$1.735 \cdot 10^{0}$	$3.571 \cdot 10^{0}$	$5.331 \cdot 10^{0}$	$6.421 \cdot 10^{0}$
0.9	$6.777 \cdot 10^{-1}$	$2.161 \cdot 10^{0}$	$5.094 \cdot 10^{0}$	$8.454 \cdot 10^{0}$	$1.095 \cdot 10^{1}$	$1.237 \cdot 10^{1}$
0.95	$2.521 \cdot 10^{0}$	$6.930 \cdot 10^{0}$	$1.333 \cdot 10^{1}$	$1.883 \cdot 10^{1}$	$2.229 \cdot 10^{1}$	$2.417 \cdot 10^{1}$

Table 4.15: *Sommerfeld number S_0^h as a function of ϵ and L/D.*

	L/D					
ϵ	1/8	1/4	1/2	1	2	4
0.1	$1.442 \cdot 10^{0}$	$1.439 \cdot 10^{0}$	$1.427 \cdot 10^{0}$	$1.386 \cdot 10^{0}$	$1.311 \cdot 10^{0}$	$1.251 \cdot 10^{0}$
0.2	$1.316 \cdot 10^{0}$	$1.316 \cdot 10^{0}$	$1.311 \cdot 10^{0}$	$1.289 \cdot 10^{0}$	$1.240 \cdot 10^{0}$	$1.198 \cdot 10^{0}$
0.3	$1.191 \cdot 10^{0}$	$1.192 \cdot 10^{0}$	$1.195 \cdot 10^{1}$	$1.191 \cdot 10^{0}$	$1.167 \cdot 10^{0}$	$1.143 \cdot 10^{0}$
0.4	$1.065 \cdot 10^{0}$	$1.068 \cdot 10^{0}$	$1.078 \cdot 10^{0}$	$1.092 \cdot 10^{0}$	$1.091 \cdot 10^{0}$	$1.083 \cdot 10^{0}$
0.5	$9.386 \cdot 10^{-1}$	$9.436 \cdot 10^{-1}$	$9.597 \cdot 10^{-1}$	$9.896 \cdot 10^{-1}$	$1.010 \cdot 10^{0}$	$1.015 \cdot 10^{0}$
0.6	$8.106 \cdot 10^{-1}$	$8.171 \cdot 10^{-1}$	$8.385 \cdot 10^{-1}$	$8.819 \cdot 10^{-1}$	$9.205 \cdot 10^{-1}$	$9.365 \cdot 10^{-1}$
0.7	$6.782 \cdot 10^{-1}$	$6.862 \cdot 10^{-1}$	$7.121 \cdot 10^{-1}$	$7.653 \cdot 10^{-1}$	$8.167 \cdot 10^{-1}$	$8.406 \cdot 10^{-1}$
0.8	$5.356 \cdot 10^{-1}$	$5.454 \cdot 10^{-1}$	$5.748 \cdot 10^{-1}$	$6.323 \cdot 10^{-1}$	$6.884 \cdot 10^{-1}$	$7.159 \cdot 10^{-1}$
0.9	$3.680 \cdot 10^{-1}$	$3.803 \cdot 10^{-1}$	$4.112 \cdot 10^{-1}$	$4.620 \cdot 10^{-1}$	$5.088 \cdot 10^{-1}$	$5.331 \cdot 10^{-1}$
0.95	$2.588 \cdot 10^{-1}$	$2.728 \cdot 10^{-1}$	$3.012 \cdot 10^{-1}$	$3.393 \cdot 10^{-1}$	$3.721 \cdot 10^{-1}$	$3.897 \cdot 10^{-1}$

Table 4.16: *Attitude angle Ψ^h (in radians) as a function of ϵ and L/D.*

As mentioned at the start of this chapter the problem of the numerical solution of the pressure in a journal bearing was already addressed by Cameron and Wood in 1947 [27] and by Sassenfeld and Walther in 1954 [95]. The results presented in Table 4.15 can directly be compared to their results. For example, Table 4.17 gives some results taken from [95]. Comparing these predictions with the results presented in Table 4.15 shows that their hand computations were already remarkably accurate.

Figure 4.8 shows a graphic representation of the data given in Table 4.15. However, instead of plotting ϵ as a function of S_0^h the figure shows $1-\epsilon$ which is the relative minimum film $H_m = h_{min}/c$ as a function of S_0^h. This graph can be used as a design graph in various

	L/D			
ϵ	1/8	1/4	1/2	1
0.1	$2.5 \cdot 10^{-3}$	$1.0 \cdot 10^{-2}$	$3.7 \cdot 10^{-2}$	$1.18 \cdot 10^{-1}$
0.2	$5.4 \cdot 10^{-3}$	$2.1 \cdot 10^{-2}$	$7.9 \cdot 10^{-2}$	$2.52 \cdot 10^{-1}$
0.3	$8.9 \cdot 10^{-3}$	$3.4 \cdot 10^{-2}$	$1.29 \cdot 10^{-1}$	$4.17 \cdot 10^{-1}$
0.4	$1.45 \cdot 10^{-2}$	$5.6 \cdot 10^{-2}$	$2.03 \cdot 10^{-1}$	$6.1 \cdot 10^{-1}$
0.5	$2.33 \cdot 10^{-2}$	$8.95 \cdot 10^{-2}$	$3.2 \cdot 10^{-1}$	$8.9 \cdot 10^{-1}$
0.6	$3.9 \cdot 10^{-2}$	$1.49 \cdot 10^{-1}$	$4.98 \cdot 10^{-1}$	$1.32 \cdot 10^{0}$
0.7	$7.35 \cdot 10^{-2}$	$2.7 \cdot 10^{-1}$	$8.6 \cdot 10^{-1}$	$2.08 \cdot 10^{0}$
0.8	$1.705 \cdot 10^{-1}$	$6.0 \cdot 10^{-1}$	$1.73 \cdot 10^{0}$	$3.55 \cdot 10^{0}$
0.9	$6.8 \cdot 10^{-1}$	$2.16 \cdot 10^{0}$	$5.1 \cdot 10^{0}$	$8.35 \cdot 10^{0}$
0.95	$2.54 \cdot 10^{0}$	$6.9 \cdot 10^{0}$	$1.34 \cdot 10^{1}$	$1.86 \cdot 10^{1}$

Table 4.17: *Sommerfeld number S_0^h for the full journal bearing as a function of ϵ and L/D presented by Sassenfeld and Walther [95].*

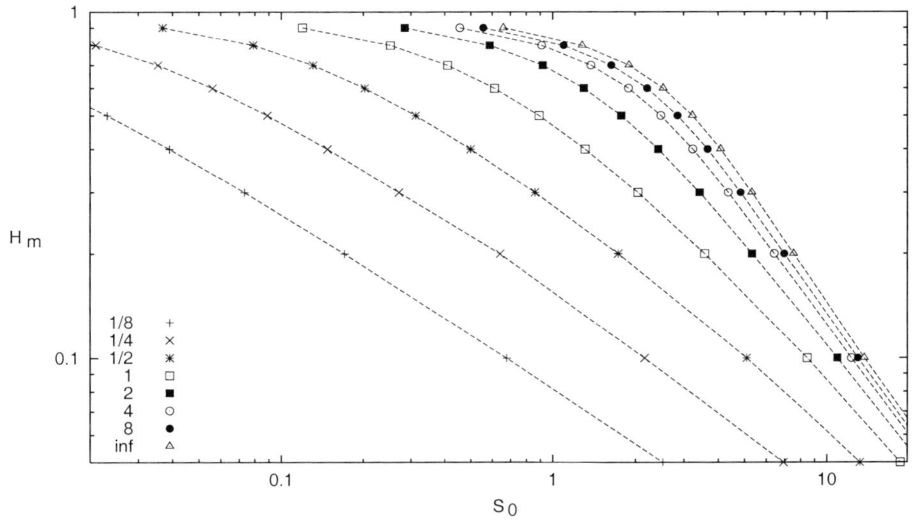

Figure 4.8: *$H_m = h_{min}/c = 1 - \epsilon$ as a function of the computed Sommerfeld number S_0^h and L/D for a full journal bearing.*

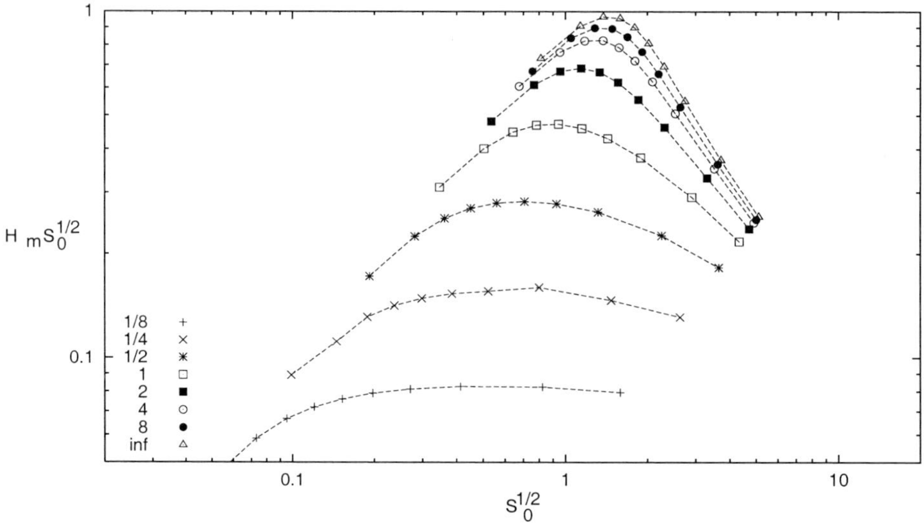

Figure 4.9: *Design graph for a full journal bearing: $H_m\sqrt{S_0}$ as a function of $\sqrt{S_0}$ and L/D.*

ways. For example, for a prescribed acceptable minimum film thickness the maximum load capacity of the bearing can be determined for a given L/D. Alternatively L/D can be chosen such that a prescribed load can be carried. However, often one can not freely choose the parameters. For example, if a bearing should be chosen for a particular application the dimensions L and D may be determined by other constraints, and the only question asked to the tribologist is to specify the optimum clearance c such that the minimum film thickness will be at its maximum, i.e. as large as possible. Such data is difficult to obtain from Figure 4.8 because the parameters on both axes contain the clearance. However, it can be obtained from the data presented in Table 4.15 if it is presented in a different way, see Blok [8]. For this purpose a graph should be made presenting the product $H_m\sqrt{S_0}$ as a function of $\sqrt{S_0}$. As can be seen easily the clearance cancels in the product $H_m\sqrt{S_0}$ so it now only appears in the parameter on the horizontal axis. Figure 4.9 shows the data of Table 4.15 presented in this way. Now the curve for a given value of L/D shows a maximum. Thus, if W, L/D and R are all specified the graph can be read as a graph giving the minimum film thickness as a function of the clearance. Naturally optimizing for the minimum film thickness is not the only possibility. One may also wish to take the friction into account. This can also be done easily as was shown by Moes and Bosma [84].

4.9 Conclusion

In this chapter the efficient solution of the pressure field in a hydrodynamically lubricated journal bearing was studied. The starting point was the efficient solver for the Poisson problem developed in the previous chapter. This solver was extended to incorporate coefficients that vary over the domain. Furthermore, the cavitation condition was imposed resulting in a complementarity problem. Introducing local relaxations, and paying attention to the transfer operators, a program was developed that solves the problem up to the level of the discretization error using FMG with a single $V(2,1)$ cycle. The solver can be applied to bearings with different L/R ratios and naturally it can also be used for the case of a partial journal bearing. Furthermore it can serve as the basis for a solver for a variety of more complex hydrodynamic lubrication problems. However, such extensions should again be considered carefully and some issues of relevance are presented in the advanced topics section below.

4.10 Advanced Topics

4.10.1 Line Relaxation

In Section 4.8 a rule was given that determines the coarsest grid to ensure efficient convergence of the coarse grid correction cycle for the journal bearing problem with different L/R:

$$(R/L)(h_\theta/h_y) \approx 1 \tag{4.50}$$

For $0 \leq \theta \leq 2\pi$ and $0 \leq Y \leq 1$ taking $n_y = 4$ as a minimum one obtains:

$$n_\theta \approx (\text{int})(8\pi\frac{R}{L}) \tag{4.51}$$

For very slender journal bearings ($L/R \ll 1$) the rule requires a rather large number of points to be used on the coarsest grid in the θ direction. This naturally leads to larger computing times. On the other hand, taking $n_\theta = 4$ as a minimum, for very large L/R a coarsest grid satisfying the above requirement can not be obtained. At this point it should be noted that this rule is not a requirement for an efficient solver in general. It only applies to the case where (Gauss-Seidel) *pointwise* relaxation is used. The smoothing behaviour of such a relaxation, and thereby the error reduction of a coarse grid correction cycle built around this relaxation, sensitively depend on the value of $(R/L)(h_\theta/h_y)$. This is illustrated in Table 4.18. For a coarsest grid with fixed number of points $n_\theta = 24$ and $n_y = 4$ this table gives the residual norm on gridlevel 7 as a function of the number of $V(2,1)$ cycles performed for a FMG algorithm. The table shows that for the case $L/R = 1$ convergence is as efficient as shown throughout this chapter.
However, the $L/R = 1/2$ case already shows a different behaviour. The first few cycles show rapid convergence, exceeding even the convergence of the $L/R = 1$ case. But after 3 V-cycles the convergence speed reduces very rapidly, and convergence stalls. Even more dramatic behaviour occurs for the case of $L/R = 1/4$, from the first cycle onwards the numerical algorithm diverges! So even if keeping the n_θ and n_y values constant allows one to limit the number of unknowns, the convergence degrades so rapidly that little is gained.

V-cycle	$L/R = 1$	$L/R = 1/2$	$L/R = 1/4$
1	$8.06 \cdot 10^{-7}$	$1.55 \cdot 10^{-6}$	$2.87 \cdot 10^{-5}$
2	$2.40 \cdot 10^{-7}$	$5.43 \cdot 10^{-7}$	$7.25 \cdot 10^{-4}$
3	$8.50 \cdot 10^{-8}$	$7.18 \cdot 10^{-8}$	$5.38 \cdot 10^{-3}$
4	$3.29 \cdot 10^{-8}$	$1.40 \cdot 10^{-7}$	$3.06 \cdot 10^{-1}$
5	$1.25 \cdot 10^{-8}$	$4.24 \cdot 10^{-8}$	$2.64 \cdot 10^{0}$
6	$4.78 \cdot 10^{-9}$	$7.38 \cdot 10^{-8}$	–
7	$1.82 \cdot 10^{-9}$	$3.12 \cdot 10^{-8}$	–

Table 4.18: *Residual norm on level 7 as a function of three different L/R ratios and of the number of $V(2,1)$ cycles, using an FMG algorithm. Coarsest grid $n_\theta = 24$ and $n_y = 4$.* $\epsilon = 0.2$.

The poor convergence displayed in Table 4.18 can be explained easily. Assuming the coefficient $\xi = H^3$ is locally sufficiently smooth the behaviour of the solver can be analysed using the anisotropic Poisson 2d problem:

$$\frac{\partial^2 u}{\partial x^2} + \frac{1}{k^2}\frac{\partial^2 u}{\partial y^2} = f(x,y) \tag{4.52}$$

By means of Local Mode Analysis it was shown in Section 2.10.2 that for $k = 1$ the smoothing behaviour of Gauss-Seidel pointwise relaxation depended on the ratio h_y/h_x. In the same way it can be shown that the smoothing behaviour of this relaxation applied to the solution of the discretized Equation (4.52) is given by Equation (2.137) with $\alpha = kh_y/h_x$.

If $k \ll 1$ ($\alpha \ll 1$) the problem exhibits what is referred to as a *weak* coupling in the x direction. As a result errors in u^h in the y direction will give rise to important residuals, whereas similar errors in the x direction will hardly affect the residual. Similarly, if $k \gg 1$ ($\alpha \gg 1$) the errors of u^h in the x direction will dominate the residuals. The consequence of weak coupling in one of the grid directions is that a pointwise relaxation scheme can not efficiently reduce error components that are smooth in the direction of the strong coupling and oscillatory in the other direction. As a result the asymptotic smoothing factor defined by Equation (3.47) will tend to unity. Naturally if the relaxation only smoothes well in one direction the error after a few relaxations will not be smooth in both directions and it can not be represented accurately on a grid that is coarse with respect to both directions. Moreover, if injection is used the remaining high frequency error will be fed into a low frequency correction which subsequently may lead to a diverging cycle.

Good error smoothing for the case of problems with strong coupling in one of the directions requires a special type of relaxation. In this case the point relaxation scheme should be replaced with a line relaxation scheme, see Section 2.10.2. Instead of scanning the grid solving the equations point by point the grid is scanned line by line each times solving all equations of one line simultaneously. The lines should be chosen in the direction of the strong coupling. For further details the reader is referred to Section 2.10.2. As an illustration Table 4.19 gives some values of the asymptotic smoothing factor $\bar{\mu}$ as a function of α for Gauss-Seidel line relaxation. The table also gives the values for point relaxation. The table clearly shows what has been explained above. For strong coupling the pointwise relaxation has poor smoothing but using line relaxation this problem can be overcome.

α	Pointwise	Linewise
1	0.50	0.45
2	0.68	0.45
4	0.89	0.45
8	0.97	0.45

Table 4.19: *Asymptotic smoothing factor of Gauss-Seidel pointwise relaxation and Gauss-Seidel line relaxation applied to the anisotropic Poisson 2d problem as a function of* $\alpha = (kh_y)/h_x$.

The results immediately translate to the journal bearing problem. This is illustrated by Table 4.20 which shows the residuals for the case $L/R = 1/4$ shown in Table 4.18 if the Gauss-Seidel pointwise relaxation is replaced by a Gauss-Seidel line relaxation. The table shows that with this relaxation indeed the efficiency of the solver is completely restored.

V-cycle	level 7
	$L/R = 1/4$
1	$7.54 \cdot 10^{-7}$
2	$6.37 \cdot 10^{-8}$
3	$5.14 \cdot 10^{-9}$
4	$4.67 \cdot 10^{-10}$
5	$4.36 \cdot 10^{-11}$
6	$4.57 \cdot 10^{-12}$
7	$7.77 \cdot 10^{-13}$

Table 4.20: *Residual norm on level 7 as a function of the number of V-cycles, using an FMG algorithm with line relaxation,* $n_\theta = 24$, $n_y = 4$.

4.10.2 Film Thickness

In the previous sections the film thickness in the y direction was assumed to be constant. This means physically that the bearing and the journal are perfectly aligned! In general the film thickness will be a function of both θ and y: $h(\theta, y)$. This generalisation of the film thickness does not affect the convergence speed of the solver, and does not require any specific attention.

A case that does deserve special care, is the case of discontinuous changes in h, like those caused by (narrow) grooves. In such a case, the coarser grids can only generate an accurate correction, when the coarse grid represents correctly the fine grid operator for smooth components. This may have consequences for the expressions to be used for $\xi^h_{i\pm1/2}$.

Alternatively one may specifically define ξ on the coarse grid such that the resulting coarse grid operator correctly approximates the global flow resistance of the fine grid operator. In that case ξ^H of Equation (4.43) should be taken as:

$$\frac{1}{\xi_I^H} = \frac{1}{4\xi_{2I-1}^h} + \frac{1}{2\xi_{2I}^h} + \frac{1}{4\xi_{2I+1}^h} \tag{4.53}$$

For more general information on the treatment of discontinuous coefficients, the reader is referred to Alcouffe et al. [1]. For more information concerning the lubrication with stepwise changes in the film thickness consult van der Stegen [97].

4.10.3 Imposed Load

What happens if instead of the eccentricity ϵ, the two components of the load W_n and W_t are given, and the eccentricity ϵ and angle Ψ are asked? The problem becomes different from the one that has previously been described. The angle Ψ can be calculated afterwards, but the eccentricity ϵ has to be initialised, the load computed, and then ϵ has to be corrected, etc. At first this process seems to require many consequent solutions, and thus to be much more computationally intensive than the problem with ϵ given. The problem with an imposed force, contains a so called global equation, which involves the pressures in all the points. As will be shown extensively in the next chapter for another global equation involving the variable h_0, it is possible to use the FMG algorithm to greatly reduce the solution time involved. Furthermore, the residuals of the force balance equation should be transferred in a FAS way. Finally, updates of the value of ϵ should be made only on the coarsest grid where convergence is fast. Using these two additions, the solution of the hydrodynamic lubrication problem for a given load, up to the level of the discretization error, can be performed in a time which is only twice as long as the time required for the solution of the problem with ϵ given.

4.10.4 Transient Effects

A further complication can be that the bearing is dynamically loaded. Assuming the load cycle is known the objective is to obtain the path of the centre of the journal as a function of time, i.e. the eccentricity as a function of time. If a solver for the problem with imposed load has been constructed it can be extended to a solver for the transient problem in the following way.

First the discrete equations have to be extended. Now the transient term $\partial(\rho h)/\partial t$ of the Reynolds equation should be taken into account. Discretising the time according to $t = t_0 + k\Delta t$, and writing the time index k as another subscript, one obtains the first order backward discretization as:

$$\frac{\partial(\rho h)}{\partial t}|_{k,i} \doteq \frac{(\rho h)|_{k,i} - (\rho h)|_{k-1,i}}{\Delta t} \tag{4.54}$$

or using a second order backward scheme:

$$\frac{\partial(\rho h)}{\partial t}|_{k,i} \stackrel{..}{=} \frac{3(\rho h)|_{k,i} - 4(\rho h)|_{k-1,i} + (\rho h)|_{k-2,i}}{2\Delta t} \tag{4.55}$$

Next the discrete problem with this additional term (made dimensionless) in the discrete Reynolds equation should be solved. It can easily be seen that this term does not change the problem of solving the pressures very much. The only complication is that now the equation should be solved at a number of timesteps where the film thickness of some previous timesteps is needed to compute the discrete transient derivative. Solving the problem at each timestep can be done with the coarse grid correction cycle presented in this chapter. An even better way is to use a so-called F-cycle for each timestep. The F-cycle is a kind of FMG cycle, preceded by a series of restrictions to the coarsest grid, after doing the time increment. The solution process is started at the coarsest grid to allow the low frequency errors induced by the time increment to converge before going to a finer grid, see Figure 4.10.

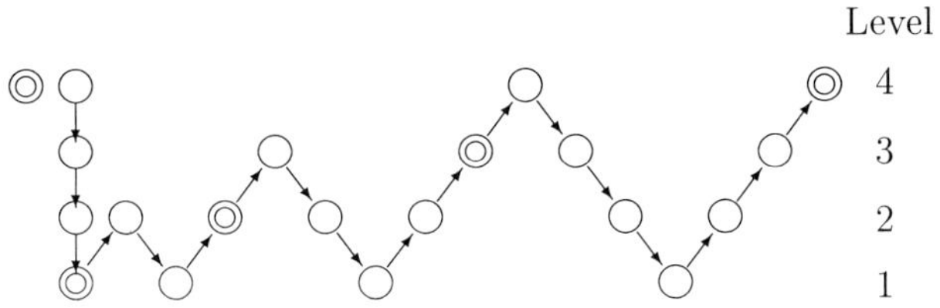

Figure 4.10: *F-cycle, the time increment occurs on the finest level between the double and the simple circle.*

Using the F-cycle, with a single $V(2,1)$ cycle per grid level is generally sufficient to solve the problem to the level of the incremental discretization error. For more detailed information the reader is referred to Brandt [20].

Chapter 5

Dry Contact

In this chapter the dry contact problem is studied as an introduction to the EHL problem. The dry contact problem has four difficulties in addition to obtaining fast convergence: the solution of the integral equation (with its non-local character), the complementarity problem: $h > 0 \,\&\, p = 0$, $h = 0 \,\&\, p > 0$ the coupling of the global variable h_0 with the force balance equation and the efficient calculation of the deformation integrals.

As was indicated in the definition of EHL in the introduction 1.3, the maximum deformation δ in an EHL contact is generally much larger than the lubricant film thickness h. Since the pressure is proportional to the maximum deformation δ, neglecting this small lubricant film thickness h hardly alters the pressure distribution. Thus, the dry contact can be regarded as an approximation to the steady state EHL problem, at least for the pressure and the deformation. It becomes an even better approximation of the lubricated problem for the limits of high load and low speed, since $h/\delta \rightarrow 0$. Thus for $\mathrm{w} \rightarrow \infty$ and/or $u_m \rightarrow 0$, the dry contact pressure distribution accurately approximates the lubricated one, and the physical and numerical character of the EHL problem will strongly resemble that of the dry contact problem. This justifies studying the dry contact problem as a final preparation before studying the EHL problem. However, it is emphasized that the techniques presented in this chapter are of significance for a wide range of problems in contact mechanics, including the evaluation of sub-surface stresses, surface temperatures and effects of frictional heating of the contact.

5.1 Equations

In physical terms the dry contact problem can be posed as follows: given two solid bodies of known geometry pressed together with a force w, what is the pressure distribution p that deforms the two bodies correctly, and that satisfies force balance.

Assuming a parabolical approximation of the two bodies with equal reduced radii of curvature in x and y: $R_y = R_x$, the gap h between the two surfaces can be expressed as:

$$h(x,y) = h_0 + \frac{x^2}{2R_x} + \frac{y^2}{2R_x} + \frac{2}{\pi E'} \int_{-\infty}^{+\infty} \int_{-\infty}^{+\infty} \frac{p(x',y')\,dx'\,dy'}{\sqrt{(x-x')^2+(y-y')^2}} \tag{5.1}$$

When the surfaces are loaded together, the gap h should remain positive or become zero.

Neglecting adhesion this implies positive local pressure (contact) or zero local pressure (no contact). In mathematical terms this leads to a complementarity problem that can be expressed as:

$$\begin{array}{lll} h(x,y) > 0, & p(x,y) = 0 & \text{no contact} \\ h(x,y) = 0, & p(x,y) > 0 & \text{contact} \end{array} \tag{5.2}$$

The applied load w should be balanced by the integral over the contact pressure, this results in the force balance equation:

$$\mathrm{w} = \int_{-\infty}^{+\infty}\int_{-\infty}^{+\infty} p(x,y)\,dx\,dy \tag{5.3}$$

This contact problem has been solved by Hertz [60]. In this specific case the contact area is a disc of radius a, so $h(x,y) = 0$ for $(x/a)^2 + (y/a)^2 \leq 1$, and the pressure is given by a semi-elliptical pressure distribution:

$$p(x,y) = \begin{cases} p_h\sqrt{1-(x/a)^2-(y/a)^2}, & \text{if } (x/a)^2+(y/a)^2 \leq 1; \\ 0, & \text{otherwise.} \end{cases} \tag{5.4}$$

and $h_0 = -a^2/R_x$, where the variables p_h and a are given by:

$$p_h = \frac{3\mathrm{w}}{2\pi a^2} \tag{5.5}$$

$$a = \sqrt[3]{\frac{3\mathrm{w}R_x}{2E'}} \tag{5.6}$$

with $2/E' = (1-\nu_1^2)/E_1 + (1-\nu_2^2)/E_2$.

Hence, for this specific case the analytical solution is known, and there is no need to solve the problem numerically. However, for more general contact problems a numerical approach is needed. This is for example the case if the shape of the undeformed surfaces differs from a paraboloide, or if the surfaces are not perfectly smooth. In this latter case their nominal shape may still be well approximated by paraboloides but the roughness of the surface or its micro-geometry introduces a term $r(x,y)$ into the right hand side of Equation (5.1).

For the numerical solution it is convenient to rewrite the equations introducing dimensionless variables. The obvious choice in this case is to use the parameters of the Hertzian solution. Introducing:

$$\begin{aligned} X &= x/a \\ Y &= y/a \\ P &= p/p_h \\ H &= hR_x/a^2 \end{aligned} \tag{5.7}$$

The problem can be written as:

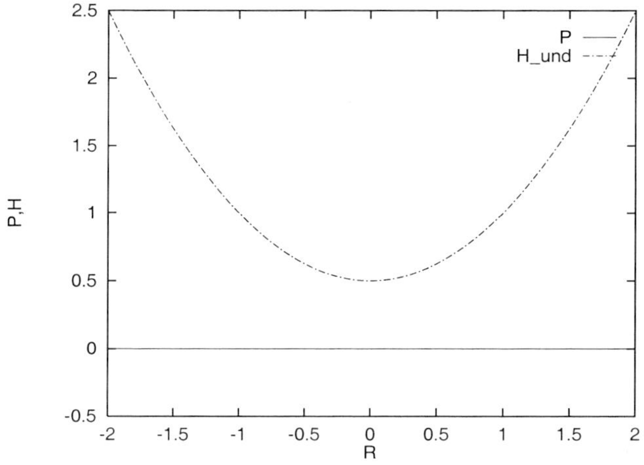

Figure 5.1: *Film thickness at the line $Y = 0$ between the two bodies before contact: $H > 0$ and $P = 0$ everywhere.*

$$\begin{array}{lll} H(X,Y) > 0, & P(X,Y) = 0 & \text{no contact} \\ H(X,Y) = 0, & P(X,Y) > 0 & \text{contact} \end{array} \tag{5.8}$$

with

$$H(X,Y) = H_0 + \frac{X^2}{2} + \frac{Y^2}{2} + \frac{2}{\pi^2} \int_{-\infty}^{+\infty} \int_{-\infty}^{+\infty} \frac{P(X',Y')\,dX'\,dY'}{\sqrt{(X-X')^2 + (Y-Y')^2}} \tag{5.9}$$

As an illustration Figure 5.1 shows the pressure and film thickness for $H_0 = 0.5$. In this case the surfaces are still fully separated and the pressure is zero everywhere. The correct value of H_0 is however the value for which force balance is satisfied:

$$\int_{-\infty}^{+\infty} \int_{-\infty}^{+\infty} P(X,Y)\,dX\,dY = 2\pi/3 \tag{5.10}$$

In that case the solution is the Hertzian solution:

$$P(X,Y) = \begin{cases} \sqrt{1 - X^2 - Y^2}, & \text{if } X^2 + Y^2 \leq 1; \\ 0, & \text{otherwise.} \end{cases} \tag{5.11}$$

and $H_0 = -1$. Figure 5.2 shows the dimensionless pressure P and film thickness H at the line $Y = 0$. In the dimensionless coordinates $H = 0$ over the disc $X^2 + Y^2 \leq 1$. For comparison the undeformed film thickness (at this value of H_0) is also given.

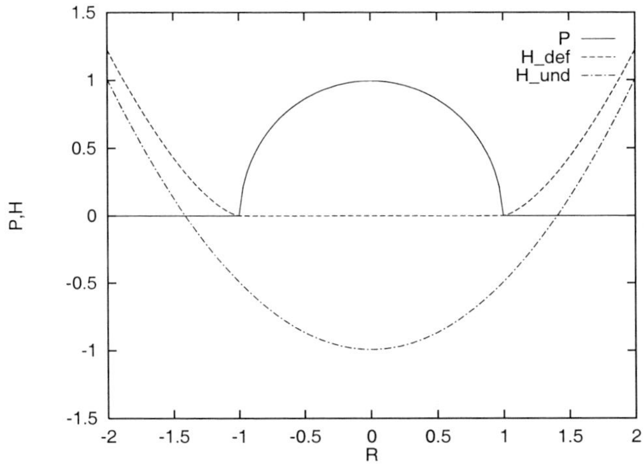

Figure 5.2: *Dimensionless Hertzian contact pressure, deformed and undeformed geometry for* $H_0 = -1$, *for the circular contact case.*

The difference between the curves indicates the total elastic deformation of the two surfaces in contact.

In the following sections it is explained how to solve the problem numerically. The fact that the analytical solution is known will be used to check the discretization error and to verify how much work is needed to obtain a solution with an error that is small compared to the discretization error.

5.2 Discrete Equations

The problem is discretized using a rectangular grid of uniform mesh size in each direction, i.e. $X_i = X_0 + ih_x$ and $Y_j = Y_0 + jh_y$. The equations will be given in the general form with h_x and h_y. However, all results presented in this chapter will be for the case $h_x = h_y$. In that case the symbol h will be used to indicate the mesh size, not to be confused with the film thickness. The discrete contact equations are given by:

$$\begin{array}{lll} H^h_{i,j} > 0, & P^h_{i,j} = 0 & \text{no contact} \\ H^h_{i,j} = 0, & P^h_{i,j} > 0 & \text{contact} \end{array} \tag{5.12}$$

where $H^h_{i,j}$ is defined by:

$$H^h_{i,j} = H_0 + \frac{X_i^2}{2} + \frac{Y_j^2}{2} + \sum_{i'}\sum_{j'} K^{hh}_{i,i',j,j'} P^h_{i',j'} \tag{5.13}$$

For a second order accurate discretization based on the approximation of the pressure by a piecewise constant function on a square $h \times h$ around $X'_{i'}, Y'_{j'}$ the coefficients $K^{hh}_{i,i',j,j'}$ are defined by:

$$K^{hh}_{i,i',j,j'} = \frac{2}{\pi^2} \int_{Y_j - h_y/2}^{Y_j + h_y/2} \int_{X_i - h_x/2}^{X_i + h_x/2} \frac{dX'dY'}{\sqrt{(X_i - X')^2 + (Y_j - Y')^2}}. \tag{5.14}$$

The coefficients can be computed analytically:

$$\begin{aligned} K^{hh}_{i,i',j,j'} = \frac{2}{\pi^2} \{ & |X_p| \text{arcsinh}\left(\frac{Y_p}{X_p}\right) + |Y_p| \text{arcsinh}\left(\frac{X_p}{Y_p}\right) \\ & - |X_m| \text{arcsinh}\left(\frac{Y_p}{X_m}\right) - |Y_p| \text{arcsinh}\left(\frac{X_m}{Y_p}\right) \\ & - |X_p| \text{arcsinh}\left(\frac{Y_m}{X_p}\right) - |Y_m| \text{arcsinh}\left(\frac{X_p}{Y_m}\right) \\ & + |X_m| \text{arcsinh}\left(\frac{Y_m}{X_m}\right) + |Y_m| \text{arcsinh}\left(\frac{X_m}{Y_m}\right) \} \end{aligned} \tag{5.15}$$

where

$$\begin{aligned} X_p &= X_{i'} - X_i + h_x/2 \quad X_m = X_{i'} - X_i - h_x/2, \\ Y_p &= Y_{j'} - Y_j + h_y/2 \quad\;\; Y_m = Y_{j'} - Y_j - h_y/2. \end{aligned}$$

For the semi-infinite half space approximation, the kernel K is independent of the position (i, j) and depends only on the distance between the point where the pressure acts (i', j') and the point where the deformation is calculated (i, j). Thus the kernel is only a function of $|i - i'|$ and $|j - j'|$ and can thus be stored in its reduced form.

The discrete force balance equation reads:

$$h_x h_y \sum_{i'} \sum_{j'} P^h_{i',j'} = \frac{2\pi}{3} \tag{5.16}$$

5.3 Relaxation

In order to achieve a stable relaxation, three aspects have to be correctly addressed: the non-local character of the equations, the treatment of the complementarity condition and the force balance equation.

5.3.1 Distributive Relaxation

To understand how to obtain a stable relaxation process it is sufficient to first study the problem without the complementarity condition. In fact this implies one studies only the contact zone. In this zone the equation to be solved is $H^h_{i,j} = 0$. However, for generality of notation the equation is written as:

$$H^h_{i,j} = H_0 + \frac{X_i^2}{2} + \frac{Y_j^2}{2} + \sum_{i'}\sum_{j'} K^{hh}_{i,i',j,j'} P^h_{i',j'} = {}_H f^h_{i,j} \tag{5.17}$$

where ${}_H f^h$ indicates the film thickness right hand side function. On a single grid, or when using Multigrid on the finest grid, this function is zero. However, in a coarse grid correction cycle on all coarser grids it will be defined by the FAS coarse grid equations, and be non-zero.

As a first step simple Jacobi and Gauss Seidel relaxation are studied. Given $\tilde{P}^h$, a new approximation $\bar{P}^h$ is computed according to:

$$\bar{P}^h_{i,j} = \tilde{P}^h_{i,j} + \omega_1 \delta^h_{i,j} \tag{5.18}$$

where ω_1 is an underrelaxation coefficient. The change $\delta^h_{i,j}$ is solved from the requirement that the equation at the point (i,j) is satisfied after making this change. This implies that it is given by:

$$\delta^h_{i,j} = \frac{r^h_{i,j}}{K^{hh}_{0,0}} \tag{5.19}$$

where $K^{hh}_{0,0} = K^{hh}_{i=i',j=j'}$. For Jacobi relaxation changes already made to points previously relaxed are not taken into account and the residual $r^h_{i,j}$ for Jacobi relaxation is given by:

$$r^h_{i,j} = {}_H f^h_{i,j} - H_0 - \frac{X_i^2}{2} - \frac{Y_j^2}{2} - \sum_{i'}\sum_{j'} K^{hh}_{|i-i'|,|j-j'|} \tilde{P}^h_{i',j'} \tag{5.20}$$

For Gauss-Seidel relaxation changes made to points already relaxed are taken into account and, assuming relaxation in lexicographic order, the residual $r^h_{i,j}$ is given by:

$$\begin{aligned} r^h_{i,j} = {}_H f^h_{i,j} - H_0 - \frac{X_i^2}{2} - \frac{Y_j^2}{2} & \\ - \sum_{i'<i}\sum_{j'} K^{hh}_{|i-i'|,|j-j'|} \bar{P}^h_{i',j'} - \sum_{i'=i}\sum_{j'<j} K^{hh}_{|i-i'|,|j-j'|} \bar{P}^h_{i',j'} & \\ - \sum_{i'=i}\sum_{j'>=j} K^{hh}_{|i-i'|,|j-j'|} \tilde{P}^h_{i',j'} - \sum_{i'>i}\sum_{j'} K^{hh}_{|i-i'|,|j-j'|} \tilde{P}^h_{i',j'} & \end{aligned} \tag{5.21}$$

Unfortunately both the Gauss-Seidel relaxation and the Jacobi relaxation are unstable. The reason for this instability is that the kernel K^{hh} is non-zero for *all* values of $|i-i'|$ and $|j-j'|$. As a result, changing the value of the pressure $\tilde{P}^h$ in a point (i',j') will affect the deformation and thus the film thickness $\bar{H}^h$ in *all* points (i,j) of the domain. Moreover, in each point (i,j) the changes applied to *all* points accumulate. Looking at the effect of an entire sweep over all the grid points, it turns out that for each point the positive effect of making the residual zero when relaxing that point is entirely outweighed by the accumulated negative effect in this point of the changes made in all other grid points. As a result the residual is larger at the end of the sweep than it was before the sweep. Because it is related to the integrals in the equation, the instability shows up as an amplification of low frequency error components.

The behaviour described above is fundamentally different from the behaviour of differential equations. In that case a local change of the pressure P, results in the change of residuals in only a few (neighbouring) points (four for the Poisson 2d problem, and in general as many points as appear in the stencil of the discrete operator). As a result the accumulated adverse effect on the residual at a point (i, j) of the changes made when relaxing the other points is smaller than the positive effect on the residual when this point itself is relaxed. From a mathematical point the behaviour can also be understood looking at the matrix of the system of equations. Due to the many off-diagonal terms the matrix is no longer diagonally dominant.

This global change in the integrals (film thickness) caused by a local variation in the function (pressure), requires a special type of relaxation to limit the accumulation of changes. This type of relaxation is referred to as *distributive relaxation* and it implies that when relaxing the equation at a given point (i, j), not only the value of the current approximation at that point is changed, but also the value of the current approximation in a few neighbouring points, for example points $(i-1, j)$, $(i+1, j)$, etc. Distributive relaxation can be applied to any problem (also to differential problems), and the changes to be applied can be solved using different criteria. For the present problem the two conditions imposed are:

(i) The equation in the point (i, j) is solved i.e. the residual $r^h_{i,j} = 0$ after the changes are applied.

(ii) The residuals in points (i', j') situated far from the point (i, j) are affected as little as possible.

The first requirement is not new as it is exactly the one generally used for a point relaxation. The point of the second requirement is that if the changes far away can be kept small the accumulation of changes as a result of one sweep can be kept small too, which will then prevent the unstable behaviour. The question to be answered is then how can one apply changes to a number of neighbouring points such that the changes in the residuals, in this case the summations will be (sufficiently) small.

From Equation (5.17) it can be seen that the change in the discrete sum $\delta H^h_{i',j'}$ in the point (i', j'), due to a pressure change $\delta^h_{i,j}$ at a point (i, j) is given by:

$$\delta H^h_{i',j'} = K^{hh}_{|i-i'|,|j-j'|}\delta^h_{i,j} \tag{5.22}$$

By definition K^{hh} is a discrete approximation to $K = 1/r$ with $r = \sqrt{(X - X')^2 + (Y - Y')^2}$. Consequently the value of $\delta H^h_{i',j'}$ decreases with increasing $|i - i'|$ and $|j - j'|$. This is illustrated in Figure 5.4. However, apparently the decrease is not fast enough to limit the accumulation of changes. Now suppose that instead of changing one value of P, two values in neighbouring points are simultaneously changed by the same amount but with opposite sign. This is graphically illustrated in Figure 5.3 and referred to as a "dipole" change. Thus, instead of changing only $P^h_{i,j}$ with $\delta^h_{i,j}$ two points are changed:

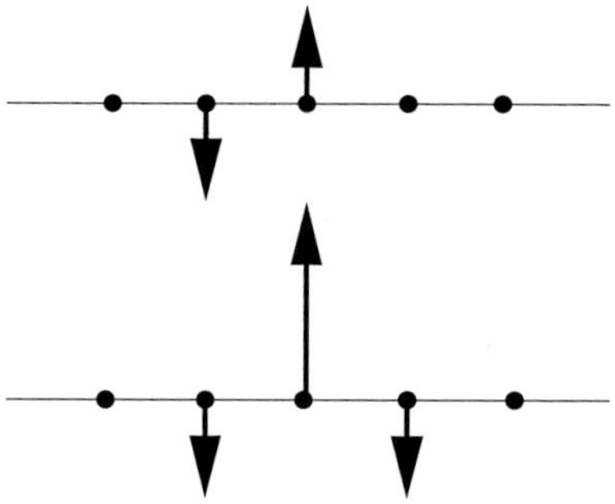

Figure 5.3: *Graphical representation using arrows of distributive relaxation, dipole (top) and tripole (bottom).*

$$\begin{array}{lll} P^h_{i-1,j} & \leftarrow & P^h_{i-1,j} & -\delta^h_{i,j} \\ P^h_{i,j} & \leftarrow & P^h_{i,j} & +\delta^h_{i,j} \end{array}$$

The change of the film thickness at a point (i', j') as a result of applying these two changes is:

$$\delta H^h_{i',j'} = (K^{hh}_{|i-i'|,|j-j'|} - K^{hh}_{|i-1-i'|,|j-j'|})\delta^h_{i,j} \propto \frac{\partial K^{hh}}{\partial |X - X'|}\delta^h_{i,j} \tag{5.23}$$

Hence, it is no longer proportional to the value of K^{hh} (and thereby K), but to the value of its derivative with respect to $|X - X'|$. With $K = 1/r$ one may expect that it decays faster with increasing $|i - i'|$ than K^{hh} itself. This is illustrated in Figure 5.4. The type of change where two points are changed with equal amount but opposite sign is also referred to as a *first order* distributive change, because its stencil $[-1, 1]$ is the stencil of a first derivative.

If applying changes with the distribution of a first derivative causes changes in the integrals proportional to the first derivative of K^{hh} for $K^{hh} \approx 1/r$ it is easy to find a distribution that causes the changes to decay even more rapidly with increasing distance: take a *higher order* distribution according to the stencil of a higher derivative. A second order distribution, also referred to as a tripole change, is graphically represented in Figure 5.3 and it entails that changes are applied to three points in the following way:

$$\begin{array}{llll} P^h_{i-1,j} & \leftarrow & P^h_{i-1,j} & -\delta^h_{i,j}/2 \\ P^h_{i,j} & \leftarrow & P^h_{i,j} & +\delta^h_{i,j} \\ P^h_{i+1,j} & \leftarrow & P^h_{i+1,j} & -\delta^h_{i,j}/2 \end{array}$$

The stencil of the distribution is $[-1, 2, 1]/2$. It is conveniently central around the point (i, j) and the change caused in the integrals at a point (i', j') as a result of this distributive change applied at a point (i, j) is given by:

$$\delta H^h_{i',j'} = (K^{hh}_{|i-i'|,|j-j'|} - \frac{1}{2}K^{hh}_{|i-1-i'|,|j-j'|} - \frac{1}{2}K^{hh}_{|i+1-i'|,|j-j'|})\delta^h_{i,j} \propto \frac{\partial^2 K^{hh}}{\partial |X - X'|^2}\delta^h_{i,j} \tag{5.24}$$

As an illustration $\delta H^h_{i',j'}$ as a function of $k = h|i - i'|$ is displayed for $j = j'$ in Figure 5.4 and indeed it decays more rapidly with increasing distance. The figure shows that by applying changes according to these distributions the changes in the residuals can be controlled such that they are small for points $|i - i'|$ far away. Alternatively, the effect of changes made at a given point can be localized by applying a suitable order of distribution.

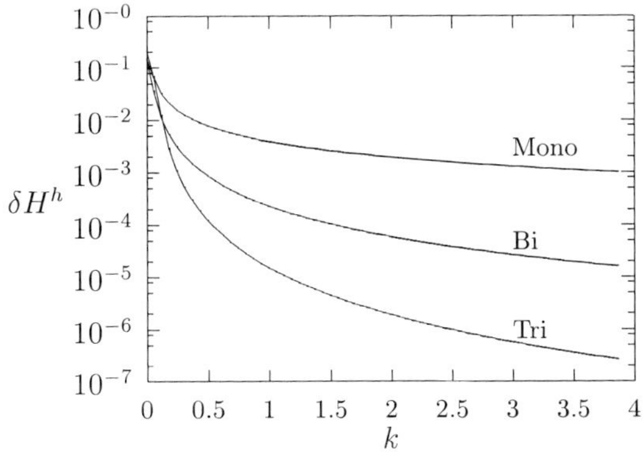

Figure 5.4: *Relative changes in the deformation integrals $\delta H^h_{i',j'}$ as a function of distance $k = h|i - i'|$ and $l = h|j - j'| = 0$.*

So far we only looked at one dimensional distributive changes. However, clearly for the dry contact problem some two dimensional distribution is needed. The distributive relaxation can be generalised directly to higher dimensions. For example the stencil of second order distributive changes that will be applied for the two dimensional dry contact problem is given by:

$$\frac{1}{4}\begin{bmatrix} & -1 & \\ -1 & 4 & -1 \\ & -1 & \end{bmatrix} \tag{5.25}$$

or in terms of the pressures:

$$\begin{aligned} &P^h_{i,j}+ \quad \delta^h_{i,j} \\ &P^h_{i+1,j}- \quad \delta^h_{i,j}/4 \\ &P^h_{i-1,j}- \quad \delta^h_{i,j}/4 \\ &P^h_{i,j+1}- \quad \delta^h_{i,j}/4 \\ &P^h_{i,j-1}- \quad \delta^h_{i,j}/4 \end{aligned} \tag{5.26}$$

The distribution proposed should limit the accumulation of changes. The only remaining question is the magnitude of the change $\delta^h_{i,j}$ to be applied. This follows directly from requirement (i) by substituting the proposed changes into the discrete equation and demanding the residual to be zero. A straightforward generalization of Equation (2.134) gives:

$$\delta^h_{i,j} = r^h_{i,j}\left(\frac{\partial H^h_{i,j}}{\partial P^h_{i,j}} - \frac{1}{4}\frac{\partial H^h_{i,j}}{\partial P^h_{i-1,j}} - \frac{1}{4}\frac{\partial H^h_{i,j}}{\partial P^h_{i+1,j}} - \frac{1}{4}\frac{\partial H^h_{i,j}}{\partial P^h_{i,j+1}} - \frac{1}{4}\frac{\partial H^h_{i,j}}{\partial P^h_{i,j-1}}\right)^{-1} \tag{5.27}$$

and using Equation (5.17) one obtains:

$$\delta^h_{i,j} = \frac{r^h_{i,j}}{K^{hh}_{0,0} - K^{hh}_{0,1}/2 - K^{hh}_{1,0}/2} \tag{5.28}$$

where $K^{hh}_{0,0} = K^{hh}_{i=i',j=j'}$, $K^{hh}_{1,0} = K^{hh}_{i=i'\pm 1,j=j'}$ and $K^{hh}_{0,1} = K^{hh}_{i=i',j=j'\pm 1}$. As the mesh size in the X direction and the Y direction are assumed to be equal: $K^{hh}_{1,0} = K^{hh}_{0,1}$ and the expression for the change $\delta^h_{i,j}$ can be simplified to:

$$\delta^h_{i,j} = \frac{r^h_{i,j}}{K^{hh}_{0,0} - K^{hh}_{1,0}} \tag{5.29}$$

The distributive relaxation can be implemented in different ways both as a Gauss Seidel process or a Jacobi type of process. A description of the implementation of the Jacobi process is as follows: let $\tilde{P}^h_{i,j}$ denote the current approximation. A new approximation $\bar{P}^h_{i,j}$ is obtained by scanning the grid points in some arbitrary order at each point computing $\delta^h_{i,j}$, and at each grid point applying this change to the current approximation to P^h at that grid point and a change $-\delta^h_{i,j}/4$ to the value of the current approximation in the neighbouring points $(i\pm 1, j)$ and $(i, j\pm 1)$. At the end of the sweep the new approximating $\bar{P}^h_{i,j}$ is obtained that satisfies:

$$\bar{P}^h_{i,j} = \tilde{P}^h_{i,j} + \delta^h_{i,j} - \frac{1}{4}(\delta^h_{i-1,j} + \delta^h_{i+1,j} + \delta^h_{i,j-1} + \delta^h_{i,j+1}) \tag{5.30}$$

Note that for the Jacobi process it is irrelevant in which order the grid points are scanned and the residuals $r^h_{i,j}$ used are not recomputed during the relaxation. This process is preferred because in that case the discrete summations have to be computed only once per relaxation sweep, for all points, and this can be done in a very efficient way as will be shown in Section 5.7.

The effect of this Jacobi relaxation can be analysed using Local Mode Analysis see Chapter 2. For $h_x = h_y$ the result is shown in Figure 5.5. From this figure it can be concluded that the relaxation is stable, and efficiently reduces high frequency components. For $\omega_1 = 1$ the asymptotic smoothing factor $\bar{\mu} = 0.47$. Using some underrelaxation an even better value can be obtained. The optimal value is $\bar{\mu} = 0.32$ for $\omega_1 = 0.9$. Summarizing, it is concluded that the distributive relaxation proposed is a good candidate to serve as the basis for a multigrid solver of the dry contact problem.

5.3.2 Complementarity

In the previous section a stable relaxation with good smoothing was presented. The solution to the stability problems observed with one point relaxations was found by applying distributive relaxation. For the contact problem a second order distributive relaxation

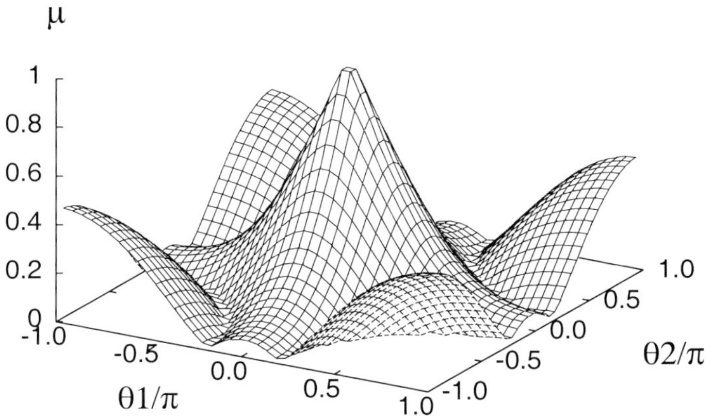

Figure 5.5: *Error amplification factor μ as a function of θ for Jacobi distributive relaxation applied to the 2 dimensional dry contact problem with $\omega = 1$.*

was described which involves applying changes to five points. The next step is to incorporate the complementarity condition. For the Hydrodynamic Lubrication problem (see Section 4.4) the cavitation condition was treated inside the relaxation process by setting each newly computed pressure immediately to zero if it was smaller than zero. The implementation of the complementarity condition for the dry contact problem is an extension of this approach. The dimensionless dry contact problem can be stated in the following manner:

$$\begin{array}{llll} \Omega_1 : & H = 0 & \& & P > 0 \\ \Omega_2 : & H > 0 & \& & P = 0 \end{array} \tag{5.31}$$

At a certain position, the gap H is either zero, and the pressure is positive (the point belongs to Ω_1), or the gap H is positive, and the pressure is zero (Ω_2). This condition excludes a positive pressure and a positive gap (no contact) which is a non-physical situation.

For simplicity of explanation first assume that a one point Jacobi or Gauss-Seidel relaxation is used. In that case the implementation of the changes such that the complementarity condition is satisfied is performed in the following way: given a current approximation to the pressure $\tilde{P}^h$, a new approximation $\bar{P}^h$ is found using the following relaxation:

$$
\begin{aligned}
\delta^h_{i,j} &= \omega_1 \frac{r^h_{i,j}}{K^{hh}_{0,0}} \\
\bar{P}^h_{i,j} &= \tilde{P}^h_{i,j} + \delta^h_{i,j} \\
\text{if} \quad \bar{P}^h_{i,j} < 0 \quad \text{then} \quad \bar{P}^h_{i,j} &= 0
\end{aligned} \tag{5.32}
$$

where ω_1 is the underrelaxation coefficient.

The extension to the five point distributive relaxation scheme can be implemented as follows:

$$
\begin{array}{llllll}
 & & & & \delta^h_{i,j} = \omega_1 \dfrac{r^h_{i,j}}{K^{hh}_{0,0} - K^{hh}_{1,0}} & \\
 & & & & \bar{P}^h_{i,j} = \tilde{P}^h_{i,j} + \delta^h_{i,j} & \\
\text{if} & \bar{P}^h_{i,j} < 0 & \text{then} & \{\bar{P}^h_{i,j} = 0 & \delta^h_{i,j} = (\bar{P}^h_{i,j} - \tilde{P}^h_{i,j})/4\} & \\
 & \text{if} & \tilde{P}^h_{i+1,j} > 0 & \text{then} & \bar{P}^h_{i+1,j} = \tilde{P}^h_{i+1,j} - \delta^h_{i,j} & \\
 & & \text{if} & \bar{P}^h_{i+1,j} < 0 & \text{then} \quad \bar{P}^h_{i+1,j} = 0 & \\
 & \text{if} & \tilde{P}^h_{i-1,j} > 0 & \text{then} & \bar{P}^h_{i-1,j} = \tilde{P}^h_{i-1,j} - \delta^h_{i,j} & \\
 & & \text{if} & \bar{P}^h_{i-1,j} < 0 & \text{then} \quad \bar{P}^h_{i-1,j} = 0 & \\
 & \text{if} & \tilde{P}^h_{i,j+1} > 0 & \text{then} & \bar{P}^h_{i,j+1} = \tilde{P}^h_{i,j+1} - \delta^h_{i,j} & \\
 & & \text{if} & \bar{P}^h_{i,j+1} < 0 & \text{then} \quad \bar{P}^h_{i,j+1} = 0 & \\
 & \text{if} & \tilde{P}^h_{i,j-1} > 0 & \text{then} & \bar{P}^h_{i,j-1} = \tilde{P}^h_{i,j-1} - \delta^h_{i,j} & \\
 & & \text{if} & \bar{P}^h_{i,j-1} < 0 & \text{then} \quad \bar{P}^h_{i,j-1} = 0 &
\end{array} \tag{5.33}
$$

This implementation has two advantages over a more straightforward implementation. The first advantage is that it accounts for the contact condition ($P \geq 0$) in the central point before implementing the changes in the four neighbouring points. Its second advantage is that it implements changes in the neighbouring points only when they are in contact ($P > 0$). This last condition limits the movement of the cavitation boundary between the contact zone and the non-contact zone. It was found that both conditions are necessary to ensure a fast convergence of the Multigrid process.

With the complementarity condition incorporated the dry contact problem can be solved for a given value of the separation H_0. As an example Figure 5.6 gives some results obtained in this way. The figure shows the residual of the equation as a function of the number of relaxations for different grids. The coarsest grid for which results are shown has $(8+1)\times(8+1)$ points and the finest grid $(128+1)\times(128+1)$ points. The figure shows that the relaxation converges, however, with decreasing mesh size the convergence speed slows down and more relaxations are needed to obtain a given error reduction. From the figure it can be seen that each time the mesh size is halved the convergence speed reduces by a factor of two. This implies that the asymptotic error reduction per relaxation is $1-O(h)$. Note that for the Poisson 2d problem the asymptotic error reduction per relaxation on a single grid was $1 - O(h^2)$. The fact that for the present problem relaxation converges faster is due to the integral character of the problem. Nevertheless, even though convergence is faster than for a differential problem the speed of convergence is still mesh size dependent and on very fine grids unacceptably slow for practical purposes. To overcome this slowness Multigrid techniques will be used. Their implementation for the present problem will be explained in Section 5.4.

So far the relaxation was applied to the solution of the problem on a single grid for a given value of the constant H_0. However, for the dry contact problem the value of H_0 is unknown and is determined by the force balance equation. Hence, for this problem H_0 is an unknown that should be solved from the force balance condition simultaneously with the solution of the pressure from the film equation. How to incorporate the iterative solution of H_0 is explained in the next section.

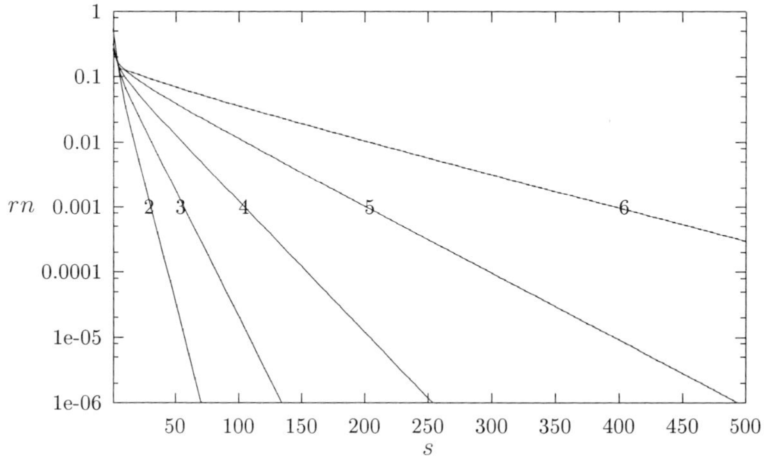

Figure 5.6: *Residual norm as a function of the number of relaxation sweeps s on a single grid for a grid with* $(2^{l-1}4+1) \times (2^{l-1}4+1)$ *points for* $l = 2, 3, 4, 5$ *and* 6. $H_0 = -1.0$, $\omega_2 = 0.2$, $-2 \leq X \leq 2$, $-2 \leq Y \leq 2$ *and* $P_{init} = 0.9P_h$.

5.3.3 Force Balance

The physical condition that there should be force balance determines the value of the unknown H_0 in the dimensionless film thickness equation (5.9), which itself represents the separation between the two remote points in the elastic bodies. By remote it is meant that the elastic deformation in these points is zero. This implies that for a solution for which the integral of the dimensionless pressure is smaller than the dimensionless applied load, the two bodies should be moved closer to one another: H_0 decreases. When the integral exceeds the load, the two bodies should be moved further apart, thus increasing H_0. However, the constant H_0 influences the dimensionless load in an indirect and complicated way: changing the value of H_0 affects all H's. Subsequently, because it changes all H's it changes all P's, and thereby the integral over the dimensionless pressures. Due to this indirect relation, given an approximation $\tilde{P}^h$ and $\tilde{H}_0$ one can not give a simple equation to determine a new value of H_0 such that the force balance condition is satisfied. This can only be done iteratively. The approach is the following. After a number of relaxations on the contact equation an approximation $\tilde{P}^h$ is obtained for a given $\tilde{H}_0$. Subsequently H_0 is corrected according to:

$$\bar{H}_0 = \tilde{H}_0 - \omega_2 \; {}_W r^h \tag{5.34}$$

where ω_2 is an underrelaxation factor and ${}_W r^h$ is the residual of the force balance equation. The force balance equation is given by (5.16) for generality it can be written as:

$$h_x h_y \sum_i \sum_j P^h_{i,j} = {}_W f^h \tag{5.35}$$

On a single grid, and when Multigrid techniques are used on the finest grid in the correction cycle the right hand side ${}_W f^h = 2\pi/3$. On all coarser grids it will generally be different and determined according to the rules of the Full Approximation Scheme FAS. Consequently its residual is given by:

$$ {}_W r^h = {}_W f^h - h_x h_y \sum_i \sum_j \tilde{P}^h_{i,j} \tag{5.36}$$

After changing the approximation to H_0 from $\tilde{H}_0$ to $\bar{H}_0$, a number of relaxations on the contact equation is performed using the new value of H_0. This process of alternating relaxations of the contact equations with changing the value of the unknown H_0 using the residual of the force balance condition, is then repeated until both the residual of the contact equation and of the force balance equation are satisfactory small.

The overall convergence behaviour is affected by two factors. Firstly, the value of ω_2 used to change the value of H_0 according to Equation (5.34) has to be tuned. This means that in the combined process of solving P^h from the contact equation and H_0 from the force balance condition the residuals of both equations converge *both* as fast as possible. A value of ω_2 that is too large will cause H_0 to oscillate, and will hinder convergence. A value of ω_2 which is too small, will also cause a reduction of the overall convergence speed. Secondly, the number of relaxations in between two H_0 changes depends on the number of points on the grid, as a certain level of convergence is required, before the next change of H_0 can be performed. In the Multigrid solver to be proposed, this latter parameter cancels as the relaxation of the force balance condition will be incorporated into the coarse grid correction cycle itself. Details regarding the effect of the choice of ω_2 on convergence will also be restricted to the case of the Multigrid results.

5.4 Coarse Grid Correction

As was shown in Figure 5.6 the relaxation on fine grids gives grid dependent slow convergence. The first step towards a Multigrid cycle is therefore to construct the coarse grid correction. Since the problem is non-linear, due to the complementarity condition, the

Full Approximation Scheme (FAS) is used, see Section 3.8.2. At this point it is emphasized that when a non-linear problem is solved the FAS formulation should be applied to *all equations*. In this case it should thus be used for the contact equation (film thickness equation), *and* for the force balance equation.

First the fine grid complementarity equation (5.12) is recalled:

$$\begin{array}{lll} H^h_{i,j} > 0, & P^h_{i,j} = 0 & \text{no contact} \\ H^h_{i,j} = 0, & P^h_{I,j} > 0 & \text{contact} \end{array} \tag{5.37}$$

For simplicity this equation will be written as:

$$H^h_{i,j} = {}_H f^h_{i,j} \tag{5.38}$$

or as:

$$H_0 + \frac{X_i^2}{2} + \frac{Y_j^2}{2} + \sum_{i'} \sum_{j'} K^{hh}_{i,i',j,j'} P^h_{i',j'} = {}_H f^h_{i,j} \tag{5.39}$$

Assuming some initial approximation to P and an approximation $\tilde{H}_0$ to H_0 the coarse grid correction cycle starts with ν_1 distributive relaxations on this equation. After these relaxations an approximation $\tilde{P}^h$, $\tilde{H}_0$, is obtained for which the residual reads:

$$r^h_{i,j} = {}_H f^h_{i,j} - \tilde{H}_0 - \frac{X_i^2}{2} - \frac{Y_j^2}{2} - \sum_{i'} \sum_{j'} K^{hh}_{i,i',j,j'} \tilde{P}^h_{i',j'} \tag{5.40}$$

Due to the smoothing effect of the relaxation the remaining error will be smooth and can be accurately represented on the coarse grid. On this grid a correction will be solved using the FAS coarse grid equation which reads:

$$\tilde{H}_0 + \frac{X_I^2}{2} + \frac{Y_J^2}{2} + \sum_{I'} \sum_{J'} K^{HH}_{I,I',J,J'} \hat{P}^H_{I',J'} = {}_H \hat{f}^H_{I,J} \tag{5.41}$$

where ${}_H \hat{f}^H$ is defined as:

$${}_H \hat{f}^H_{I,J} = [I_h^H r^h_{.,.}]_{I,J} + \tilde{H}_0 + \frac{X_I^2}{2} + \frac{Y_J^2}{2} + \sum_{I'} \sum_{J'} K^{HH}_{I,I',J,J'} [I_h^H \tilde{P}^h_{.,.}]_{I',J'} \tag{5.42}$$

Please note that on the finest grid in the contact area the film thickness equation reduces to $H^h = 0$. However, this is no longer true on the coarse grids! After an approximation $\tilde{\underline{P}}^H$ has been obtained on the coarse grid, the fine grid approximation $\tilde{\underline{P}}^h$ is corrected according to Equation (5.43):

$$\bar{\underline{P}}^h = \tilde{\underline{P}}^h + I_H^h(\tilde{\underline{P}}^H - I_h^H \tilde{\underline{P}}^h) \tag{5.43}$$

Because of the complementarity character of the contact problem, special attention should be given to the choice of the restriction and the interpolation operators. The restriction of the residuals uses injection, to avoid mixing residuals from the contact and non-contact zone. The interpolation of the pressure correction is followed by the condition that the resulting pressure should be positive. If not the pressure is set to zero.

So far we only considered the solution of the contact equation for a fixed value of H_0. Next the incorporation of the relaxation of the force balance equation changing H_0 in the cycle is explained. As mentioned above, the force balance equation must also be treated using FAS:

$$h_x h_y \sum_i \sum_j P^h_{i,j} = {}_W f^h \qquad \text{with} \qquad {}_W f^h = 2\pi/3 \qquad \text{on the finest grid} \tag{5.44}$$

and after an approximation $\tilde{P}^h$ is obtained, its residual is given by:

$${}_W r^h = {}_W f^h - h_x h_y \sum_i \sum_j \tilde{P}^h_{i,j} \tag{5.45}$$

Please note that the force balance equation is a scalar equation associated with a single unknown H_0. As the residual ${}_W r^h$ of the force balance equation is a scalar, the transfer to a coarser grid is trivial: $I^H_h \, {}_W r^h = {}_W r^h$! Thus the coarse grid right hand side function ${}_W f^H$ is defined on all coarser grids:

$${}_W f^H = H_x H_y \sum_I \sum_J \tilde{P}^H_{I,J} + {}_W r^h = H_x H_y \sum_I \sum_J \tilde{P}^H_{I,J} + {}_W f^h - h_x h_y \sum_i \sum_j \tilde{P}^h_{i,j} \tag{5.46}$$

The next question is now at which points in the cycle the value of H_0 should be changed using the residual of the force balance condition. The answer is that this should only be done on *the coarsest grid* in the cycle. This implies that the residual of the force balance equation is transferred all the way down to the coarsest grid, *without any modification of the H_0 value taking place.* The numerical solution process changes the value of H_0 only on the very coarsest grid (level 1). Changing H_0 induces a smooth (low frequency) change in the solution, resulting in important additional residuals of the contact equation in all points. Since convergence is generally slow on finer grids, reducing these residuals on finer grids would require many relaxation sweeps. However, on the coarsest grid convergence is so fast (see Figure 5.6), that only a few relaxation sweeps are required to reduce the residuals of the contact equation (5.40) to small values after changing the value of H_0. Since the changes in the solution have a low frequency, they can be quite well approximated on the coarsest grid. The correction is performed according to:

$$\bar{H}_0 = \tilde{H}_0 - \omega_2 \, {}_W r^h \tag{5.47}$$

where ω_2 is an underrelaxation coefficient and ${}_W r^h$ is the residual of the force balance equation (5.45), computed on the *coarsest* grid.

The value of ω_2 has to be tuned such that the residuals of the contact equation and the residual of the force balance equation converge *both* as fast as possible. A value of ω_2 that is too large or too small will degrade convergence of the coarse grid correction cycle.

5.5 Cycle Performance

In the previous section the coarse grid equations have been derived. In this section the performance of the constructed coarse grid correction cycle will be evaluated as a function of the coefficients ω_1 and ω_2 used in the relaxation of the pressure and the relaxation of the force balance condition. The coarse grid correction cycle will be used with a coarsest grid of $(4+1) \times (4+1)$ points and results will be presented as a function of the finest grid used in the cycle. The domain used in the calculations extended over $-2.0 \leq X \leq 2.0$ and $-2.0 \leq Y \leq 2.0$. As an initial approximation $P^h = 0.999p_h$ is used where p_h is Hertzian pressure solution given by Equation (5.11). Indeed, this implies that a first approximation is used that, on fine grids, is already very close to the exact solution to the problem. On coarse grids the choice of the first approximation is not critical. However, on fine grids an accurate first approximation is needed for the process to converge as will be explained in the following section. Notice that this is by no means a restriction of the solver because an excellent tool is available to obtain a good first approximation: the Full Multigrid algorithm. However, in this section the aim is to illustrate cycle performance only, and therefore an accurate first approximation had to be chosen.

First the influence of the value of ω_1 is studied, while freezing the rigid body motion ($H_0 = -1$, $\omega_2 = 0$). Tables 5.1-5.3 show the residual norm of the contact equation as a function of the gridlevel and the number of $V(2,1)$ cycles performed for $\omega_1 = 1.0$, $\omega_1 = 0.8$ and $\omega_1 = 0.6$ respectively. From these three tables it can be concluded that the value of ω_1 is not critical. Using a moderate underrelaxation coefficient like $\omega_1 = 0.6$ a very stable convergence is obtained, independent of the level and the number of V-cycles. Note that, based on the contact region alone (the linear problem), the Fourier analysis predicted an optimal value of $\omega_1 = 0.9$ with a potential error reduction of a factor of 20 per cycle. The fact that this factor is not reached is attributed to the effects of the intergrid transfer, and the free boundary.

rn	number of V-cycles						
level	1	2	3	4	5	6	7
2	$4.7 \cdot 10^{-2}$	$1.4 \cdot 10^{-2}$	$4.1 \cdot 10^{-3}$	$1.2 \cdot 10^{-3}$	$3.6 \cdot 10^{-4}$	$1.1 \cdot 10^{-4}$	$3.3 \cdot 10^{-5}$
3	$9.7 \cdot 10^{-3}$	$2.3 \cdot 10^{-3}$	$6.6 \cdot 10^{-4}$	$1.7 \cdot 10^{-4}$	$4.6 \cdot 10^{-5}$	$1.2 \cdot 10^{-5}$	$3.2 \cdot 10^{-6}$
4	$2.9 \cdot 10^{-3}$	$1.1 \cdot 10^{-3}$	$2.8 \cdot 10^{-4}$	$8.0 \cdot 10^{-5}$	$2.3 \cdot 10^{-5}$	$6.6 \cdot 10^{-6}$	$1.9 \cdot 10^{-6}$
5	$3.0 \cdot 10^{-4}$	$6.2 \cdot 10^{-5}$	$1.6 \cdot 10^{-5}$	$4.4 \cdot 10^{-6}$	$1.2 \cdot 10^{-6}$	$3.1 \cdot 10^{-7}$	$8.2 \cdot 10^{-8}$
6	$4.6 \cdot 10^{-4}$	$1.7 \cdot 10^{-4}$	$4.8 \cdot 10^{-5}$	$1.8 \cdot 10^{-5}$	$6.3 \cdot 10^{-6}$	$2.2 \cdot 10^{-6}$	$7.9 \cdot 10^{-7}$
7	$6.1 \cdot 10^{-4}$	$2.7 \cdot 10^{-4}$	$2.8 \cdot 10^{-5}$	$8.0 \cdot 10^{-7}$	$9.0 \cdot 10^{-7}$	$2.2 \cdot 10^{-7}$	$3.8 \cdot 10^{-8}$

Table 5.1: *Residual norm as a function of the level and the number of V-cycles,* $H_0 = -1.0$, $\omega_1 = 1.0$, $\omega_2 = 0$, *initial pressure is* $0.999p_h$.

Next, choosing $\omega_1 = 0.6$ the influence of the value of ω_2 on the convergence behaviour of the cycle is studied. Again $V(2,1)$ cycles are used. Table 5.4 shows the residual of the contact equation and of the force balance equation as a function of the grid level, the number of V-cycles, and the value of ω_2. From this table it can be concluded that slow convergence of the residual norm can be caused in two different ways. For a value of ω_2 which is too small, the force balance equation will be *undercorrected*. This causes

rn	number of V-cycles						
level	1	2	3	4	5	6	7
2	$1.3 \cdot 10^{-2}$	$5.6 \cdot 10^{-4}$	$2.7 \cdot 10^{-5}$	$1.3 \cdot 10^{-6}$	$6.4 \cdot 10^{-8}$	$3.1 \cdot 10^{-9}$	$1.5 \cdot 10^{-10}$
3	$4.8 \cdot 10^{-3}$	$1.8 \cdot 10^{-4}$	$6.1 \cdot 10^{-6}$	$2.1 \cdot 10^{-7}$	$7.5 \cdot 10^{-9}$	$2.8 \cdot 10^{-10}$	$1.1 \cdot 10^{-11}$
4	$8.9 \cdot 10^{-4}$	$3.5 \cdot 10^{-5}$	$1.4 \cdot 10^{-6}$	$5.8 \cdot 10^{-8}$	$3.0 \cdot 10^{-9}$	$1.3 \cdot 10^{-10}$	$7.5 \cdot 10^{-12}$
5	$2.9 \cdot 10^{-4}$	$1.7 \cdot 10^{-5}$	$1.2 \cdot 10^{-6}$	$7.9 \cdot 10^{-8}$	$5.5 \cdot 10^{-9}$	$3.7 \cdot 10^{-10}$	$2.5 \cdot 10^{-11}$
6	$1.6 \cdot 10^{-4}$	$7.0 \cdot 10^{-6}$	$1.0 \cdot 10^{-6}$	$7.8 \cdot 10^{-8}$	$8.9 \cdot 10^{-9}$	$7.8 \cdot 10^{-10}$	$8.0 \cdot 10^{-11}$
7	$1.9 \cdot 10^{-4}$	$1.9 \cdot 10^{-5}$	$3.4 \cdot 10^{-6}$	$4.7 \cdot 10^{-7}$	$7.2 \cdot 10^{-8}$	$1.1 \cdot 10^{-8}$	$1.6 \cdot 10^{-9}$

Table 5.2: *Residual norm as a function of the level and the number of V-cycles,* $H_0 = -1.0$, $\omega_1 = 0.8$, $\omega_2 = 0$, *initial pressure is* $0.999p_h$.

rn	number of V-cycles						
level	1	2	3	4	5	6	7
2	$4.8 \cdot 10^{-3}$	$2.4 \cdot 10^{-4}$	$1.6 \cdot 10^{-5}$	$1.1 \cdot 10^{-6}$	$7.2 \cdot 10^{-8}$	$4.7 \cdot 10^{-9}$	$3.1 \cdot 10^{-10}$
3	$2.0 \cdot 10^{-3}$	$1.5 \cdot 10^{-4}$	$1.3 \cdot 10^{-5}$	$1.2 \cdot 10^{-6}$	$1.3 \cdot 10^{-7}$	$1.7 \cdot 10^{-8}$	$2.3 \cdot 10^{-9}$
4	$4.3 \cdot 10^{-4}$	$4.1 \cdot 10^{-5}$	$4.3 \cdot 10^{-6}$	$4.7 \cdot 10^{-7}$	$5.3 \cdot 10^{-8}$	$6.4 \cdot 10^{-9}$	$8.2 \cdot 10^{-10}$
5	$1.1 \cdot 10^{-4}$	$9.9 \cdot 10^{-6}$	$1.1 \cdot 10^{-6}$	$1.3 \cdot 10^{-7}$	$1.5 \cdot 10^{-8}$	$1.9 \cdot 10^{-9}$	$2.4 \cdot 10^{-10}$
6	$8.5 \cdot 10^{-5}$	$8.4 \cdot 10^{-6}$	$1.1 \cdot 10^{-6}$	$1.5 \cdot 10^{-7}$	$2.2 \cdot 10^{-8}$	$3.2 \cdot 10^{-9}$	$4.7 \cdot 10^{-10}$
7	$1.0 \cdot 10^{-4}$	$6.5 \cdot 10^{-6}$	$7.4 \cdot 10^{-7}$	$8.5 \cdot 10^{-8}$	$1.1 \cdot 10^{-8}$	$1.3 \cdot 10^{-9}$	$1.8 \cdot 10^{-10}$

Table 5.3: *Residual norm as a function of the level and the number of V-cycles,* $H_0 = -1.0$, $\omega_1 = 0.6$, $\omega_2 = 0$, *initial pressure is* $0.999p_h$.

the convergence of the force balance equation to lag behind, and generates continuous changes of the H_0 value. These continued changes will eventually cause a reduction of the overall convergence speed. This type of convergence can be called creeping convergence.

When ω_2 is too large, the force balance equation tends to be *overcorrected* causing the H_0 value to oscillate. Small oscillations are not very harmful, but larger ones tend to hinder the overall convergence as much as undercorrection. This type of convergence can be called oscillating convergence. An extreme example is seen in the table for $\omega_2 = 0.6$. The H_0 value is overcorrected such that the residual of the force balance equation changes sign, but its value hardly reduces. Even larger ω_2 will cause the solution process to diverge.

A value of $\omega_2 = 0.3$ for the smooth circular contact problem was found to be a good compromise, causing something close to critical convergence. The optimal value of ω_2 can also be obtained explicitly by studying the residual of the force balance equation $_Wr^h$ as a function of H_0, as is done in Figure 5.7. The derivative $\partial_W r^h/\partial H_0$ corresponds to the contact stiffness and a value close to 3.0 was found. The exact value can be obtained using a Hertzian analysis, as the load is proportional to the rigid body displacement to the power 3/2. For the residual of the dimensionless force balance equation one finds $_Wr = 2\pi/3(-H_0)^{3/2}$ yielding a dimensionless stiffness of π around $H_0 = -1.0$. The ω_2 value is the inverse of this stiffness, and should thus have an optimum value of $0.318\cdots$. However, for problems involving the contact between extremely rough surfaces, a value of $\omega_2 = 0.2$ yields a very robust convergence behaviour.

From a comparison of Tables 5.1 and 5.4 or from a comparison of the lines $\omega_2 = 0.0$ and

number of V-cycles						
1	2	3	4	6	8	10
$\omega_2 = 0.00$						
$1.0 \cdot 10^{-4}$	$6.5 \cdot 10^{-6}$	$7.4 \cdot 10^{-7}$	$8.5 \cdot 10^{-8}$	$1.3 \cdot 10^{-9}$	$2.4 \cdot 10^{-11}$	$4.2 \cdot 10^{-13}$
$-2.3 \cdot 10^{-6}$	$1.3 \cdot 10^{-4}$	$1.5 \cdot 10^{-4}$	$1.5 \cdot 10^{-4}$	$1.5 \cdot 10^{-4}$	$1.5 \cdot 10^{-4}$	$1.5 \cdot 10^{-4}$
$\omega_2 = 0.05$						
$1.0 \cdot 10^{-4}$	$7.0 \cdot 10^{-6}$	$1.1 \cdot 10^{-6}$	$4.1 \cdot 10^{-7}$	$2.3 \cdot 10^{-7}$	$1.6 \cdot 10^{-7}$	$1.2 \cdot 10^{-7}$
$-3.2 \cdot 10^{-5}$	$7.9 \cdot 10^{-5}$	$8.2 \cdot 10^{-5}$	$7.1 \cdot 10^{-5}$	$5.0 \cdot 10^{-5}$	$3.6 \cdot 10^{-5}$	$2.5 \cdot 10^{-5}$
$\omega_2 = 0.1$						
$1.0 \cdot 10^{-4}$	$7.3 \cdot 10^{-6}$	$1.2 \cdot 10^{-6}$	$4.1 \cdot 10^{-7}$	$1.6 \cdot 10^{-7}$	$7.3 \cdot 10^{-8}$	$3.4 \cdot 10^{-8}$
$-6.2 \cdot 10^{-5}$	$3.9 \cdot 10^{-5}$	$3.8 \cdot 10^{-5}$	$2.8 \cdot 10^{-5}$	$1.3 \cdot 10^{-5}$	$6.2 \cdot 10^{-6}$	$2.9 \cdot 10^{-6}$
$\omega_2 = 0.2$						
$1.0 \cdot 10^{-4}$	$7.2 \cdot 10^{-6}$	$9.5 \cdot 10^{-7}$	$1.5 \cdot 10^{-7}$	$9.0 \cdot 10^{-9}$	$1.0 \cdot 10^{-9}$	$1.4 \cdot 10^{-10}$
$-1.2 \cdot 10^{-4}$	$-1.3 \cdot 10^{-5}$	$7.4 \cdot 10^{-8}$	$6.3 \cdot 10^{-7}$	$1.3 \cdot 10^{-7}$	$1.8 \cdot 10^{-8}$	$2.5 \cdot 10^{-9}$
$\omega_2 = 0.3$						
$1.0 \cdot 10^{-4}$	$6.5 \cdot 10^{-6}$	$6.4 \cdot 10^{-7}$	$6.0 \cdot 10^{-8}$	$6.0 \cdot 10^{-10}$	$7.8 \cdot 10^{-12}$	$1.4 \cdot 10^{-13}$
$-1.8 \cdot 10^{-4}$	$-2.7 \cdot 10^{-5}$	$-3.4 \cdot 10^{-6}$	$-4.1 \cdot 10^{-7}$	$-5.8 \cdot 10^{-9}$	$-8.6 \cdot 10^{-11}$	$-1.3 \cdot 10^{-12}$
$\omega_2 = 0.4$						
$1.1 \cdot 10^{-4}$	$5.0 \cdot 10^{-6}$	$1.0 \cdot 10^{-6}$	$4.5 \cdot 10^{-8}$	$4.6 \cdot 10^{-9}$	$3.3 \cdot 10^{-10}$	$2.1 \cdot 10^{-11}$
$-2.4 \cdot 10^{-4}$	$-3.8 \cdot 10^{-6}$	$-7.4 \cdot 10^{-6}$	$8.3 \cdot 10^{-7}$	$6.9 \cdot 10^{-8}$	$4.6 \cdot 10^{-9}$	$3.0 \cdot 10^{-10}$
$\omega_2 = 0.5$						
$1.1 \cdot 10^{-4}$	$2.9 \cdot 10^{-6}$	$2.8 \cdot 10^{-6}$	$1.1 \cdot 10^{-6}$	$3.9 \cdot 10^{-7}$	$1.3 \cdot 10^{-7}$	$4.0 \cdot 10^{-8}$
$-3.0 \cdot 10^{-4}$	$5.7 \cdot 10^{-5}$	$-4.8 \cdot 10^{-5}$	$2.5 \cdot 10^{-5}$	$8.2 \cdot 10^{-6}$	$2.6 \cdot 10^{-6}$	$8.4 \cdot 10^{-7}$
$\omega_2 = 0.6$						
$1.1 \cdot 10^{-4}$	$2.9 \cdot 10^{-6}$	$6.8 \cdot 10^{-6}$	$5.3 \cdot 10^{-6}$	$4.2 \cdot 10^{-6}$	$3.3 \cdot 10^{-6}$	$2.5 \cdot 10^{-6}$
$-3.6 \cdot 10^{-4}$	$1.6 \cdot 10^{-5}$	$-1.6 \cdot 10^{-4}$	$1.4 \cdot 10^{-4}$	$1.1 \cdot 10^{-4}$	$8.3 \cdot 10^{-5}$	$6.5 \cdot 10^{-5}$

Table 5.4: *Residual norm rn and force balance residual on level 7, as a function of ω_2 and the number of V-cycles, $\omega_1 = 0.6$, initial pressure is $0.999p_h$, initial $H_0 = -1.0$.*

$\omega_2 = 0.3$ of Table 5.4 it can be concluded that the convergence of the force balance equation occurs at the same time as the convergence of the contact problem. The additional equation does not significantly reduce the overall convergence speed.

5.6 Full MultiGrid

The convergence of the V-cycles not only depends on the values of ω_1 and ω_2, the initial solution and the number of points on the finest grid are also important. The simplest way of understanding this dependence is through an analysis of relaxations only. When starting from an initial zero pressure solution, the relaxation process will seek to adjust the pressure in a certain point (i, j) as to correct the local residual. This process is repeated in all points. However, the elastic deformation depends on all pressures, it is an integral over all points, see Equation (5.9), and the correction in every individual point will be

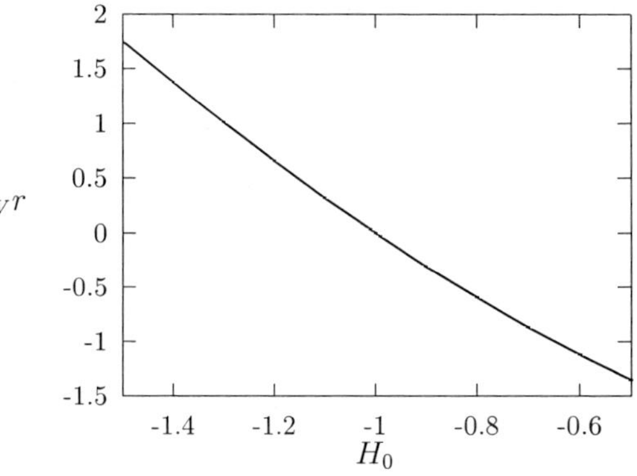

Figure 5.7: *Force balance residual as a function of* H_0.

far too large. The overcorrection thus made will be proportional to the total number of points as the influence coefficient is proportional to $1/h^2$. Underrelaxation can be used to stabilize the relaxation process, but smaller and smaller values are needed on finer and finer grids, slowing down subsequent convergence. In order to ensure a stable process, an initial solution close to the final solution can also be chosen. This reduces the difference in the film thickness equation and thus the overcorrection. For this reason an initial solution close to the analytical solution was chosen in the previous section.

On a very coarse grid though, the deformation is determined by the pressure in only a few points, and the relaxation process is stable, independent of the initial solution. As the relaxation process is continued on finer and finer grids, it is important to have initial solutions which are closer and closer to the final solution. This is exactly the opportunity offered by the Full MultiGrid (FMG) Scheme in which the final solution on a coarse grid is interpolated and used as an initial solution on the next finer grid, etc. The final errors using an FMG process with one and two V(2,1) cycles are given in Table 5.5 using zero as an initial approximation for the pressure distribution and the rigid body separation H_0 on grid level 1.

From this table it can be observed that the FMG process converges even on level 8. Hence, the first approximation obtained by interpolation of the solution from a coarser grid is sufficiently accurate to ensure convergence. The residuals on each level at the end of the cycle(s) however, are not reduced a great deal from one level to the next, and a more detailed analysis is required to answer the question whether convergence to the level of the discretization error occurs using a single V(2,1) cycle. This is done in the next table in which the approximate error norm on all levels is computed.

From the approximate error norm in Table 5.6 it can indeed be concluded that the results obtained using FMG with a single V(2,1) cycle are accurate up to the level of the discretization error, with the possible exception of the level 2 result. Furthermore, it can be observed from this table that the aen^H does not reduce with a factor of 4.0 from a level to the next finer one. The reduction factor lies somewhere between 2 and 3. The discrete solution is second order accurate at each point. However, the analytical solution, and thus the solution that is approximated on finer and finer grids, is the Hertzian solution, which

	1 V-cycle			2 V-cycles		
level	rn	wr	ΔH_0	rn	wr	ΔH_0
1	$1.0 \cdot 10^{-4}$	$-1.2 \cdot 10^{-0}$	$+3.7 \cdot 10^{-1}$	$1.0 \cdot 10^{-4}$	$-1.2 \cdot 10^{-0}$	$+7.2 \cdot 10^{-3}$
2	$4.1 \cdot 10^{-1}$	$-2.1 \cdot 10^{-1}$	$+6.9 \cdot 10^{-2}$	$7.7 \cdot 10^{-2}$	$-5.7 \cdot 10^{-3}$	$-3.1 \cdot 10^{-3}$
3	$3.4 \cdot 10^{-2}$	$-1.2 \cdot 10^{-2}$	$+1.3 \cdot 10^{-3}$	$9.7 \cdot 10^{-4}$	$-3.7 \cdot 10^{-4}$	$-5.6 \cdot 10^{-4}$
4	$2.4 \cdot 10^{-3}$	$-3.3 \cdot 10^{-3}$	$+2.5 \cdot 10^{-4}$	$2.3 \cdot 10^{-4}$	$-3.2 \cdot 10^{-4}$	$+2.5 \cdot 10^{-4}$
5	$1.0 \cdot 10^{-3}$	$-1.1 \cdot 10^{-3}$	$+4.5 \cdot 10^{-5}$	$4.0 \cdot 10^{-5}$	$-4.8 \cdot 10^{-5}$	$+1.1 \cdot 10^{-4}$
6	$1.7 \cdot 10^{-4}$	$-7.7 \cdot 10^{-4}$	$+8.7 \cdot 10^{-5}$	$1.6 \cdot 10^{-5}$	$-8.7 \cdot 10^{-5}$	$+6.7 \cdot 10^{-5}$
7	$9.4 \cdot 10^{-5}$	$-4.1 \cdot 10^{-4}$	$+6.4 \cdot 10^{-5}$	$3.7 \cdot 10^{-6}$	$-3.1 \cdot 10^{-5}$	$+5.0 \cdot 10^{-5}$
8	$4.3 \cdot 10^{-5}$	$-1.7 \cdot 10^{-4}$	$+6.8 \cdot 10^{-5}$	$2.2 \cdot 10^{-6}$	$-9.9 \cdot 10^{-6}$	$+6.2 \cdot 10^{-5}$

Table 5.5: *Residual norm (rn), residual of the force balance equation (wr) and the solid body approach H_0, where $\Delta H_0 = 1 + H_0$ as a function of the level using FMG with one and two V(2,1) cycles, using zero for the initial approximations of P and H_0.*

level	1 V-cycle	2 V-cycles
1	$1.0 \cdot 10^{-1}$	$1.1 \cdot 10^{-1}$
2	$2.0 \cdot 10^{-2}$	$8.1 \cdot 10^{-3}$
3	$3.3 \cdot 10^{-3}$	$4.1 \cdot 10^{-3}$
4	$1.4 \cdot 10^{-3}$	$1.5 \cdot 10^{-3}$
5	$3.9 \cdot 10^{-4}$	$4.0 \cdot 10^{-4}$
6	$1.6 \cdot 10^{-4}$	$1.4 \cdot 10^{-4}$
7	$8.6 \cdot 10^{-5}$	$7.4 \cdot 10^{-5}$

Table 5.6: *Approximate error norm aen as a function of the level using FMG with one and two V(2,1) cycles, using zero for the initial approximations of P and H_0.*

has a singularity in the pressure derivative at $X^2 + Y^2 = 1$. At each point the solution is $O(h^2)$ accurate. However, when the average discretization error is measured it is strongly influenced by the points closest to this singularity. At each finer grid, new points are added that are closer to the singularity. The overall result is that the convergence behaviour of an average quantity is $O(h^{3/2})$ instead of $O(h^2)$. Asymptotically the approximate error should therefore show a reduction with a factor of $2^{3/2} = 2.83$ each time the mesh size is refined. This explains the reduction factor of less than 4.0 in this table. For further details the reader is referred to Brandt and Venner [24].

At this point the Multigrid solver for the dry contact problem seems to be complete. It contains the same characteristic elements as the Multigrid solvers for the differential problems considered in the previous chapters. The slowness of convergence is overcome by the use of coarser grids, and to obtain an accurate first approximation the FMG algorithm is used. Together this ensures that the problem can be solved up to the level of the discretization error in a FMG algorithm with only a few cycles per gridlevel. However, the efficiency of the solver is still far removed from the efficiency of the differential solvers discussed in the previous chapters. This can be seen from analysing the amount of work invested to obtain a solution. The total work of a FMG algorithm is a constant times the amount of work of a single relaxation. For a differential problem the work of a relaxation is $O(N)$ if there are N points on the grid. Thus in that case a FMG algorithm yields a solution in $O(N)$ operations which is the optimal efficiency. For the present problem one relaxation will consist of the (re)computation of the discrete summations representing the elastic deformation to obtain the current film thickness values, see Equation (5.13), followed by an update of the pressure in all points. Obviously this latter step requires $O(N)$ operations. However, the computation of the discrete summation in one grid point already requires a summation over all grid points and thus $O(N)$ operations. Consequently its computation in *all* grid points will require $O(N^2)$ operations! It is this computation of the elastic deformation summations that will determine the amount of work of a single relaxation, and thus the efficiency of the entire solver. Consequently, the solver developed so far has a work count of $O(N^2)$ and measured against the optimal work count of $O(N)$ it is still inefficient. This raises the question if one really needs to invest $O(N^2)$ operations to obtain these integrals, or can one obtain them in a more efficient way. In the next section it will be shown that by allowing an additional error that is at most comparable to the discretization error that is made anyway, the summations can be computed in $O(N\ln(N))$ operations. Incorporating this *fast evaluation* algorithm into the solver will then result in an algorithm solving the dry contact problem in an amount of work (computing time) of $O(N\ln(N))$. As $\ln(N)$ increases only slowly with N this is sufficiently close to the $O(N)$ optimal efficiency.

5.7 Multilevel Multi-Integration

The integral to be analysed is given in (5.1), or written in a dimensionless form in (5.9). Using a general notation it can be written as:

$$w(x) = \int_{\Omega} K(x,y)u(y)\,dy, \quad x \in \Omega \subseteq I\!R^d \tag{5.48}$$

where the function $u(y)$ and the kernel $K(x,y)$ are given, d is the dimension of the domain and $x = (x_1, ..., x_d)$ and $y = (y_1, ..., y_d)$. In terms of the notation used for the two dimensional dry contact problem $x = (x_1, x_2) = (X, Y)$ and $y = (y_1, y_2) = (X', Y')$

It is an integral over all points of the integration domain Ω, to be calculated in all points of the domain Ω. This is called a multi-integral or integral transform. The evaluation of such integrals is a common task in mathematics, physics, and engineering and can be a task by itself, or be needed in the process of (numerically) solving a (system of) integrodifferential equation(s) in which u is the unknown function. Note that this is exactly the case for the dry contact problem. Discretizing (5.48) on a grid with mesh size h implies that the continuous transform is replaced by a matrix multiplication, or multi-summation of the form:

$$\underline{w}^h = K^{hh}\underline{u}^h \tag{5.49}$$

where $\underline{u}^h$ and $\underline{w}^h$ are vectors and K^{hh} is a dense matrix. Obviously, a certain discretization error is made in this process. In this form the task also appears in problems with gravitating masses, vortex schemes, coulombic molecular forces, and other many-body long-range interactions. If the matrix K^{hh} has arbitrary entries there is no way in which the evaluation can be done faster than by straightforward multiplication. At least if one insists on introducing only a controlled error. However, in most physical problems the matrix K^{hh} or *the discrete kernel* has special properties that can be used to reduce the cost of the multisummation $K^{hh}\underline{u}^h$. Examples include far field expansions, [48, 88, 94] hierarchical solvers in many body simulations [3, 6, 30, 47], and FFT based schemes [91] (for the solution of the integral equation). Also in the past decade wavelet techniques have become popular [2, 7] where the complexity reduction is obtained by representation on a suitable set of increasingly coarse base functions.

In this section a simple and general approach referred to as *multilevel matrix multiplication* or *multilevel multi-integration* (abbreviated to MLMI) will be explained. This approach was outlined by Brandt [15, sect. 8.6.] and first implemented in [18] for smooth and asymptotically smooth kernels. An overview including the extension to oscillatory kernels can be found in [19]. The algorithm uses the smoothness properties of the discrete kernel K^{hh} which makes it more general than most other approaches.

Assuming a grid with N points for $\underline{u}^h$ and the integral to be evaluated in N points too, the work count of the evaluation up to an accuracy ϵ can be reduced from $O(N^2)$ to $O(N \ln(1/\epsilon)/\ln\ln(1/\epsilon))$ for smooth kernels, and to $O(N \ln(1/\epsilon))$ for asymptotically smooth kernels, where potential-type kernels ($\ln|x-y|$, $((x_1-y_1)^2+(x_2-y_2)^2)^{-1/2}$) are an example of this latter category. Typically, in the framework of solving the integral equation, or an integro-differential equation in which the integral appears, the allowed error ϵ will be the error that is made in discretising Equation (5.48).

In this section the basic algorithm will be explained. This algorithm is used to evaluate the discretized multi-integral on a uniform grid as described by Brandt and Lubrecht [18]. For clarity the same notation is used. The algorithm can be, and has been, applied to a variety of problems in contact mechanics, such as the computation of the elastic deformation integrals for which it is used in this book, the computation of the subsurface stress field, see [74] and the computation of the contact temperature, and the heat partition function, see Bos and Moes [9, 10, 11]. Recent developments aimed at its application to non-uniform grids are described by Brandt and Venner [23, 24].

5.7.1 Outline

In the classical Multigrid techniques coarser grids are employed to improve the convergence speed (of low frequency error components). In this particular case of the calculation of multi-integrals, our aim is again to utilize coarser grids to decrease the computing time without significantly reducing the accuracy of the integrals. Now, how can this be achieved ? Which function can be accurately represented on the coarser grid? In classical Multigrid techniques the smooth error components are solved on the coarser grids. Surely the function u or its discrete equivalent u^h is a likely candidate, however, there is no good reason to assume that it is smooth, it may be very non-smooth. Therefore the only realistic candidate is the kernel K, or its discrete equivalent K^{hh}. In many cases the kernel is smooth over at least a large part of the domain. For the moment it is sufficient to state that by smooth it is meant that at a given point the value of K^{hh}

can be obtained accurately from the value of K^{hh} in a number of points in its vicinity by means of a sufficiently high order interpolation. It is this smoothness property that can be exploited to obtain a fast evaluation algorithm. In the following sections first the case is considered where the kernel satisfies the smoothness requirement over the entire domain. Next, the extension of the algorithm to a class of kernels with a wider interest will be treated: singular smooth kernels, which stem from problems in contact mechanics (stress, deformation) and heat transfer. In that case the kernel is smooth except for a small region around a singularity.

5.7.2 Discretization Multi-Integral

The domain Ω is discretized using a uniform grid with mesh size h in each dimension. For simplicity the x grid (evaluation grid) and the y grid (integration grid) are assumed to coincide. This simply implies that the integrals are to be evaluated at the same points at which the function u is given as is the case for the dry contact problem. The discretized integral (5.48) is defined as:

$$w^h(x_i^h) = w_i^h \stackrel{\text{def}}{=} \int_\Omega K(x_i^h, y)\hat{u}^h(y)\, dy = h^d \sum_j K_{i,j}^{hh} u_j^h \tag{5.50}$$

where $\hat{u}^h$ is a piecewise polynomial function of degree $2s - 1$ and $\hat{u}^h(y_j^h) = u_j^h$. The coefficients $K_{i,j}^{hh}$ are calculated such that Equation (5.50) holds. The dimension of the domain is d and the factor h^d is introduced to ensure that both K^{hh} and u^h are of comparable magnitude on grids with different mesh sizes. For each x the discretization error in w^h (which is the truncation error τ^h of the discrete integral) made in this process will be:

$$|\tau^h| < h^{2s}|K|\,||u^{(2s)}|| \tag{5.51}$$

where $|K|$ stands for an average of the absolute value of K^{hh} over the domain and $||u^{(2s)}||$ denotes the maximum of the $2s^{th}$ derivative of u.

Coarser grids will be introduced to facilitate the fast evaluation. For simplicity, only one coarser grid (in x and y) is introduced in the present description. The coarse grid mesh size H is $H = 2h$. The indices on this coarse grid will be denoted by capitals. The two grids are arranged such that $x_{2I}^h = x_I^H$ and $y_{2J}^h = y_J^H$, as was done previously. In later sections we will use more than two grids and refer to them as levels, starting with the coarsest level which will be called level 1, as has been done in the previous chapters. Information between the grids will be transferred using transfer operators. Firstly, the interpolation yielding the value w_i^h on the fine grid from the values of w^H in a number of coarse grid points will be denoted by:

$$w_i^h = [\mathit{\Pi}_H^h w_{\cdot}^H]_i \tag{5.52}$$

where the symbol $\mathit{\Pi}_H^h$ is used to distinguish the operator from the usual interpolation operator applied in a Multigrid algorithm as it is generally of a higher order. The order of interpolation will be denoted as $2p$ as only even orders are considered. The dot in Equation (5.52) indicates that the interpolation is done with respect to the I index. This is trivial when it is applied to w_I^H which is a function of only one index. However, the interpolation will also be applied to the discrete kernel in which case one should distinguish if interpolation takes place with respect to the I index or the J index. For the transfer of a fine grid to the coarse grid the transpose of the interpolation operator will be used denoted by $(\Pi_H^h)^T$.

The algorithm is described in detail in the following sections. The notation used in these sections applies to a d dimensional problem. However, at first reading it may be helpful to visualize a one dimensional problem.

5.7.3 Smooth Kernel

In this section, the kernel K is assumed to exhibit the required smoothness properties everywhere. If the integral is discretized on a uniform grid the discrete kernel K^{hh} will have the same smoothness behaviour as K. First assume that the kernel $K(x, y)$ is smooth with respect to the variable y. In that case the value of K^{hh} in a grid point y_j^h can approximated accurately by means of interpolation from its values in points y_J^H. Defining:

$$K_{i,J}^{hH} \stackrel{\text{def}}{=} K_{i,2J}^{hh} \tag{5.53}$$

one can define an approximate kernel $\tilde{K}_{i,j}^{hh}$ such that:

$$K_{i,j}^{hh} = \tilde{K}_{i,j}^{hh} + O(\epsilon) \tag{5.54}$$

where:

$$\tilde{K}_{i,j}^{hh} = [\Pi_H^h K_{i,.}^{hH}]_j \tag{5.55}$$

Using the approximate kernel it follows that up to an error of $O(\epsilon)$ Equation (5.50) can be approximated by:

$$\begin{aligned} w_i^h \simeq \tilde{w}_i^h &\stackrel{\text{def}}{=} h^d \sum_j \tilde{K}_{i,j}^{hh} u_j^h = h^d \sum_j [\Pi_H^h K_{i,.}^{hH}]_j u_j^h \\ &= h^d \sum_J K_{i,J}^{hH} [(\Pi_H^h)^T u_.^h]_J = H^d \sum_J K_{i,J}^{hH} u_J^H \end{aligned} \tag{5.56}$$

where:

$$u_J^H \stackrel{\text{def}}{=} 2^{-d} [(\Pi_H^h)^T u_.^h]_J \tag{5.57}$$

Equation (5.56) expresses that, once K^{hh} is replaced by the interpolated kernel $\tilde{K}^{hh}$ the summation of the fine grid quantity u_j^h multiplied with $\tilde{K}_{i,j}^{hh}$, can be rewritten into a summation of a coarse grid quantity u_J^H multiplied with a coarse grid kernel $K_{i,J}^{hH}$. The change from a fine grid summation (index j) to a coarse grid summation (index J) is made possible because the interpolated kernel uses only coarse grid values $K_{i,J}^{hH}$, and by the definition of the coarse grid summation quantity u_J^H using the adjoint of the interpolation operator Π_H^h: $\Pi_h^H = (\Pi_H^h)^T$.

The operation in Equation (5.56) is best illustrated using an example. For a one dimensional problem with a grid with $0 \leq j \leq n$ and $0 \leq J \leq n/2$ the interpolated kernel will be given by the following vector:

$$[\Pi_H^h K_{i,.}^{hH}] = \frac{1}{2}[2K_{i,0}^{hH}, K_{i,0}^{hH} + K_{i,1}^{hH}, 2K_{i,1}^{hH}, K_{i,1}^{hH} + K_{i,2}^{hH}, \cdots, \\ 2K_{i,n/2-1}^{hH}, K_{i,n/2-1}^{hH} + K_{i,n/2}^{hH}, 2K_{i,n/2}^{hH}] \quad (5.58)$$

thus the multiplication can be written as:

$$\sum_j [\Pi_H^h K_{i,.}^{hH}]_j u_j^h = \frac{1}{2}[2K_{i,0}^{hH} u_0^h + (K_{i,0}^{hH} + K_{i,1}^{hH})u_1^h + 2K_{i,1}^{hH} u_2^h + (K_{i,1}^{hH} + K_{i,2}^{hH})u_3^h \\ + \cdots + 2K_{i,n/2-1}^{hH} u_{n-2}^h + (K_{i,n/2-1}^{hH} + K_{i,n/2}^{hH})u_{n-1}^h + 2K_{i,n/2}^{hH} u_n^h] \quad (5.59)$$

but this summation can be rewritten, regrouping the values of u^h as:

$$\sum_j [\Pi_H^h K_{i,.}^{hH}]_j u_j^h = \frac{1}{2}[K_{i,0}^{hH}(2u_0^h + u_1^h) + K_{i,1}^{hH}(u_1^h + 2u_2^h + u_3^h) + \cdots + \\ + K_{i,n/2-1}^{hH}(u_{n-3}^h + 2u_{n-2}^h + u_{n-1}^h) + K_{i,n/2}^{hH}(u_{n-1}^h + 2u_n^h)] \quad (5.60)$$

thus this fine grid summation can be rewritten as a coarse grid summation:

$$\sum_j [\Pi_H^h K_{i,.}^{hH}]_j u_j^h = K_{i,0}^{hH} u_0^H + K_{i,1}^{hH} u_1^H + \cdots + K_{i,n/2-1}^{hH} u_{n/2-1}^H + K_{i,n/2}^{hH} u_{n/2}^H \quad (5.61)$$

this can be written as:

$$h \sum_j [\Pi_H^h K_{i,.}^{hH}]_j u_j^h = H \sum_J K_{i,J}^{hH} u_J^H \quad (5.62)$$

using the definition:

$$u_J^H \stackrel{\text{def}}{=} 2^{-d}[(\Pi_H^h)^T u^h]_J \quad (5.63)$$

Note that the definition (5.57) ensures that $u_{2J}^h \simeq u_J^H$ if u^h is a smooth function. Because u^H is computed using the transpose of the interpolation, its computation is also referred to as *anterpolation*. Note that, even though it is an operation on u^h yielding a coarse grid quantity u^H, no assumptions regarding u^h have been made. The anterpolation is simply the result of the definition of a coarse grid quantity using the transpose of the interpolation (5.56) (in the example rewriting Equation (5.59) into Equation (5.60)). The only assumption made is that K^{hh} is smooth as a function of y such that replacing $K_{i,j}^{hh}$ by $\tilde{K}_{i,j}^{hh}$ introduces an $O(\epsilon)$ that can be made arbitrarily small if the order of interpolation $2p$ is sufficiently large.

Generally, the kernel K, and thus on a uniform grid K^{hh} has similar smoothness properties in x and y. Previously, it was shown that if the kernel is smooth as a function of y the evaluation of the discrete summation over a fine grid j to obtain w_i^h for each grid point i can be replaced by a summation over a coarse grid J. Next it will be shown that if the kernel is smooth as a function of x it is not necessary to compute w_i^h in each grid point. Rather, it is sufficient to compute it only in those points which coincide with the coarse grid points. In the other points it can be obtained accurately using interpolation from its values in points i that coincide with coarse grid points. If the kernel is smooth with respect to x too, the value of K^{hh} at a given point i can be obtained accurately from interpolation using its value in coarse grid points I only:

$$K_{i,j}^{hh} = \hat{K}_{i,j}^{hh} + O(\epsilon) \tag{5.64}$$

where:

$$\hat{K}_{i,j}^{hh} = [\hat{\mathit{\Pi}}_H^h K_{.,j}^{Hh}]_i \tag{5.65}$$

with $K_{I,j}^{Hh} = K_{2I,j}^{hh}$.
Assuming similar smoothness properties in x and y, $\hat{\mathit{\Pi}}_H^h = \mathit{\Pi}_H^h$ can be used. Together with Equation (5.56) which was the result of the previous step, one obtains that up to an error $O(\epsilon)$:

$$w_i^h \simeq [\mathit{\Pi}_H^h w_{.}^H]_i \tag{5.66}$$

where:

$$w_I^H \stackrel{\text{def}}{=} \tilde{w}_{2I}^h = H^d \sum_J K_{I,J}^{HH} u_J^H \tag{5.67}$$

and $K_{I,J}^{HH}$ is defined as $K_{I,J}^{HH} = K_{2I,2J}^{hh}$.
Combining the results of both steps, it follows that if the kernel is smooth as a function of x and y, the task of calculating (5.50) for all points i, summing over all points j, can be replaced by the transfer of u^h to the coarse grid according to (5.57), the calculation of a coarse grid multi-summation (5.67), and an interpolation of the result from the coarse grid to the fine grid (5.56). An error $O(\epsilon)$, controlled by the choice of the interpolation order $2p$, is made in the process. The entire algorithm is summarized in the box on the next page.

The total work involved in the computation of the summations on grid h can now be determined. First analyse the *anterpolation* of u^h to the coarse grid. Assuming uniform grids and a coarse grid with $H = 2h$ this operation can be done using only $O(2pN)$ operations if N is the total number of grid points on the fine grid. In the same way the interpolation of w^H from the coarse grid to the fine grid can be done in $O(2pN)$ operations. For a one dimensional problem, it is easy to see that the work count of interpolation and anterpolation is $O(2pN)$. However, this is also true for the general case of a d dimensional problem, provided that the interpolation and anterpolation are carried out one dimension at the time, which is the most efficient way. Note that one dimension at the time refers to the $x_1, x_2, \ldots$ directions (X, and Y for the two dimensional dry contact problem) for interpolation and to the $y_1, y_2, \ldots$ directions (X' and Y' for the dry contact) for the anterpolation.

Multi-Integration: two grids *Smooth Kernel*

- **Anterpolation**

  For each point J compute u_J^H according to:

$$u_J^H \stackrel{\text{def}}{=} 2^{-d}[(\mathit{\Pi}_H^h)^T u_{\cdot}^h]_J \tag{5.68}$$

- **Coarse grid summation**

  For each point I compute w_I^H according to:

$$w_I^H = H^d \sum_J K_{I,J}^{HH} u_J^H \tag{5.69}$$

  where $K_{IJ}^{HH} = K_{2I,2J}^{hh}$

- **Interpolation**

  For each point i compute w_i^h using:

$$w_i^h = [\mathit{\Pi}_H^h w_{\cdot}^H]_i \tag{5.70}$$

Finally, the summation on the coarse grid will require $O(2^{-2d}N^2)$ operations. If N is large the number of coarse grid points $N/2^d$ may still be large. Consequently, the work involved in the coarse grid summation may still be the dominating factor. However, as the kernel was assumed to be smooth over the entire domain, and the coarse grid summation has exactly the same form as the fine grid summation the algorithm can be applied recursively. Thus coarser and coarser grids are used until a grid is reached on which the number of points is $O(\sqrt{N})$. On this grid the summation is then actually performed, requiring $O(N)$ operations. In that case the work of each step in the algorithm, and thereby also the total work invested to obtain the summations, is $O(N)$ operations. In [18] results are presented for an example problem with a smooth kernel. Here the problem with a smooth kernel is only used as an introduction to the case where the kernel is asymptotically smooth. However, before directing our attention to this case, some final remarks are made. Firstly, if the kernel is smooth over the entire domain, the step by step approach going to coarser and coarser grids is not needed. In that case one can directly make the step to a grid with $\sqrt{N}$ points. However, the step by step approach is needed anyway for the asymptotically smooth kernel case that is discussed next. Secondly, the only parameter in the algorithm is the order of transfer $2p$. This parameter controls the accuracy of the result and should be chosen such that a prescribed accuracy limit is not exceeded. If the kernel is smooth on the scale of the domain $2p = O(\ln(1/\epsilon)/\ln(c/H))$ is needed to make an error of $O(\epsilon)$, where c is some constant. Hence, for a given coarse grid H, in order to obtain a more accurate solution, an increase of the interpolation order $2p$ is required. Furthermore, in order to maintain the accuracy, the interpolation order $2p$ should also be increased with increasing H. This implies that for the optimal algorithm the order of transfer should be increased going to coarser and coarser grids. However, in practice there is little loss in efficiency using the same (sufficiently high) order of transfer for all the coarser grids. Finally, when dealing with multi-summations obtained by discretizing a multi-integral, the natural choice for the prescribed accuracy limit will be the discretization error. The condition determining the transfer order, will be a condition of the form:

$$|e^H| < |\tau^h| \tag{5.71}$$

where $|e^H|$ is the error introduced in the fast evaluation on grid H which is a function of H and $2p$ and $|\tau^h|$ the error made anyway by discretizing the equation which is a function of the fine grid mesh size h.

5.7.4 Singular Smooth Kernels

In the previous section, the kernel K was assumed to be sufficiently smooth over the entire domain Ω; for a more quantitative description of the smoothness required, the reader is referred to [18]. Expressed in words, this requirement implies that the kernel can be interpolated from coarser grid using $2p = O(\ln(1/\epsilon))$ order interpolation, yielding an error of only $O(\epsilon)$. A number of kernels of practical interest, however, do not fulfil this requirement; for instance, the potential-type kernels $K(x,y) = \ln|x-y|$ and $K(x,y) = |x-y|^{-1}$ are non-smooth (singular) in the neighbourhood of $x = y$. Fortunately, their smoothness increases rapidly with increasing distance $r = |x-y|$. Since the kernel is smooth in a large portion of the domain, it is possible to maintain the previously outlined approach. However, to keep the additional error, made in the coarse grid integration process, below the required level, some additional work in the neighbourhood of the singularity will be performed in the form of corrections. First an exact expression that will replace Equation (5.56) will be derived for the case that the fine grid point i belongs also to the coarse grid (even points, $i = 2I$).

$$w_i^h = h^d \sum_j K_{i,j}^{hh} u_j^h = h^d \sum_j \tilde{K}_{i,j}^{hh} u_j^h + h^d \sum_j (K_{i,j}^{hh} - \tilde{K}_{i,j}^{hh}) u_j^h$$

$$= h^d \sum_j [\mathit{II}_H^h K_{i,\cdot}^{hH}]_j u_j^h + h^d \sum_j (K_{i,j}^{hh} - \tilde{K}_{i,j}^{hh}) u_j^h$$

$$= w_I^H + h^d \sum_j (K_{i,j}^{hh} - \tilde{K}_{i,j}^{hh}) u_j^h \tag{5.72}$$

In this derivation, Equations (5.65), (5.56) and (5.67) have been used.
Now it can be shown that the correction term $(K_{i,j}^{hh} - \tilde{K}_{i,j}^{hh}) \to 0$ as $|i-j| \to \infty$. Remember that $\tilde{K}^{hh}$ is obtained by interpolation from K^{hh} itself and that K^{hh} becomes smoother with increasing $|i-j|$. To be more precise, for a $2p$ order interpolation from a coarse grid with $H = 2h$:

$$(K_{i,j}^{hh} - \tilde{K}_{i,j}^{hh}) = \begin{cases} 0, & \text{if } j = 2J; \\ O(h^{2p}K^{(2p)}(\xi)), & \text{otherwise.} \end{cases} \tag{5.73}$$

Where $K^{(2p)}(\xi)$ is the $2p$'th derivative of K at some intermediate point. Thus, whenever this derivative of K becomes small, the correction term will become small and can be neglected. Clearly, this is no longer true for $i \simeq j$ in the case of the singular smooth kernels mentioned above and thus we will have to carry out the corrections in a neighbourhood of $i = j$ ($|j-i| \leq m$ or $i-m \leq j \leq i+m$ in 1-d). The precise shape of this neighbourhood in higher dimensions is discussed in [18]. Equation (5.72) can therefore be simplified to:

$$w_i^h \simeq w_I^H + h^d \sum_{|j-i|\leq m} (K_{i,j}^{hh} - \tilde{K}_{i,j}^{hh})u_j^h \tag{5.74}$$

For points i that do not belong to the coarse grid (odd points, $i = 2I+1$ for a one dimensional problem), the value of w^h is obtained by interpolation from the coarse grid values w_I^H which followed from the introduction of the approximate kernel $\hat{K}^{hh}$ defined as:

$$\hat{K}_{i,j}^{hh} = [I\!I_H^h K_{.,j}^{Hh}]_i \tag{5.75}$$

where $\hat{K}_{I,j}^{Hh} = K_{2I,j}^{hh}$. Using this approximate kernel we can derive an expression similar to (5.72), for the integrals in these points (index $i = 2I+1$ for a one dimensional problem):

$$\begin{aligned} w_i^h &= h^d \sum_j K_{i,j}^{hh} u_j^h = h^d \sum_j \hat{K}_{i,j}^{hh} u_j^h + h^d \sum_j (K_{i,j}^{hh} - \hat{K}_{i,j}^{hh})u_j^h \\ &= h^d \sum_j [I\!I_H^h K_{.,j}^{Hh}]_i u_j^h + h^d \sum_j (K_{i,j}^{hh} - \hat{K}_{i,j}^{hh})u_j^h \\ &\simeq [I\!I_H^h w_{.}^H]_i + h^d \sum_j (K_{i,j}^{hh} - \hat{K}_{i,j}^{hh})u_j^h \end{aligned} \tag{5.76}$$

Far from the singularity the correction terms become small and are once again neglected. Equation (5.76) then reduces to:

$$w_i^h \simeq [I\!I_H^h w_{.}^H]_i + h^d \sum_{|j-i|\leq m} (K_{i,j}^{hh} - \hat{K}_{i,j}^{hh})u_j^h \tag{5.77}$$

Assuming that $K(x,y)$ has similar smoothness properties in x and y, and identical interpolation operators are used in (5.65) and (5.75), the correction term in (5.76) is similar to (5.73). Since the interpolation is carried out with respect to the i index, the correction term will be non-zero for all values of j:

$$(K_{i,j}^{hh} - \hat{K}_{i,j}^{hh}) = O(h^{2p}K^{(2p)}(\xi)) \qquad (\forall j, i = 2I+1) \tag{5.78}$$

This will result in larger errors (approximately three times as large) in the points i that do not coincide with coarse grid points I as compared to the points $i = 2I$ (a factor of two comes from Equations (5.77) and (5.78), a factor of one comes from the approximation in

Equation (5.76)). The resulting algorithm for the fast evaluation of the integral transform with an asymptotically smooth kernel is summarized in the box below.

Multi-Integration: two grids *Singular Smooth Kernel*

- **Anterpolation**

  For each point J compute u_J^H according to:

$$u_J^H \stackrel{\text{def}}{=} 2^{-d}[(\mathit{\Pi}_H^h)^T u_.^h]_J \tag{5.79}$$

- **Coarse grid summation**

  For each point I compute w_I^H according to:

$$w_I^H = H^d \sum_J K_{I,J}^{HH} u_J^H \tag{5.80}$$

- **Coarse grid correction**

  For each point I add a correction:

$$w_I^H \leftarrow w_I^H + h^d \sum_{|2I-j|\leq m} (K_{2I,j}^{hh} - \tilde{K}_{2I,j}^{hh})u_j^h \tag{5.81}$$

- **Interpolation**

  For each point i compute w_i^h using:

$$w_i^h = [\mathit{\Pi}_H^h w_.^H]_i \tag{5.82}$$

- **Fine grid correction**

  For each point i add a correction:

$$w_i^h \leftarrow [\mathit{\Pi}_H^h w_.^H]_i + h^d \sum_{|i-j|\leq m} (K_{i,j}^{hh} - \hat{K}_{i,j}^{hh})u_j^h \tag{5.83}$$

The total work of the algorithm will now be $O((2p+\bar{m})N)$ where $\bar{m}$ is the total number of correction points. The values of $2p$ and $\bar{m}$ should be determined from an optimization, minimizing the work per grid point given the requirement that the error introduced by the fast evaluation should be smaller than the discretization error. For a detailed example of such an analysis the reader is referred to [18]. For the one dimensional problem one obtains $2p \propto \ln(1/h)$ and $m \propto 2p$ giving a total work involved in the evaluation of $O(N\ln(N))$ if N is the total number of points on the grid. The dependence of $2p$ on the fine grid mesh size is to be expected. On a finer grid the discretization error will be smaller, hence, to obtain a result with an evaluation error smaller than the discretization error, requires that the evaluation should be carried out more accurately.

For a d dimensional problem the situation is a bit more complex because the correction region is a d dimensional region and $\bar{m} = \Pi_{k=1}^{d} m_k$. However, if the kernel has exactly the same smoothness behaviour in each dimension, and the algorithm is implemented as an extension of the one dimensional problem, see Section 5.7.5, one can use $m = O(\ln(1/h))$ only in the direction of interpolation and a fixed small number independent of the mesh size in the other dimensions.

5.7.5 Implementation

Before demonstrating the efficiency of the Multilevel Multi-Integration algorithm for the fast evaluation of the film thickness in the dry contact problem first in this section some hints regarding its implementation in code are given. These hints (or tricks), also help to understand the way in which the multi-integration is programmed in the dry contact program given in Appendix G. The central aim in writing the program was to create a code as simple as possible. The second requirement was to keep the code readable, the third one was to keep it short. Considerations like computing time have been neglected with respect to these three considerations.

Ghost Points

The interpolation $\mathit{I\!I}_H^h$ of the kernel from coarse grid points to fine grid points that do not coincide with coarse grid points is a straightforward process in the interior of the domain. The only difference compared to the interpolation operators discussed in Chapter 3 is that the stencils are wider due to the higher order. Near the boundary of the domain, however this higher order interpolation causes complications as non central interpolation is required for a few points near the boundary. Their number depends on the order of the interpolation used. High order interpolation would require a relatively large number of points with special interpolation operators. This would result in additional and complicated lines in the program, and thus the possibilities to make errors are numerous. Besides, non-central interpolations are by definition less accurate than central operators.

The use of non-central operators in the interpolation can be avoided if all grids are extended with a fixed number of *ghost points*. Their number depends on the degree of the interpolation. Second order interpolation requires zero ghost points, fourth order requires two ghost points, sixth order requires four ghost points, etc. The ghost points will increase the total number of unknowns, but since our goal is to calculate on very fine grids, they will hardly affect the total number of unknowns. In these ghost points, at least on all coarser grids, the summations are also computed. Their introduction allows the use of the central interpolation in all points as is shown in Figures 5.8 and 5.9. In these two figures the start (edge) of the domain is depicted. The ghost points are represented by open circles, the 'real' points by closed circles. The 'real' domain starts at the dotted line. The first full circle has index zero. A similar extension with ghost points occurs at the end of the domain. Please note that the domain including the ghost points covers different geometrical domains on every level. Note that the use of these ghost points is justified by the fact that the kernel is usually well defined over the entire extended domain too, and all interpolations in the algorithm are interpolations of the kernel.

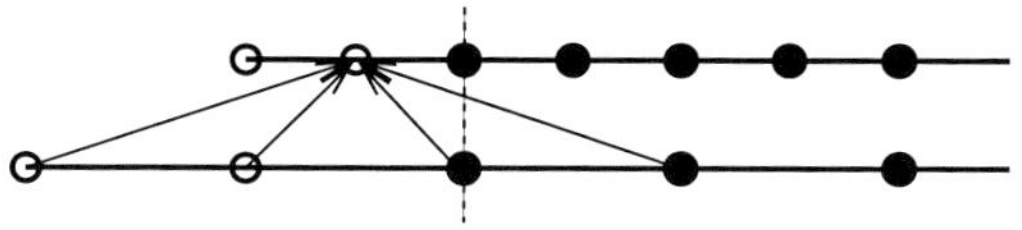

Figure 5.8: *Fourth order interpolation with ghost points.*

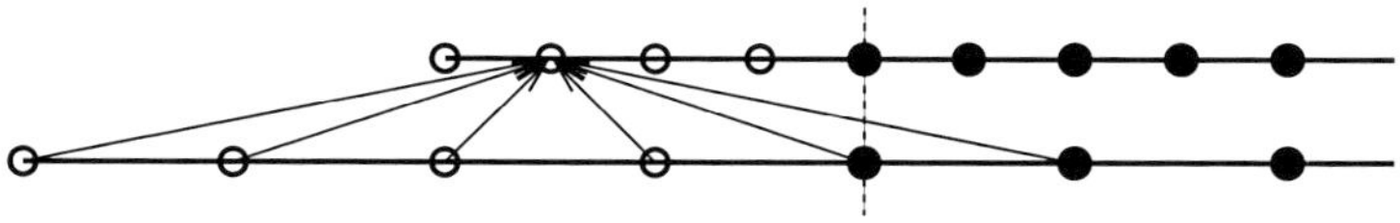

Figure 5.9: *Sixth order interpolation with ghost points.*

The values of the variables in the ghost points on the finest grid are defined to be zero. On coarser grids, the restriction $u^H = \Pi_h^H u^h$ will generally create non-zero values in ghost points. The use of the ghost points is hidden in the program, as is explained in detail in Appendix A. In higher dimensions the ghost points cover domains that are straight extensions of the one dimensional problem.

Next, a few remarks about the implementation of the fine to coarse transfer of u^h to obtain u^H. In the code in Appendix G the transfer is implemented using the stencils as they appear in the transpose of the interpolation matrix. However, a simpler and cheaper way is to use the interpolation stencil itself. Instead of multiplying the coarse grid value with a coefficient and giving it to the fine grid point as is done when interpolating one now reverses the approach, adding the fine grid value, multiplied with the coefficient associated with the coarse grid point, to each of the coarse grid points appearing in the stencil. As can be verified easily, after all fine grid points have been scanned, each coarse grid point has accumulated the correct result. Programmed in this way it is easy to see that the anterpolation requires exactly $2p$ operations per fine grid point, and thus in total $2pN$ operations.

Finally, different orders of transfers have been programmed in separate routines. This has been done for clarity. To reduce the program length they can easily be replaced by two single routines, one for interpolation and one for anterpolation, each having the required order as an input parameter. The coefficients of the interpolation can easily be precomputed using the Lagrange formula and stored in a small array.

Higher Dimensions

In spite of the use of the ghost points, the high order interpolation and restriction operators already generate complicated interpolation and correction routines in one dimension. In higher dimensions these routines would become very complicated and long, and therefore difficult to debug. Besides, interpolation and anterpolation from a uniform grid to a uniform grid are most efficiently performed one dimension at the time. Therefore, it was decided to program the two dimensional problem using the one dimensional routines twice. A first interpolation takes place with respect to x, the second interpolation takes place with respect to y. After the first interpolation a correction routine with respect to x corrects for the non-smoothness of the kernel. After the second interpolation with respect to y a second correction routine with respect to y removes the large errors due to the singular kernel. This means that two interpolation and two correction routines are required. However, in both cases the second routine is a trivial permutation of the first one, so it can be written and debugged very rapidly.

This approach has a second disadvantage since the integrals are calculated and corrected on intermediate grids with $h_x = h$ and $h_y = H = 2h$. Although other solutions are possible, these intermediate grids are implemented to form part of the general grid structure, and as such they contribute to the overall storage requirement of the program. The additional storage (50%) and the double routines were considered to be a small price to pay for the gained simplicity and clarity. In the case of a three dimensional problem, the approach can be straightforwardly extended, requiring three interpolation and restriction routines (with respect to x, y and z) and two intermediate levels. Once again the additional programming effort and the additional storage space 78% are considered a small price to pay for the gained simplicity.

An additional advantage of programming the multi-dimensional integration in this way is the choice of the correction regions. When coarsening in say x direction one can use $m = O(\ln(1/h))$ for the x direction and a small constant in the other dimensions, thus giving $\bar{m} = O(\ln(1/h))$ rather than $\bar{m} = O((\ln(1/h))^d)$.

Multilevel Multi-Integration Results

In this section some results obtained with the *multilevel multi-integration* or *fast evaluation* algorithm will be presented in detail. As a model problem the computation of the (dimensionless) film thickness is taken for the 2 dimensional dry contact problem. P^h is taken as the Hertzian pressure solution given by Equation (5.11). The domain extends over $[-2, 2] \times [-2, 2]$ and the coarsest grid used contained $(4 + 1) \times (4 + 1)$ points. The fast evaluation has been used with a fixed transfer order on all grids. The number of correction points was taken:

$$m_1 = 3 + \ln(n) \tag{5.84}$$

in the direction of interpolation with n the number of points in one dimension on the grid at which the integrals are to be evaluated, and

$$m_2 = 2 \tag{5.85}$$

in the direction perpendicular to the direction of interpolation. To monitor the performance, two quantities have been computed. Let k denote the gridlevel at which the evaluation is needed, and l the coarsest grid used in the algorithm, i.e. the grid at which the actual multi-summation is performed. The total error in the result is defined as:

$$E^{k,l} = h^2 \sum_i \sum_j |H_{i,j}^{k,l} - H(x_i, y_j)| \tag{5.86}$$

The summation in this definition is taken over all points in the contact circle $X_i^2+X_j^2 \leq 1$ where the analytical solution is given by $H(x_i, y_j) = 0$. The superscript k, l indicates the level on which the evaluation is needed k and level l at which the coarse grid multi-summation is performed. Consequently, $E^{k,k}$ is the result of summation on a single grid and a measure of the discretization error, and $K^{k,l}$ with $l < k$ the error in the result obtained using the multilevel multi-integration with $k - l$ coarsening steps (in each dimension). The multiplication with h^2 in the summation is done to obtain a grid independent value where h refers to the mesh size of the target grid (level k).
The second quantity monitored is the incremental error defined as:

$$EI^{k,l} = h^2 \sum_i \sum_j |H_{i,j}^{k,l} - H_{i,j}^{k,l+1}| \tag{5.87}$$

This incremental error is only defined for $l \leq k - 1$ and is a measure of the error in the level k result introduced by the coarser level l.

Tables 5.7-5.9 show the results obtained using sixth order transfers. Table 5.7 gives the total error, Table 5.8 the incremental error and Table 5.9 the computing time as obtained on a SGI O2 R5000 computer. Please note that the actual speed of the computer is irrelevant. What counts is the relative gain (ratio) in computing time. This ratio should be virtually machine independent. In all tables the results marked with an asterisk indicate the cases where the actual multisummation is carried out on a grid with $\sqrt{N}$ points.

First examine the results for $l = k$ in the first column of Table 5.7. Each time the mesh size is decreased by a factor 2 one would expect this error to reduce by a factor 4 in accordance with the second order discretization. However, the table shows the actual reduction is approximately a factor 3. This is due to the singularity in the Hertzian pressure solution at $X^2 + Y^2 = 1$, as explained in Section 5.6. The computing times in the first column of Table 5.9 show that each time the mesh size is refined, the computing time needed to obtain the result increases by a factor 16. As each finer grid has about 4 times the number of nodes this column illustrates the $O(N^2)$ work count of the evaluation using classical single grid summation.

k	$l = k$	$l = k - 1$	$l = k - 2$	$l = k - 3$	$l = k - 4$	$l = k - 5$
1	$2.86 \cdot 10^{-1}$					
2	$1.58 \cdot 10^{-1}$	$1.58 \cdot 10^{-1}$				
3	$4.68 \cdot 10^{-2}$	$4.67 \cdot 10^{-2}$	$*4.68 \cdot 10^{-2}$			
4	$1.56 \cdot 10^{-2}$	$1.53 \cdot 10^{-2}$	$1.53 \cdot 10^{-2}$	$1.54 \cdot 10^{-2}$		
5	$3.07 \cdot 10^{-3}$	$3.00 \cdot 10^{-3}$	$2.95 \cdot 10^{-3}$	$*3.02 \cdot 10^{-3}$	$3.13 \cdot 10^{-3}$	
6	$9.50 \cdot 10^{-4}$	$9.05 \cdot 10^{-4}$	$8.38 \cdot 10^{-4}$	$7.84 \cdot 10^{-4}$	$8.63 \cdot 10^{-4}$	$9.76 \cdot 10^{-4}$
7	$2.47 \cdot 10^{-4}$	$2.30 \cdot 10^{-4}$	$1.99 \cdot 10^{-4}$	$1.53 \cdot 10^{-4}$	$*1.23 \cdot 10^{-4}$	$2.78 \cdot 10^{-4}$
8	$8.93 \cdot 10^{-5}$	$8.53 \cdot 10^{-5}$	$7.77 \cdot 10^{-5}$	$6.41 \cdot 10^{-5}$	$4.55 \cdot 10^{-5}$	$5.88 \cdot 10^{-5}$

Table 5.7: *Error in integral on level k, where the integration is actually carried out on level l ($2p = 6$).*

Next consider the cases with $l < k$. From Table 5.7 it can be seen that the error for all cases up to eight levels ($k = 8$) and five levels deep ($l = k - 5$) is at most a little larger than the error for $l = k$. Therefore, the error introduced as a result of the evaluation using

k	$l=k$	$l=k-1$	$l=k-2$	$l=k-3$	$l=k-4$	$l=k-5$
1	$2.86 \cdot 10^{-1}$					
2	$1.58 \cdot 10^{-1}$	$4 \cdot 10^{-6}$				
3	$4.64 \cdot 10^{-2}$	$4 \cdot 10^{-4}$	$*8 \cdot 10^{-5}$			
4	$1.56 \cdot 10^{-2}$	$3 \cdot 10^{-4}$	$2 \cdot 10^{-4}$	$1 \cdot 10^{-4}$		
5	$3.07 \cdot 10^{-3}$	$7 \cdot 10^{-5}$	$5 \cdot 10^{-5}$	$*7 \cdot 10^{-5}$	$1 \cdot 10^{-4}$	
6	$9.50 \cdot 10^{-4}$	$5 \cdot 10^{-5}$	$7 \cdot 10^{-5}$	$7 \cdot 10^{-5}$	$2 \cdot 10^{-4}$	$1 \cdot 10^{-4}$
7	$2.47 \cdot 10^{-4}$	$2 \cdot 10^{-5}$	$3 \cdot 10^{-5}$	$5 \cdot 10^{-5}$	$*6 \cdot 10^{-5}$	$2 \cdot 10^{-4}$
8	$8.93 \cdot 10^{-5}$	$4 \cdot 10^{-6}$	$8 \cdot 10^{-6}$	$1 \cdot 10^{-5}$	$2 \cdot 10^{-5}$	$5 \cdot 10^{-5}$

Table 5.8: *Incremental error caused by the integration level l. The column $l = k$ displays the discretization error ($2p = 6$).*

k	$l=k$	$l=k-1$	$l=k-2$	$l=k-3$	$l=k-4$	$l=k-5$
5	00 : 00 : 11	00 : 00 : 01	00 : 00 : 01	*00 : 00 : 01	00 : 00 : 01	
6	00 : 02 : 39	00 : 00 : 14	00 : 00 : 03	00 : 00 : 02	00 : 00 : 02	00 : 00 : 02
7	00 : 32 : 39	00 : 02 : 47	00 : 00 : 17	00 : 00 : 07	*00 : 00 : 06	00 : 00 : 06
8	11 : 22 : 40	00 : 32 : 47	00 : 03 : 08	00 : 00 : 35	00 : 00 : 25	00 : 00 : 24

Table 5.9: *Computing times (hours : minutes : seconds) for the integral on level k, where the integration is actually carried out on level l ($2p = 6$).*

k	$l=k$	$l=k-1$	$l=k-2$	$l=k-3$	$l=k-4$	$l=k-5$
1	$2.86 \cdot 10^{-1}$					
2	$1.58 \cdot 10^{-1}$	$1.58 \cdot 10^{-1}$				
3	$4.68 \cdot 10^{-2}$	$4.63 \cdot 10^{-2}$	$*4.64 \cdot 10^{-2}$			
4	$1.56 \cdot 10^{-2}$	$1.66 \cdot 10^{-2}$	$1.70 \cdot 10^{-2}$	$1.73 \cdot 10^{-2}$		
5	$3.07 \cdot 10^{-3}$	$3.14 \cdot 10^{-3}$	$3.19 \cdot 10^{-3}$	$*3.17 \cdot 10^{-3}$	$3.22 \cdot 10^{-3}$	
6	$9.50 \cdot 10^{-4}$	$9.94 \cdot 10^{-4}$	$1.06 \cdot 10^{-3}$	$1.12 \cdot 10^{-3}$	$1.10 \cdot 10^{-3}$	$1.15 \cdot 10^{-3}$
7	$2.47 \cdot 10^{-4}$	$2.57 \cdot 10^{-4}$	$2.73 \cdot 10^{-4}$	$2.97 \cdot 10^{-4}$	$*3.12 \cdot 10^{-4}$	$3.05 \cdot 10^{-4}$
8	$8.93 \cdot 10^{-5}$	$8.98 \cdot 10^{-5}$	$9.06 \cdot 10^{-5}$	$9.18 \cdot 10^{-5}$	$9.19 \cdot 10^{-5}$	$8.45 \cdot 10^{-5}$
9	$1.88 \cdot 10^{-5}$	$1.90 \cdot 10^{-5}$	$1.95 \cdot 10^{-5}$	$2.03 \cdot 10^{-6}$	$2.14 \cdot 10^{-5}$	$*2.17 \cdot 10^{-5}$

Table 5.10: *Error in integral on level k, where the integration is actually carried out on level l ($2p = 8$).*

k	$l=k$	$l=k-1$	$l=k-2$	$l=k-3$	$l=k-4$	$l=k-5$
1	$2.86 \cdot 10^{-1}$	-				
2	$1.58 \cdot 10^{-1}$	$3 \cdot 10^{-5}$				
3	$4.68 \cdot 10^{-2}$	$6 \cdot 10^{-4}$	$*1 \cdot 10^{-4}$			
4	$1.56 \cdot 10^{-2}$	$1 \cdot 10^{-3}$	$4 \cdot 10^{-4}$	$3 \cdot 10^{-4}$		
5	$3.07 \cdot 10^{-3}$	$7 \cdot 10^{-5}$	$6 \cdot 10^{-5}$	$*9 \cdot 10^{-5}$	$6 \cdot 10^{-5}$	
6	$9.50 \cdot 10^{-4}$	$5 \cdot 10^{-5}$	$7 \cdot 10^{-5}$	$6 \cdot 10^{-5}$	$1 \cdot 10^{-4}$	$6 \cdot 10^{-5}$
7	$2.47 \cdot 10^{-4}$	$1 \cdot 10^{-5}$	$2 \cdot 10^{-5}$	$2 \cdot 10^{-5}$	$*3 \cdot 10^{-5}$	$9 \cdot 10^{-5}$
8	$8.93 \cdot 10^{-5}$	$5 \cdot 10^{-7}$	$1 \cdot 10^{-6}$	$2 \cdot 10^{-6}$	$6 \cdot 10^{-6}$	$3 \cdot 10^{-5}$
9	$1.88 \cdot 10^{-5}$	$2 \cdot 10^{-7}$	$5 \cdot 10^{-7}$	$1 \cdot 10^{-6}$	$2 \cdot 10^{-6}$	$*6 \cdot 10^{-6}$

Table 5.11: *Incremental error caused by the integration level l. The column $l = k$ displays the discretization error ($2p = 8$).*

k	$l=k$	$l=k-1$	$l=k-2$	$l=k-3$	$l=k-4$	$l=k-5$
5	00 : 00 : 15	00 : 00 : 03	00 : 00 : 01	*00 : 00 : 01	00 : 00 : 01	
6	00 : 03 : 00	00 : 00 : 16	00 : 00 : 04	00 : 00 : 02	00 : 00 : 02	00 : 00 : 02
7	00 : 32 : 39	00 : 03 : 03	00 : 00 : 21	00 : 00 : 08	*00 : 00 : 06	00 : 00 : 06
8	11 : 22 : 40	00 : 47 : 46	00 : 03 : 46	00 : 00 : 42	00 : 00 : 32	00 : 00 : 26
9	173 : 47 : 27	11 : 22 : 40	00 : 47 : 51	00 : 05 : 54	00 : 03 : 22	*00 : 03 : 10

Table 5.12: *Computing times (hours : minutes : seconds) for the integral on level k, where the integration is actually carried out on level l (using eight order transfers).*

coarser grids must be smaller than or at most comparable to the discretization error. This is confirmed by the results of the incremental error presented in Table 5.8. Summarizing it is concluded that with $2p = 6$ up to level 8 the coarsening process can be repeated all the way down to a grid with $O(\sqrt{N})$ points at the expense of an error that is at most comparable to the discretization error.

From Table 5.9 it can be seen that the computing time reduction obtained by the fast evaluation is huge. For example using the integration five levels deep, the level eight computing time of eleven hours and twenty two minutes is reduced to twenty four seconds, a reduction of more than a factor of one thousand! However, already on much coarser grids, relevant computing time reductions are obtained. Finally, comparing the computing times for different levels k with $l \approx k/2$ it can be seen that each time the mesh size is refined, the computing time increases only by a factor 4. As explained in the previous section the algorithm has in principle an $O(N \ln(N))$ complexity. However, the table shows that the factor $\ln(N)$ is hardly recognizable in the results.

In Section 5.7.4 it was pointed out that in order to ensure an error smaller than the discretization error the order of transfer should be chosen $2p = O(\ln(1/h))$. This implies that with decreasing mesh size, a higher order of transfer is needed at a certain point. Tables 5.10-5.12 present the results obtained for $2p = 8$. In this case the fast integration can be used up to even finer grids. Even on level 9 which has about a million points the integration can be taken down to a grid with $\sqrt{N}$ points at the expense of an error smaller than the discretization error. The computing time reduction is again huge. Comparing Table 5.9 with 5.12 shows that the increase of computing time for the eight order transfers is small.

5.8 Incorporating MLMI into the FMG Solver

In order to obtain the optimal $O(N \ln(N))$ efficiency for the dry contact problem, the Multigrid solver discussed in the beginning of this chapter has to be combined with the fast integration routines described previously. Since both use a very similar Multigrid structure, the fast solver was implemented on the grid using the ghost points. Whenever the fast solver requires the integrals to be calculated, the current pressure distribution is transferred to coarser grids, and the integral is calculated using these coarser grids. This implementation raises a number of questions about interference of coarse grid variables like P and K which have different values and purposes in the convergence acceleration part and the fast integration part. When thinking carefully about it, it can be seen that the two can use the same arrays, since they do not interfere. Concerning the kernel, it is necessary though to clarify how the kernel on the coarser grids is defined. In order to maintain a coherent argument, it was decided to redefine the kernel, whenever the coarse grid correction cycle reaches a coarser grid, using the kernel definition, applied to this grid.

5.9 Conclusion

In this chapter Multigrid techniques have been applied to efficiently solve the dry contact problem. First, the same technique as introduced in the previous chapters was used to accelerate convergence, paying attention to the complementarity condition. Because of the non-local character of the deformation equation, it was necessary to introduce distributive relaxation.

The imposed load introduced a second equation, which was relaxed only on the coarsest grid. Residuals of this equation are transfered down from the finest grid to the coarsest grid, such that the coarse grid changes in H_0 do indeed satisfy the fine grid force balance equation. It was shown that the system of equations could be solved almost as efficiently as the problem with imposed rigid body approach H_0. The resulting FMG solution algorithm was shown to yield solutions to the problem with an error that is small compared to the discretization error using only a few V-cycles per refinement.

Finally *Multilevel Multi-Integration* was described. This is a separate Multigrid algorithm designed for the fast evaluation of multi-integrals. In this case it is used for the elastic deformation integrals which otherwise, i.e. if evaluated on a single grid with classical summation, would require $O(N^2)$ operations if N is the number of nodes. It was shown that using the Multilevel Multi-Integration the work count of the evaluation can be reduced to $O(N \ln(N))$ at the expense of an error that is small compared to the discretization error that is made anyway. Incorporated into the FMG algorithm the result is an algorithm that provides an efficient and stable solver for the dry contact problem.

Fast Fourier Transform techniques [26] can be used to calculate the deformation integrals yielding very similar gains in computing time. In order to obtain an efficient solver of the contact problem, however, these techniques have to be coupled to an efficient solution technique. The advantage of the method described in this chapter is that the two techniques (fast integration and fast solution) use the same data structure and are thus relatively compact when integrated. Another advantage of the fast integration is that it does not rely on periodicity, as does FFT. However, this last disadvantage can be easily removed, by introducing a larger domain, albeit at a computing time penalty.

5.10 Advanced Topics

The techniques presented in this chapter can be used to efficiently solve a wide variety of problems in contact mechanics. Some of these problems and some advanced insights are briefly addressed below.

5.10.1 Wavy Surfaces

The development of a numerical solver for the dry contact between parabolical bodies is not a very useful occupation, even if one tries to develop a solver that is as efficient as possible. The reason for this is that the solution of the problem has been known for more than a century: [60]. If the numerical solution method is to be useful, it has to be able to solve non-Hertzian contacts. The simplest way of changing the contact is to introduce a 'roughness' term. The term roughness is used in a general sense: including form errors (components with a wavelength comparable or larger than the contact radius), waviness (components with a wavelength around one tenth of the contact radius) and roughness (components with a wavelength around one hundredth of the contact radius). These definitions are not strict, but will be useful for the rest of this section to identify the type of problem.

In the case of form errors, the overall geometry changes significantly, effectively changing the contact area, and the pressure distribution. As long as contact occurs only in one specific zone, the standard contact solver will converge efficiently. However, it is possible that the force balance equation is very perturbed, and therefore many V-cycles are required for convergence. An FMG approach, perhaps combined with W-cycles will in that case reduce the total number of cycles required.

In the case of waviness, the geometry changes only locally, which results in a pressure distribution that looks like a perturbed Hertzian distribution. Whenever the perturbations are small compared to the Hertzian pressure, the standard solver will converge efficiently. However, when the pressure variations become so large that the contact is split into many, small, distinct contact areas, the standard settings will no longer allow convergence. The underrelaxation factors have to be decreased: 0.2 for the relaxation of the pressures, and 0.2 for the relaxation of H_0 work fine. This will cause a slower, but more robust convergence.

In the case of roughness, the same measures as described for the waviness tend to work fine. For both wavy and rough contact problems the FMG approach should be used with care. Whenever the surface is very wavy or rough, it is possible that the coarser grid solutions will give a bad starting solution for the finer grid. It is then recommended to use a truncated FMG cycle, starting with a Hertzian solution on an intermediate level (say level 4) that is sufficiently fine to show the outline of the solution, see Figure 5.10. The extreme case results in simple V or W-cycles.

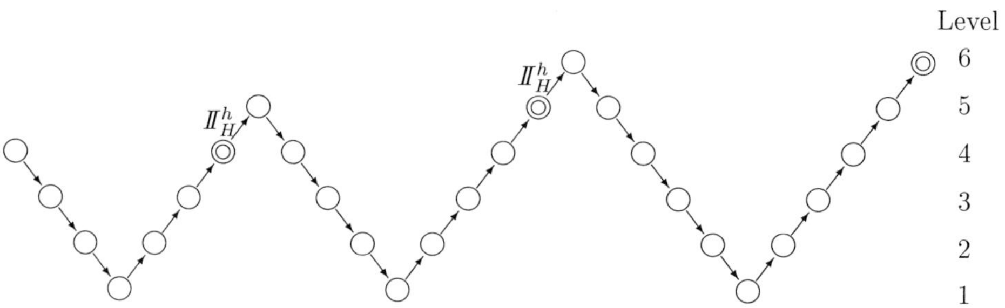

Figure 5.10: *Truncated FMG cycle starting on level 4.*

It is important to understand that the coarser grids will still be effective in the error reduction process (since the errors are smooth) even though these coarser grids can no longer accurately represent the solution. An extreme case occurs for the following example: all the pressures on the coarsest grid are always zero. This means that the coarsest grid no longer contributes to the solution process, which takes place on the one but coarsest grid. Some results for such a case are given below.

The roughness is introduced as follows in the dimensionless film thickness equation:

$$H(X,Y) = H_0 + \frac{X^2}{2} + \frac{Y^2}{2} - \mathcal{R}(X,Y) + \frac{2}{\pi^2}\int_{-\infty}^{+\infty}\int_{-\infty}^{+\infty}\frac{P(X',Y')\,dX'\,dY'}{\sqrt{(X-X')^2+(Y-Y')^2}} \quad (5.88)$$

For this example the roughness was modeled as a product of two cosine functions, with identical wavelengths in the x and y direction.

$$\mathcal{R}(X,Y) = \mathcal{A}\cos(\frac{2\pi X}{\lambda})\cos(\frac{2\pi Y}{\lambda}) \quad (5.89)$$

In this particular case $\mathcal{A} = 0.05$ and $\lambda = 0.2$ was used. Figure 5.11 shows the computed pressure at the line $Y = 0$ as a function of X. The contact area is split in approximately one hundred distinct contact patches, but convergence is hardly affected, see Dumont [36].

5.10.2 Tangential Stresses

The calculation of the deformation integrals used only the normal stress field. But in real applications tangential stresses will also exists i.e. due to frictional forces. The deformations (and subsurface stresses see next section) arising from these tangential stresses can be computed in a similar way, see Kalker [64, 65]. The kernels describing the effects are also singular smooth kernels, and the fast integration can be directly extended to include these effects.

5.10.3 Subsurface Stresses

The calculation of the subsurface stress field in a semi-infinite half-space, is a task similar to the calculation of the deformation integrals. The stress tensor in a point involves the

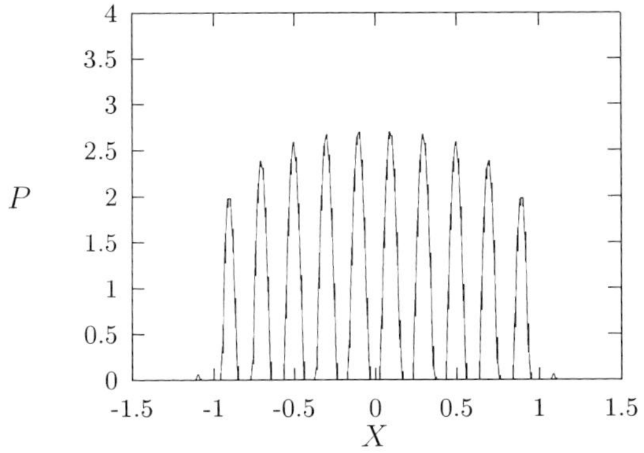

Figure 5.11: *Dimensionless dry contact pressure* $P(X, Y = 0)$ *for a wavy circular contact case.*

contribution from each point on the surface. The tensor has six independent components and thus six different kernels are required, see Kalker [64, 65]. Furthermore, these kernels change with the depth z. The simplest way of approaching this problem is through the calculation of the stress tensor at planes of constant depth z. In that case the problem becomes completely equivalent to the fast calculation of the deformation integrals, and the techniques described previously can be directly applied. When all components of the stress tensor at a certain depth are calculated, the depth is increased and the process starts anew. As the kernels become smoother and smoother (less singular) with increasing depth, integration on coarser and coarser grids can be used to limit the total computing time, without any loss of accuracy, i.e. fine pressure grid information is only required close to the surface to obtain high accuracy stress results. For increasing depths z, coarser pressure grids can be used *maintaining* the same accuracy level for the stress results. Furthermore, the stress grid can also coarsened without any accuracy loss, as the stress field becomes smoother and smoother. Both types of coarsening result in an additional gain in computing time.

5.10.4 Fast Integration, Recent Developments

The fast integration described in the previous sections permits the calculation of the integrals in N points in $O(N \ln(N))$ operations instead of $O(N^2)$ operations, without a significant loss of accuracy. The fast integration calculates the integrals on coarser grids and partially corrects them on finer grids. The gain in computing is based on the smoothness properties of the discrete kernel K^{hh}. Therefore, the continuous kernel has to have certain smoothness properties, but the grid has to be uniform as well. Local grid refinements as described in the last section of Chapter 3 are thus not possible. To overcome this problem Brandt and Venner [23, 24] have developed another method which only uses the smoothness properties of the continuous kernel K.

5.10.5 Physical Interpretation

It is possible to understand in a physical way why the MultiLevel Multi-Integration can obtain such a reduction in calculational complexity (time) without any loss of accuracy. To see this it is useful to recall the Saint-Venants principle:

"altering a pressure distribution without affecting the resultant force results in a new stress field which is only significantly different in the neighbourhood of the changes"

This implies that one can replace the pressure distribution over a certain area by its resultant force, without changing the deformation in points far away. In other words, to calculate the deformation in far away points, the pressure distribution can be lumped together without any significant reduction in accuracy.

The MLMI uses this principle in a recursive way, computing the contribution to the deformation in a point due to the concentrated forces far away using a coarse grid. The closer the points are to the point of interest, the finer the grids used in the calculation. Or conversely, the farther away the points are, the more they are lumped together. Graphically this is represented by Figure 5.12.

Figure 5.12: *Calculation on grid with increasing meshsize to obtain integral in the centre (arrow).*

In order to be able to continue calculating on equidistant grids, with mesh size ratio 0.5, the calculation is performed on small parts of the different grids, graphically this is represented by Figure 5.13.

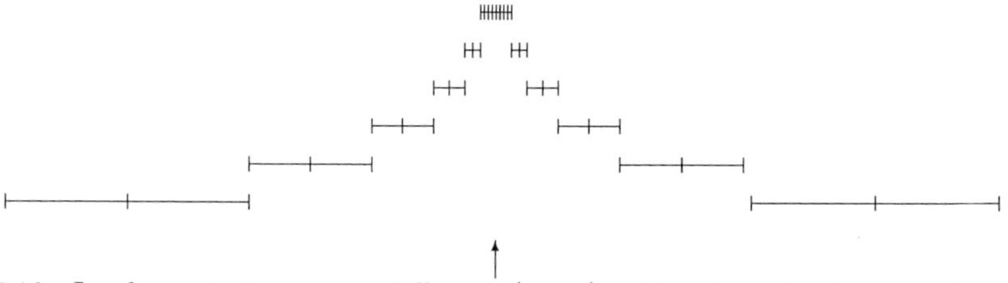

Figure 5.13: *Implementation using different (open) grids of constant mesh size to obtain the integral in the centre (arrow).*

Because of the interpolated kernel used in the summation, the domains on which the integrals are calculated are not as clear cut as in this figure. Consequently, the fine grid contribution is computed as a fine grid correction over the coarse grid integral.

This means that the coarse grid integral is computed over the entire domain, and thus some calculations are done twice, resulting in a slightly less rapid but much simpler implementation. This implementation is graphically represented in Figure 5.14.

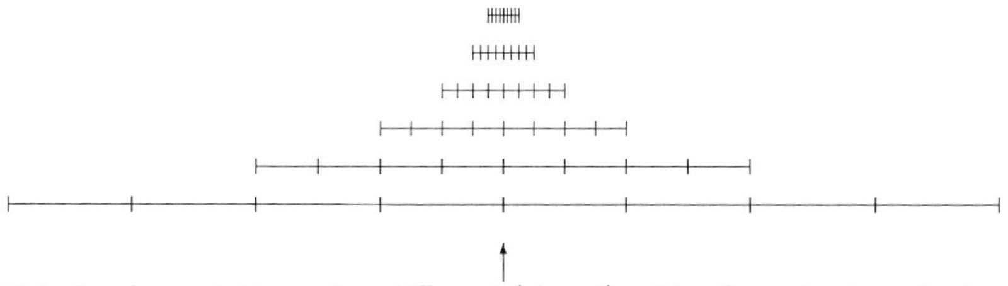

Figure 5.14: *Implementation using different (closed) grids of constant mesh size to obtain the integral in the centre (arrow).*

In the MLMI the calculation is carried out over the entire coarse grid, but corrections are performed over increasingly smaller sub-domains, on increasingly finer grids. The size of the correction domains, as well as the order of the corrections are determined by the precision required. In this chapter, the precision is required to be close to the discretization error on the finest grid, but other criteria can be used.

The implementation presented above can be easily extended to problems in higher dimensions as long as intermediate grids are used, so coarsening and refining occurs one dimension at the time.

Chapter 6

ElastoHydrodynamic Lubrication

In the final chapter of this book the different characteristic elements of the solvers for the two dimensional Poisson problem (Chapter 3), the Hydrodynamic Lubrication problem discussed in Chapter 4, and the Dry Contact problem discussed in Chapter 5 are all brought together in a Multigrid solver for the EHL circular contact problem.

6.1 Introduction

In the previous chapter the dry contact between elastic bodies was considered. From the results it may be clear that even if sufficient lubricant is supplied to form a film separating the surfaces, the shape and thickness of this film will be strongly influenced by the deformation of the contacting elements. This situation is typical for lubricated concentrated contacts, where concentrated refers to the condition that the radii of curvature of the two surfaces are of opposite sign (contraform) or of the same sign but with a large difference in value. Practical examples are the contacts between rolling element and the inner and outer raceway in rolling element bearings, the contacts between gear teeth, and the contacts between a cam and follower in a cam tappet mechanism.

In the second half of the twentieth century the study of the film formation and the performance of these contacts has evolved into a separate discipline referred to as Elasto-Hydrodynamic Lubrication, abbreviated to EHL. Characteristic for EHL contacts are the extreme conditions of a very thin lubricant film, high pressure (up to 3.0 GPa for contacts between steel surfaces) and high shear rates. Consequently, the elastic deformation is not the only characteristic element in EHL studies. Under these extreme conditions the lubricant behaviour can no longer be characterized with a constant viscosity and density. Pressure dependence of the viscosity and the density and also non-Newtonian lubricant behaviour play an important role in EHL contacts.

For a detailed account of the history of EHL the reader is referred to Dowson [35], [37] and Gohar [45]. Only in asymptotic and much simplified cases an estimate of a characteristic value such as the central or minimum film thickness could be obtained analytically. To obtain the complete pressure profile and film shape, a numerical approach was needed; simultaneously solving pressure and film thickness from the Reynolds equation, the film equation including elastic deformation, and the force balance condition. Classical work in this sense is the work of Dowson and Higginson [34] (incorporated in [35]) for the line contact problem and of Hamrock and Dowson [55] and [56] for the point contact problem. The film thickness formulas presented in those works, which were obtained from the numerical results, are still widely used today.

Unfortunately the solution algorithms for the EHL problem turned out to be rather time consuming. Cpu times tended to be $O(N^3)$ if N is the number of nodes on the grid. This seriously limited the number of nodes that could be used and as a result very dense grids could not be utilised.

A major step forward was formed by the introduction of multigrid techniques by Lubrecht [71, 72, 73] as an alternative method for the numerical solution of EHL problems. Their further development is recorded in Venner [99] and Wijnant [110]. These works have resulted in algorithms that enable the solution of line and point contact problems in an amount of work of $O(N\ln(N))$ if N is the total number of grid points. This low work count facilitates the use of dense grids, using work stations and even personal computers. Moreover, the algorithms are stable for a wide range of load conditions. The efficiency with which the basic equations could now be solved created room for more complex problems to be studied. The numerical simulations for these problems have led to a significantly increased understanding of mechanisms and phenomena that can play a role in EHL contacts. Some of these aspects will be discussed in the advanced techniques section at the end of this chapter.

This chapter describes how an efficient solver for the steady state EHL circular contact problem can be obtained by combining the lessons of the previous chapters. The resulting algorithm forms an excellent basis for the study of more complex EHL problems.

6.2 Equations

For completeness the equations describing the EHL point contact problem as given in Chapter 1 are repeated here. The Reynolds equation for the steady state case reads:

$$\frac{\partial}{\partial x}(\frac{\rho h^3}{12\eta}\frac{\partial p}{\partial x}) + \frac{\partial}{\partial y}(\frac{\rho h^3}{12\eta}\frac{\partial p}{\partial y}) - u_m\frac{\partial(\rho h)}{\partial x} = 0 \tag{6.1}$$

with $p = 0$ on the boundaries and the cavitation condition $p \geq 0$ everywhere in the domain. The viscosity η is given by either the Barus equation [5]:

$$\eta(p) = \eta_0 \exp(\alpha p) \tag{6.2}$$

where η_0 is the atmospheric viscosity and α is the pressure viscosity coefficient, or by the empirical relation proposed by Roelands [93]:

$$\eta(p) = \eta_0 \exp((\ln(\eta_0) + 9.67)(-1 + (1 + \frac{p}{p_0})^z)) \tag{6.3}$$

where z is the pressure viscosity index, typically $z = 0.6$, and $p_0 = 1.96 \cdot 10^8$ is a constant. The density is either assumed to be constant, $\rho = \rho_0$, or it is given by the Dowson and Higginson relation [35]:

$$\rho(p) = \rho_0 \frac{5.9 \cdot 10^8 + 1.34\,p}{5.9 \cdot 10^8 + p} \tag{6.4}$$

where ρ_0 is the atmospheric density and p is given in Pa. The equation for the gap height or film thickness was given as:

$$h(x,y) = h_0 + \frac{x^2}{2R_x} + \frac{y^2}{2R_y} + \frac{2}{\pi E'} \int_{-\infty}^{+\infty} \int_{-\infty}^{+\infty} \frac{p(x',y')\,dx'\,dy'}{\sqrt{(x-x')^2+(y-y')^2}} \tag{6.5}$$

where R_x and R_y are the reduced radii of curvature of the surfaces in x and y direction respectively, and E' is the reduced modulus of elasticity. For the circular contact studied in this chapter $R_y = R_x$. The constant h_0 is determined by the force balance condition:

$$\mathrm{w} = \int_{-\infty}^{+\infty} \int_{-\infty}^{+\infty} p(x',y')\,dx'\,dy' \tag{6.6}$$

In physical terms $-h_0$ represents the *mutual approach* of two remote points in the solids. By remote it is meant that the elastic deformation at these points is negligible.

From these equations one can solve the pressure and film thickness as a function of the operating conditions. In particular, one is interested in characteristic solution parameters such as the minimum film thickness:

$$h_m = \min\ h(x,y) \tag{6.7}$$

or the central film thickness which is defined as the film thickness in the centre of the contact, i.e. at $(x,y) = (0,0)$:

$$h_c = h(x=0, y=0) \tag{6.8}$$

Alternatively, it can be defined as the film thickness at the location where $(\partial p/\partial x) = 0$ and $(\partial p/\partial y) = 0$.

$$h_c = h(x,y)|_{(\partial p/\partial x)=(\partial p/\partial y)=0} \tag{6.9}$$

For moderately to highly loaded cases there is no difference between these two definitions. In those cases the pressure resembles the Hertzian dry contact pressure with the maximum occurring at the centre of the contact. However, for low load cases the maximum of the pressure profile occurs on the centreline but in front of the point $x = 0$.

6.3 Dimensionless Equations

Assuming that the Barus viscosity-pressure equation is used and that the lubricant behaves incompressible, the dimensional equations given above feature 8 parameters, u_m, η_0, ρ_0, α, R_x, R_y, E' and w. This implies that the minimum film thickness or the central film thickness are a function of these 8 parameters. However, these parameters are not all independent. To reduce the number of independent parameters, dimensionless variables can be introduced. An optimal choice is the set of dimensionless variables that reduces the number of parameters to a minimum. This is for example the case if the equations are made dimensionless using the Hertzian dry contact parameters, and the viscosity and density at ambient pressure. For the circular contact ($R_x = R_y$) the dimensionless variables are defined as:

$$\begin{array}{rclcrcl} X & = & x/a & & Y & = & y/a \\ P & = & p/p_h & & H & = & hR_x/a^2 \\ \bar{\eta} & = & \eta/\eta_0 & & \bar{\rho} & = & \rho/\rho_0 \end{array} \tag{6.10}$$

Substitution in Equation (6.1) gives:

$$\frac{\partial}{\partial X}\left(\xi\frac{\partial P}{\partial X}\right)+\frac{\partial}{\partial Y}\left(\xi\frac{\partial P}{\partial Y}\right)-\frac{\partial(\bar{\rho}H)}{\partial X}=0 \tag{6.11}$$

for $X \in [X_a, X_b]$ and $Y \in [-Y_a, Y_a]$
where:

$$\xi = \frac{\bar{\rho}H^3}{\bar{\eta}\bar{\lambda}} \quad \text{with} \quad \bar{\lambda} = \frac{12u_m\eta_0R_x^2}{a^3p_h}$$

The boundary conditions are $P(X_a,Y) = P(X_b,Y) = P(X,-Y_a) = P(X,Y_a) = 0$. The cavitation condition imposes $P(X,Y) \geq 0$.
For an incompressible lubricant $\bar{\rho} \equiv 1$. For a compressible lubricant one obtains:

$$\bar{\rho} = \frac{5.9 \cdot 10^8 + 1.34p_hP}{5.9 \cdot 10^8 + p_hP} \tag{6.12}$$

The Barus viscosity-pressure equation (6.2) in dimensionless form reads:

$$\bar{\eta} = \exp(\bar{\alpha}P) \tag{6.13}$$

with $\bar{\alpha} = \alpha p_h$.
The Roelands equation in dimensionless variables is given by:

$$\bar{\eta} = \exp((\ln(\eta_0) + 9.67)(-1 + (1 + \frac{p_h}{p_0}P)^z)) \tag{6.14}$$

Substitution of the dimensionless variables in Equation (6.5) gives the following dimensionless film thickness equation:

$$H(X,Y) = H_0 + \frac{X^2}{2} + \frac{Y^2}{2} + \frac{2}{\pi^2}\int_{-\infty}^{+\infty}\int_{-\infty}^{+\infty}\frac{P(X',Y')\,dX'dY'}{\sqrt{(X-X')^2+(Y-Y')^2}} \tag{6.15}$$

with H_0 determined by the dimensionless force balance condition:

$$\int_{-\infty}^{+\infty}\int_{-\infty}^{+\infty} P(X,Y)\,dXdY = \frac{2\pi}{3} \tag{6.16}$$

If the Barus viscosity-pressure equation is used and the lubricant is assumed to behave incompressible only two parameters remain, $\bar{\alpha}$ and $\bar{\lambda}$. This implies that the dimensionless minimum film thickness H_m and the dimensionless central film thickness H_c (and any other characteristic solution value) will be a function of $\bar{\alpha}$ and $\bar{\lambda}$ only. This greatly reduces the number of computations to be performed in parameter studies.

6.4 Dimensionless Parameters

In the past different sets of dimensionless parameters have been proposed to characterize EHL contacts. It does not really matter much which set is used, as long as it uses the minimum number and no redundant parameters are introduced.

To present the results in design charts or survey diagrams, a convenient set of dimensionless variables for EHL problems was proposed by Blok and co-workers, see [35]. For the line contact they were pointed out by Moes [83] and for the point contact by Moes and Bosma [84]. Their derivation can be found in [86]. They are now often referred to as the Moes dimensionless parameters and for the circular contact they are defined as:

$$M = \frac{\mathrm{w}}{E'R_x^2}\left(\frac{\eta_0 u_s}{E'R_x}\right)^{-3/4} \tag{6.17}$$

and:

$$L = \alpha E'\left(\frac{\eta_0 u_s}{E'R_x}\right)^{1/4} \tag{6.18}$$

and the associated dimensionless film thickness parameter is:

$$H^M = \frac{h}{R_x}\left(\frac{\eta_0 u_s}{E'R_x}\right)^{-1/2} \tag{6.19}$$

Design charts are created giving the dimensionless minimum or central film thickness defined in this way as a function of M and L. For EHL contacts different regimes of asymptotic behaviour can be distinguished. In terms of M and L these regimes are listed in Table 6.1.

regime	M	L
rigid isoviscous	*small*	0
rigid piezoviscous	*small*	*large*
elastic isoviscous	*large*	0
elastic piezoviscous	*large*	*large*

Table 6.1: *Regimes of EHL solutions.*

The parameters M, L and H^M are a convenient way to present the results in design charts. However, from a numerical solution point of view the dimensionless equations as given above are better suited as all variables remain roughly in the range $[0, 1]$. Therefore, M and L will be used in this work only to characterize a given load situation and to present design graphs. Computed pressure profiles and film shapes will all be given in terms of the dimensionless variables P and H as defined in the previous section.

For convenience, the relations between $\bar{\alpha}$, $\bar{\lambda}$ and H as they appear in the dimensionless equations and M, L, and H^M are given:

$$\begin{aligned}
\bar{\lambda} &= \left(\frac{128\pi^3}{3M^4}\right)^{1/3} \\
\bar{\alpha} &= \frac{L}{\pi}\left(\frac{3M}{2}\right)^{1/3} \\
H^M &= H\sqrt{\frac{6\pi}{\bar{\lambda}}}
\end{aligned} \tag{6.20}$$

At this point it is noted that three parameters are often used in the literature, to characterize a given contact. For the point contact problem they are referred to as the Hamrock and Dowson parameters [55] defined as:

$$\begin{aligned}
W &= \frac{\mathrm{w}}{E'R_x^2} \\
U &= \frac{\eta_o u_m}{E'R_x} \\
G &= \alpha E'
\end{aligned} \tag{6.21}$$

and the associated dimensionless film thickness is defined as $H^{HD} = h/R_x$. The disadvantage of using these parameters is that it is not the minimum set of parameters. One reason mentioned in favour of these parameters is that they represent parameters that one generally varies independently in experiments, i.e. load, speed and lubricant. However, this advantage does not outweigh the extra work needed in parametric studies. Also, in terms of these parameters the different asymptotic regimes are not so clearly defined.

Finally, the situation changes if the lubricant is considered compressible and/or the Roelands pressure viscosity relation is used. In addition to M, L or $\bar{\alpha}$, $\bar{\lambda}$ two parameters of the group α, z and η_0 have to be given whereas the third parameter is determined by the relation:

$$\frac{\alpha p_0}{z} = \ln(\eta_0) + 9.67 \tag{6.22}$$

where $p_0 = 1.96 \cdot 10^8$ as mentioned before.

6.5 Discrete Equations

The equations are discretized on a uniform grid with mesh size h in X and Y direction on a rectangular domain $X_a \leq X \leq X_b$, $-Y_a, \leq Y \leq Y_a$. A second order accurate discretization of the Poiseuille terms in Reynolds equation is obtained in the same way as described in Section 4.2 for the hydrodynamic problem. This gives the following discrete Reynolds equation for the point (i, j) of the grid:

$$\frac{\xi^h_{i-1/2,j}P^h_{i-1,j} - (\xi^h_{i-1/2,j} + \xi^h_{i+1/2,j})P^h_{i,j} + \xi^h_{i+1/2,j}P^h_{i+1,j}}{h^2} + \frac{\xi^h_{i,j-1/2}P^h_{i,j-1} - (\xi^h_{i,j-1/2} + \xi^h_{i,j+1/2})P^h_{i,j} + \xi^h_{i,j+1/2}P^h_{i,j+1}}{h^2} - (\bar{\rho}H)^h_x = 0 \quad (6.23)$$

with $P^h_{i,j} = 0$ for points on the boundary, and the cavitation condition $P^h_{i,j} \geq 0$. The coefficients $\xi^h_{i\pm1/2,j}$ and $\xi^h_{i,j\pm1/2}$ are defined by:

$$\begin{aligned} \xi^h_{i\pm1/2,j} &= (\xi^h_{i,j} + \xi^h_{i\pm1,j})/2 \\ \xi^h_{i,j\pm1/2} &= (\xi^h_{i,j} + \xi^h_{i,j\pm1})/2 \end{aligned} \quad (6.24)$$

with:

$$\xi^h_{i,j} = \frac{\bar{\rho}(P^h_{i,j})(H^h_{i,j})^3}{\bar{\eta}(P^h_{i,j})\bar{\lambda}} \quad (6.25)$$

$(\bar{\rho}H)^h_x$ in Equation (6.23) denotes the discrete 'wedge' term. If $H^h_{i,j}$ were a given function independent of the pressure any second order discretization would be suitable, e.g. a second order central discretization such as:

$$(\bar{\rho}H)^h_x \doteq \frac{\bar{\rho}^h_{i+1,j}H^h_{i+1,j} - \bar{\rho}^h_{i-1,j}H^h_{i-1,j}}{2h} \quad (6.26)$$

where $\bar{\rho}_{i\pm1,j} = \bar{\rho}(P^h_{i\pm1,j})$. However, in EHL this discretization can only be applied for low loads. As the central term for the point (i,j) does not appear in the stencil with increasing load, and thus dependence of H on P, numerous iterative processes will be unstable when applied to this discretization. Therefore, an upstream (one-sided) discretization is preferred. A second order upstream discretization is given by:

$$(\bar{\rho}H)^h_x \doteq \frac{1.5\bar{\rho}^h_{i,j}H^h_{i,j} - 2\bar{\rho}^h_{i-1,j}H^h_{i-1,j} + 0.5\bar{\rho}^h_{i-2,j}H^h_{i-2,j}}{h} \quad (6.27)$$

This discretization can be used for all points $i \geq 2$. At the first line near the boundary $(i = 1)$ it can be replaced by a first order discretization:

$$(\bar{\rho}H)^h_x \doteq \frac{\bar{\rho}^h_{i,j}H^h_{i,j} - \bar{\rho}^h_{i-1,j}H^h_{i-1,j}}{h} \quad (6.28)$$

Alternatively one can compute the film thickness in one extra point $(i = -1)$ and use the second order discretization throughout.

As was shown in Section 5.2, (5.14), the film thickness equation can be discretized as:

$$H^h_{i,j} = H_0 + \frac{X_i^2}{2} + \frac{Y_j^2}{2} + \sum_{i'}\sum_{j'} K^{hh}_{i,i',j,j'}\ P^h_{i',j'} \quad (6.29)$$

where, for a second order discretization based on an approximation of the pressure by a piecewise constant function on a square $h \times h$ around at $X'_{i'}, Y'_{j'}$ the coefficients $K^{hh}_{i,i',j,j'}$ are given by:

$$K^{hh}_{i,i',j,j'} = \frac{2}{\pi^2} \int_{Y_j-h/2}^{Y_j+h/2} \int_{X_i-h/2}^{X_i+h/2} \frac{dX'dY'}{\sqrt{(X_i - X')^2 + (Y_j - Y')^2}} \tag{6.30}$$

and can be computed analytically:

$$\begin{aligned} K^{hh}_{i,i',j,j'} = \frac{2}{\pi^2}\{ &|X_p|\text{arcsinh}\left(\frac{Y_p}{X_p}\right) + |Y_p|\text{arcsinh}\left(\frac{X_p}{Y_p}\right) \\ &-|X_m|\text{arcsinh}\left(\frac{Y_p}{X_m}\right) - |Y_p|\text{arcsinh}\left(\frac{X_m}{Y_p}\right) \\ &-|X_p|\text{arcsinh}\left(\frac{Y_m}{X_p}\right) - |Y_m|\text{arcsinh}\left(\frac{X_p}{Y_m}\right) \\ &+|X_m|\text{arcsinh}\left(\frac{Y_m}{X_m}\right) + |Y_m|\text{arcsinh}\left(\frac{X_m}{Y_m}\right)\} \end{aligned} \tag{6.31}$$

with:

$$\begin{aligned} X_p &= X_i - X_{i'} + h/2 \\ X_m &= X_i - X_{i'} - h/2 \\ Y_p &= Y_j - Y_{j'} + h/2 \\ Y_m &= Y_j - Y_{j'} - h/2 \end{aligned} \tag{6.32}$$

On a uniform grid the coefficients $K^{hh}_{i,i',j,j'}$ are a function of $|i - i'|$ and $|j - j'|$ only. Thus they can be pre-computed and stored in a single array of $n_x \times n_y$ points. In the following sections therefore also $K^{hh}_{k,l}$ will be used, designating $K^{hh}_{i,i',j,j'}$ with $|i - i'| = k$ and $|j - j'| = l$.

Finally the discretized force balance equation is derived, it is the same equation as obtained for the dry contact problem, (5.16):

$$h^2 \sum_i \sum_j P^h_{i,j} = \frac{2\pi}{3} \tag{6.33}$$

6.6 Model Problem

In order to design a stable relaxation scheme that efficiently reduces high frequency error components, it is necessary to understand the nature of the equations. Due to the exponential viscosity-pressure relation the coefficient ξ in Equation (6.11) varies many orders of magnitude over the domain. In the inlet region $\xi >> 1$ whereas in the Hertzian contact region $\xi << 1$. With the change of ξ the character of the problem changes. This can be illustrated by a simple model problem that is obtained when the force balance equation is neglected, the coefficient ξ is taken constant, and $\bar{\rho} = 1$ is assumed:

$$\xi \left(\frac{\partial^2 P}{\partial X^2} + \frac{\partial^2 P}{\partial Y^2} \right) - \frac{\partial H}{\partial X} = 0 \tag{6.34}$$

with:

$$H(X,Y) = H_0 + \frac{X^2}{2} + \frac{Y^2}{2} + \frac{2}{\pi^2} \int_{-\infty}^{+\infty} \int_{-\infty}^{+\infty} \frac{P(X',Y')\,dX'dY'}{\sqrt{(X-X')^2 + (Y-Y')^2}} \tag{6.35}$$

for an arbitrary value of H_0. $P = 0$ is used as boundary condition. This model problem has little physical meaning, as is illustrated by the fact that the value of H_0 is irrelevant as it cancels in the equation. However, the problem is characteristic for the local behaviour of the EHL problem. For large values of ξ the differential aspects as represented by the second order derivatives of the pressure determine the behaviour. For small values of ξ the $\partial H / \partial X$ term dominates the behaviour, and, because the film thickness is given by an integral equation, the problem behaves as an integral problem. Understanding how to solve this model problem for large and small values of ξ respectively, forms the key to understanding how to construct an efficient solver for the complete EHL problem. Consistent with the discretization of the full problem the discretized model problem is given by:

$$\begin{aligned} \xi \frac{P^h_{i-1,j} - 2P^h_{i,j} + P^h_{i+1,j}}{h^2} + \xi \frac{P^h_{i,j-1} - 2P^h_{i,j} + P^h_{i,j+1}}{h^2} - & \\ \frac{1.5H^h_{i,j} - 2H^h_{i-1,j} + 0.5H^h_{i-2,j}}{h} &= 0 \end{aligned} \tag{6.36}$$

with:

$$H^h_{i,j} = H_0 + \frac{X_i^2}{2} + \frac{Y_j^2}{2} + \sum_{i'} \sum_{j'} K^{hh}_{i,i',j,j'} P^h_{i',j'} \tag{6.37}$$

where the coefficients $K^{hh}_{i,i',j,j'}$ are given by (6.31).

6.6.1 Relaxation for Large ξ Values

For large values of ξ Equation (6.34) is dominated by the first two terms. Hence, it behaves essentially the same as the Poisson 2d problem. For this problem it was already shown that Gauss-Seidel one point relaxation is an efficient smoother when H^h is given in all points. For the present problem this relaxation can be described as follows. Let $\tilde{P}^h_{i,j}$ denote the current approximation to $P^h_{i,j}$ and $\tilde{H}^h_{i,j}$ the associated approximation to $H^h_{i,j}$ computed using (6.37). One Gauss-Seidel type relaxation then consists of two steps. First compute a new approximation $\bar{P}^h_{i,j}$ in each grid point according to:

$$\bar{P}^h_{i,j} = \tilde{P}^h_{i,j} + \omega \delta^h_{i,j} \tag{6.38}$$

where ω is the relaxation factor, and $\delta^h_{i,j}$ is the change that should be applied to $\tilde{P}^h_{i,j}$ to solve the equation at point (i,j) and can be written as (see Chapter 2):

$$\delta^h_{i,j} = r^h_{i,j} \left(\frac{\partial (L^h \underline{P}^h)_{i,j}}{\partial P^h_{i,j}} \right)^{-1} \tag{6.39}$$

with $r^h_{i,j}$ the dynamic residual defined as:

$$\begin{aligned} r^h_{i,j} &= -\xi \frac{\bar{P}^h_{i-1,j} - 2\tilde{P}^h_{i,j} + \tilde{P}^h_{i+1,j}}{h^2} \\ &\quad -\xi \frac{\bar{P}^h_{i,j-1} - 2\tilde{P}^h_{i,j} + \tilde{P}^h_{i,j+1}}{h^2} \\ &\quad + \frac{1.5\tilde{H}^h_{i,j} - 2\tilde{H}^h_{i-1,j} + 0.5\tilde{H}^h_{i-2,j}}{h} \end{aligned} \tag{6.40}$$

and $\partial (L^h \underline{P}^h)_{i,j} / \partial P^h_{i,j}$ the derivative of the discrete operator working on $\underline{P}^h$ at the point (i,j) with respect to $P^h_{i,j}$:

$$\left. \frac{\partial (L^h \underline{P}^h)_{i,j}}{\partial P^h_{i,j}} \right|_{i,j} = -\frac{4\xi}{h^2} - \frac{1.5K^{hh}_{0,0} - 2K^{hh}_{1,0} + 0.5K^{hh}_{2,0}}{h} \tag{6.41}$$

Next, given the new approximation $\bar{P}^h_{i,j}$ in each point, a new approximation $\bar{H}^h_{i,j}$ to $H^h_{i,j}$ is computed in each point using Equation (6.37). Notice that it is not a full Gauss-Seidel relaxation. When relaxing the pressure, the changes already made in points relaxed previously are taken into account in the first two terms of the equation, but not in the $H^h_{i,j}$ terms, i.e. the discrete wedge term in the dynamic residual defined by Equation (6.40) only uses $\tilde{H}^h_{i,j}$ values which were computed with the *old* values $\tilde{P}^h_{i,j}$. Updating $\tilde{H}^h_{i,j}$ for the changes applied to points (X_i, Y_j) previously relaxed is feasible but tends to be computationally expensive, unless it would be sufficient to update only a few neighbouring points. Besides, as the Multilevel Multi-integration described in the previous chapter provides a fast and efficient way to obtain *all* values of $\bar{H}^h_{i,j}$ at once, a scheme as described above is preferred.

The action of relaxation upon a Fourier component of the solution can be analysed by means of local mode analysis, see Section 2.8. The behaviour depends on the parameter ξ/h^2. For large values of ξ/h^2 the error amplification graph is the same as given in Figure 2.9. The asymptotic smoothing factor is $\bar{\mu} = 1/2$ for $\omega = 1$. However, with decreasing ξ/h^2 the relaxation gradually becomes unstable. This is illustrated in Figure 6.1 which gives $\mu(\theta_1, \theta_2)$ for $\xi/h^2 = 0.5$. The figure shows that certain low frequency components are amplified. For this particular case applying the changes with an underrelaxation factor $\omega = 0.5$ will stabilize the scheme at the expense of a poor smoothing of high frequency components. For smaller values of ξ/h^2 the amplification of low frequency components becomes stronger and an increasingly smaller value of ω is needed to stabilize the scheme.

Alternatively, the relaxation can be applied in line relaxation manner. Instead of scanning the grid point by point it is scanned line by line solving simultaneously the equations of one line. Assuming line relaxation in the X direction for a given line j the

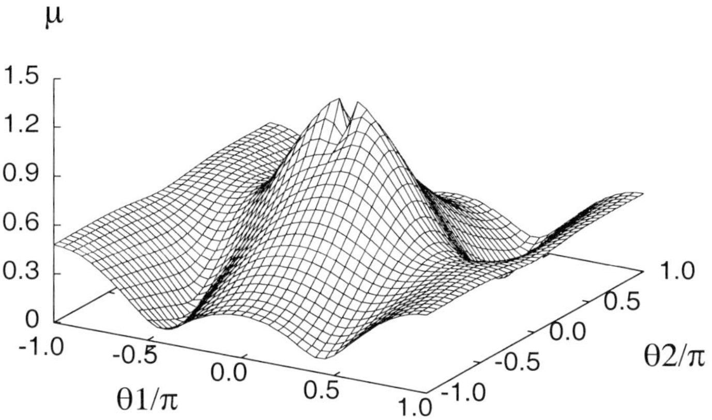

Figure 6.1: *Error amplification factor $\mu(\theta_1, \theta_2)$ for Gauss-Seidel pointwise relaxation applied to the model problem for $\xi/h^2 = 0.5$.*

changes to be applied to $\tilde{P}^h_{i,j}$ for $0 < i < n_x$ of that line are simultaneously solved from a system of equations, see Section 2.10.2:

$$A^j \underline{\delta}^h_j = \underline{r}^h_j \tag{6.42}$$

where $\underline{\delta}^h_j$ is a vector of changes $\delta^h_{i,j}$ and $\underline{r}^h_j$ a vector with the current residuals $r^h_{i,j}$. Both are vectors of n_x elements. The matrix coefficients $A^j_{i,k}$ are defined by:

$$A^j_{i,k} = \frac{\partial(L^h \underline{P}^h)_{i,j}}{\partial P^h_{k,j}} \tag{6.43}$$

for $0 < k < n_x$ and $0 < i < n_x$. This gives:

$$A^j_{i,k} = -\frac{1.5K^{hh}_{|i-k|,0} - 2K^{hh}_{|i-k-1|,0} + 0.5K^{hh}_{|i-k-2|,0}}{h} \tag{6.44}$$

for $|i-k| > 1$ and

$$\begin{aligned} A^j_{i,i} &= -\frac{4\xi}{h^2} - \frac{1.5K^{hh}_{0,0} - 2K^{hh}_{1,0} + 0.5K^{hh}_{2,0}}{h} \\ A^j_{i,i-1} &= \frac{\xi}{h^2} - \frac{1.5K^{hh}_{1,0} - 2K^{hh}_{0,0} + 0.5K^{hh}_{1,0}}{h} \quad \text{if} \;\; i > 1 \\ A^j_{i,i+1} &= \frac{\xi}{h^2} - \frac{1.5K^{hh}_{1,0} - 2K^{hh}_{2,0} + 0.5K^{hh}_{3,0}}{h} \quad \text{if} \;\; i < n_x - 1 \end{aligned} \tag{6.45}$$

$A^j_{i,i-1}$ and $A^j_{i,i+1}$ are zero for $i = 1$ and $i = n_x - 1$ respectively. Next the system of Equations (6.42) should be solved. For the two dimensional Poisson problem discussed in Chapter 2 the system of equations to be solved for each line was a tridiagonal system which can be solved quickly. This would also apply here if H did not depend on P. However, due to this dependence the matrix A^j is a full matrix and its solution would normally require $O(n_x^3)$ operations. The work count of one relaxation sweep which involves $n_y - 1$ lines would be $O(n_y n_x^3)$ and assuming $n_x = n_y = \sqrt{N}$ one relaxation would cost $O(N^2)$ operations and thus be very expensive. However, to obtain the line relaxation efficiency it is generally not needed to solve the system exactly. In practice it is often sufficient to take into account only the terms in the summations related to the direct neighbours of a point i. This is justified because $K^h_{k,0}$ decreases with increasing distance. In particular it is sufficient to solve a hexadiagonal system defining all $A^j_{i,k} = 0$ for $k < i - 3$ and $k > i + 2$. This hexadiagonal system can then be solved quickly in $O(n_x)$ operations using Gaussian elimination. In that case the work count of the relaxation is $O(N)$ as for a pointwise scheme.

After the system of equations for the line j is solved the changes $\delta^h_{i,j}$ are simultaneously applied to all points of the line with some relaxation factor ω_{gs}:

$$\bar{P}^h_{i,j} = \tilde{P}^h_{i,j} + \omega_{gs}\delta^h_{i,j} \tag{6.46}$$

Subsequently, the system of equations for the next line is constructed and solved. This process is repeated for each line of the grid. Finally using the new approximation $\bar{P}^h$ a new approximation $\bar{H}^h$ to H^h can be computed.

The behaviour of the relaxation can again be analysed with Local Mode Analysis. The conclusion is the same as for the pointwise relaxation. For large ξ/h^2 it is stable with excellent smoothing: $\bar{\mu} = 0.44$. With decreasing ξ/h^2 the process becomes unstable due to the amplification of low frequency components. The process can be stabilized using underrelaxation but not for the small values of ξ/h^2 that may occur in EHL contacts. In fact, the small values of ω needed would completely remove the smoothing properties of the scheme.

Summarizing, in this section Gauss-Seidel point- and line-relaxation applied to the model problem were explained. In both cases the stability of the relaxation is determined by the ratio ξ/h^2. For large values of ξ/h^2 both schemes are stable and provide good smoothing. Hence, a multigrid solver using either of these schemes will yield an efficient solver. Note that this is only true if ξ/h^2 remains large on all the grids that are used. With decreasing ξ/h^2 both schemes become unstable due to the amplification of smooth error components. The amplification of low frequency components is related to the changes occurring in $H^h_{i,j}$ as a result of the changes applied to all pressure values. As $H^h_{i,j}$ is defined by a summation over all pressures, it will accumulate the changes made to the pressures in the entire field. One can account for the changes by using updated values of $\tilde{H}^h_{i,j}$ in the dynamic residual, i.e. updated for changes $\omega_{gs}\delta^h_{i,j}$ applied to points (i,j) already relaxed. This does help to stabilize the scheme, however, the improvement is only marginal. Even combined with underrelaxation it does not result in a relaxation scheme that is stable and has acceptable smoothing behaviour for the small values of ξ/h^2 anticipated to occur in EHL contacts.

6.6.2 Relaxation for Small ξ Values

In the previous section it was explained that the Gauss-Seidel point- and line-relaxation are unstable for small values of ξ/h^2. This instability is due to the accumulation of changes in the summation involved in the definition of H^h. To be precise: a change $\delta^h_{i,j}$ applied in point (i,j) will cause a change to $\tilde{H}^h_{i',j'}$ that is proportional to $K^{hh}_{i,i',j,j'}$. As $K \propto 1/r$ with r being the distance between the points (i,j) and (i',j'), the change in $\tilde{H}^h_{i',j'}$ will be proportional to $\delta^h_{i,j}/r$. The change in $\tilde{H}^h_{i',j'}$ as a result of the entire relaxation sweep will be the summation of all such changes. It is this accumulation occurring when re-computing the film thicknesses, that causes the instability of the entire process. Chapter 5 shows that the cure for this kind of instability is *distributive relaxation*. When relaxing a point (i,j) changes are not only applied to the point (i,j) but also to a few of its neighbours in such a way that the local Equation (6.34) is solved, as usual, but that the resulting changes in the integral (and thus in $\tilde{H}^h_{i,j}$) are essentially local. Using such a distributive relaxation may indeed take care of the stability problem. However, it still does not provide efficient smoothing. For very small ξ values Equation (6.34) reduces to:

$$\frac{\partial H}{\partial X} \approx 0 \tag{6.47}$$

which is an equation in X direction only. The only coupling between equations in the Y direction is the coupling in the elastic deformation integrals. Consequently, as for the anisotropic 2D Poisson problem (Section 2.10.2), errors that are smooth in the X direction and oscillatory in the Y direction can not be efficiently reduced by the relaxation. However, good smoothing can be obtained if line relaxation is applied.

A distributive Jacobi line relaxation is described as follows. Given an approximation $\tilde{P}^h$ and the associated values of $\tilde{H}^h$, for each line j changes $\delta^h_{i,j}$, to be applied distributively, are solved from:

$$A^j \underline{\delta}^h_j = \underline{r}^h_j \tag{6.48}$$

where $\underline{\delta}^h_j$ is a vector of changes $\delta^h_{i,j}$ and $\underline{r}^h_j$ a vector with the residuals $r^h_{i,j}$. Both are vectors of $n_x - 1$ elements. The matrix coefficients $A^j_{i,k}$ are defined by:

$$A^j_{i,k} = \frac{\partial(L^h \underline{P}^h)_{i,j}}{\partial P^h_{k,j}} - \frac{1}{4}\left(\frac{\partial(L^h \underline{P}^h)_{i,j}}{\partial P^h_{k+1,j}} + \frac{\partial(L^h \underline{P}^h)_{i,j}}{\partial P^h_{k-1,j}} + \frac{\partial(L^h \underline{P}^h)_{i,j}}{\partial P^h_{k,j+1}} + \frac{\partial(L^h \underline{P}^h)_{i,j}}{\partial P^h_{k,j-1}}\right) \tag{6.49}$$

for $0 < k < n_x$ and $0 < i < n_x$. For details regarding the system of equations the reader is referred to Appendix C. This system of equations is constructed and solved for each line j. Subsequently, these changes are applied distributively in the same way as for the pointwise distributive relaxation applied to the dry contact problem discussed in the previous chapter. The new approximation at $\bar{P}^h_{i,j}$ is given by:

$$\bar{P}^h_{i,j} = \tilde{P}^h_{i,j} + \omega_{ja}(\delta^h_{i,j} - (\delta^h_{i+1,j} + \delta^h_{i-1,j} + \delta^h_{i,j-1} + \delta^h_{i,j+1})/4); \tag{6.50}$$

With respect to the actual implementation of the relaxation it is noted that again the matrix A^j is a full matrix but, as is shown in Appendix C, it can be truncated to a hexadiagonal system which can be solved quickly as mentioned before.

The action of this line relaxation on the error can be analysed using Local Mode Analysis. Figure 6.2 shows the error amplification factor $\mu(\theta_1, \theta_2)$. The figure shows that for small θ_1, θ_2 the relaxation is stable. Moreover, it efficiently reduces high frequency components. The asymptotic smoothing rate $\bar{\mu} = 0.41$. The relaxation is also stable for large values of ξ/h^2, but it is much less efficient than the Gauss-Seidel line relaxation in terms of smoothing behaviour.

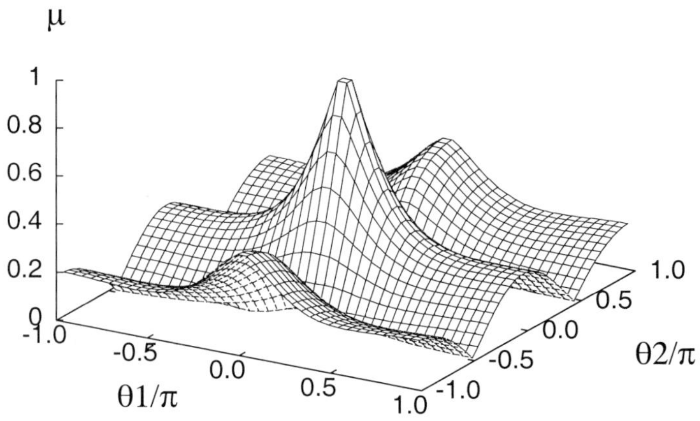

Figure 6.2: *Error amplification factor* $\mu(\theta_1, \theta_2)$ *for distributive Jacobi line relaxation applied to the model problem for* $\xi/h^2 = 0.0$.

6.6.3 Relaxation for Varying ξ Values

Neither of the relaxation processes described in the previous section is suited to efficiently solve the model problem for all values of ξ. The Gauss-Seidel line relaxation is an excellent smoother for large ξ/h^2 but is unstable for small ξ/h^2. On the other hand the distributive line relaxation is a good smoother for very small ξ/h^2 but rapidly loses efficiency with increasing ξ/h^2. The objective is to obtain an efficient multigrid solver. In that case even though ξ is a constant the ratio ξ/h^2 will depend on the grid that is used. Therefore, a good choice to obtain an efficient coarse grid correction cycle seems to be to use different relaxations on different grids.

- On grids where $\xi/h^2 > \xi_{limit}$ the Gauss-Seidel line relaxation is used.
- On grids where $\xi/h^2 \leq \xi_{limit}$ the Jacobi distributive line relaxation is used.

Where ξ_{limit} is a switch parameter to be defined. From practical tests it was found that an efficient cycle can be obtained using $\xi_{limit}\approx 0.3$ and some underrelaxation in both the processes. Typically $\omega_{ja}=0.6$ and $\omega_{gs}=0.8$.

In the EHL problem the coefficient ξ varies over the grid. The next step towards a relaxation is therefore to consider the following model problem:

$$\frac{\partial}{\partial X}\left(\xi\frac{\partial P}{\partial X}\right)+\frac{\partial}{\partial Y}\left(\xi\frac{\partial P}{\partial Y}\right)-\frac{\partial H}{\partial X}=0 \tag{6.51}$$

with $P = 0$ on the boundaries and H given by Equation (6.5) for different functions $\xi(X,Y)$. Note that the constant H_0 is still irrelevant for this problem. To approximate the behaviour for the full circular contact problem $\xi(X,Y)$ can be chosen as:

$$\xi(X,Y)=\begin{cases}(X+Y)^6, & \text{if} \quad X^2+Y^2\geq 1;\\ 0, & \text{otherwise.}\end{cases} \tag{6.52}$$

When discretized on a uniform grid one obtains exactly the discrete Reynolds equation (6.23) if $\bar{\rho}=1$ is substituted. A suitable relaxation process for this model problem is now obtained by combining the two relaxations on a single grid. This is justified by the fact that relaxation, as far as its effect on the error is concerned, is a local process.

The combined relaxation scheme can be described as follows. For a given approximation $\tilde{P}^h$ and associated approximation $\tilde{H}^h$ a new approximation $\bar{P}^h$ is obtained by scanning the grid line by line, for each line j constructing a system of equations of the form (6.48). The matrix coefficients for row i are defined by Equation (6.43) for the Gauss-Seidel line relaxation scheme if the local value of $\xi^h_{i,j}/h^2>\xi_{limit}$, and by Equation (6.49) for the distributive relaxation otherwise. For further details the reader is referred to Appendix C.

The resulting system of equations is truncated to a hexadiagonal system and solved. Subsequently, the change $\delta^h_{i,j}$ is added to the current approximation if at the point (i,j) the coefficients $\xi/h^2>\xi_{limit}$ and it is also distributed to its four direct neighbours if $\xi/h^2\leq \xi_{limit}$. Each type of change can be (and will be) applied with a certain underrelaxation factor. Using the same relaxation factors as for the problem with constant ξ this type of hybrid relaxation yields an efficient multilevel solver for Equation (6.51) with (6.52) and (6.15).

6.7 Relaxation of the EHL Problem

In the previous sections a stable relaxation process with good smoothing properties for Equation (6.23) with $\bar{\rho}=1$ was developed. In this equation the variable H is defined by Equation (6.29) and ξ is a known function of X and Y. The next steps towards an efficient relaxation of the EHL problem, are to incorporate the non-linearity in the definition of ξ, the cavitation condition $P\geq 0$ and to include the relaxation of the force balance equation. In addition $\partial H/\partial X$ should be replaced by $\partial(\bar{\rho}H)/\partial X$. This last change is trivial and requires only some minor changes in the coefficients of the system of equations that is solved for each line, see Appendix C.

The non-linearity is simply introduced by replacing Equation (6.52) with (6.25). However, when computing the change to be applied at each point the variation of ξ is not taken into account in the relaxation. As a result the changes have to be applied using some underrelaxation.

Cavitation was treated in the hydrodynamic problem (Chapter 4) as follows; after a change $\delta^h_{i,j}$ is applied the new pressure $\bar{P}^h_{i,j}$ is checked. If its value is smaller than zero it is set to zero. In this way when relaxing the next neighbour, cavitation of the previous point is taken into account. However, for a line relaxation process, the situation is a bit more complicated as all changes for a line are solved simultaneously. The simplest way to incorporate cavitation would be to apply all changes to a given line and set negative values to zero when needed. However, as the changes of the line are solved simultaneously this implies that within a single line the change for a point i is computed as if a point $i-1$ is not cavitated at all. As a result convergence of the relaxation may stall due to the fact that a few points around the cavitation boundary continuously switch back and forth between cavitated and pressurized. The alternative is to create the system of equations for a line taking into account cavitation. Whenever one of the neighbours of a point i has zero pressure, only the central term of the local system is used. As a result the change for this point is computed as it would be done for a pointwise Jacobi relaxation.

Next the force balance equation is added to the system. How to relax the force balance equation was explained in Section 5.4. After a number of relaxations on a single grid, the present value of the variable $\tilde{H}_0$ is updated according to:

$$\bar{H}_0 = \tilde{H}_0 + \omega_{H_0}({}_W f^h - h^2 \sum_{i,j} \tilde{P}^h_{i,j}) \tag{6.53}$$

where ω_{H_0} is some small constant and ${}_W f^h$ is the right hand side of the force balance equation. For a single grid ${}_W f^h = 2\pi/3$. The better the first approximation to H_0 the quicker the convergence. However, the nature of the convergence of the force balance equation is determined by the constant ω_{H_0} too. If this value is too large and the changes occur too often the solution can not digest the change of H_0 and the residual of the force balance equation will oscillate, which will also hamper convergence of the Reynolds equation. If ω_{H_0} is too small, convergence will be too slow. In general the convergence behaviour will be a damped oscillation around zero.

The importance of the force balance equation should not be underestimated. In the model problem with constant ξ, the integration constant H_0 cancels from the problem and thereby the force balance equation. This implies that for the EHL problem the value of H_0 is in fact determined by the *boundary layer* that connects the region of large and small ξ. Hence, the value of H_0 is strongly coupled to the coefficients ξ in this region.

The developed relaxation can now be applied to the solution of the EHL problem on a single grid. As an illustration its convergence behaviour for two load conditions is illustrated below. Table 6.2 specifies the load cases in terms of different sets of dimensionless parameters. Table 6.3 gives values of the parameters for a contact between steel surfaces that would result in these conditions. Also given are the maximum Hertzian pressure and the Hertzian contact radius.

Figure 6.3 shows the error reduction as a function of the number of relaxations for case 1. Results are presented for four grids with decreasing mesh size covering the domain

Parameter	case 1	case 2
M	20	200
L	10	10
$\bar{\alpha}$	9.89	21.31
$\bar{\lambda}$	0.20	$9.35 \cdot 10^{-2}$
W	$1.73 \cdot 10^{-7}$	$1.73 \cdot 10^{-6}$
U	$8.85 \cdot 10^{-12}$	$8.85 \cdot 10^{-12}$
G	4972	4972

Table 6.2: *Values of the different sets of dimensionless parameters characterizing the two load cases considered, and of the Hertzian dry contact parameters.*

Parameter	Value		Dimension
	case 1	case 2	
α	$2.2 \cdot 10^{-8}$		$[N^{-1}m^2]$
η_0	$40.0 \cdot 10^{-3}$		$[Nm^{-2}s]$
R_x	$1.6 \cdot 10^{-2}$		$[m]$
E'	$2.26 \cdot 10^{11}$		$[N^{-1}m^2]$
u_s	1.6		$[ms^{-1}]$
w	10.0	100.0	$[N]$
p_h	$0.45 \cdot 10^9$	$0.97 \cdot 10^9$	$[Nm^{-2}]$
a	$1.02 \cdot 10^{-4}$	$2.20 \cdot 10^{-4}$	$[m]$

Table 6.3: *Values of the parameters for the contact between steel surfaces for the two load cases.*

$X \in [-4.5, 1.5]$ and $Y \in [-3.0, 3.0]$. The coarsest grid for which results are presented has $(16+1) \times (16+1)$ points. The next finer grid $(32+1) \times (32+1)$ points etc. For each case the Hertzian pressure profile was taken as a first approximation to P^h and the first approximation to H^h was obtained from Equation (6.29) using the first approximation to H_0 that is given as input. For the present case $H_0 = -0.53$ was used. The value of H_0 was changed according to Equation (6.53) each tenth relaxation with a value $\omega_{H_0} = 0.1$.

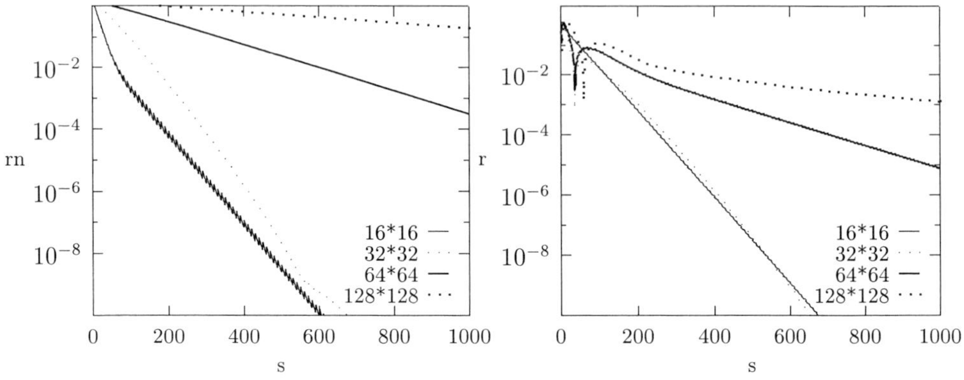

Figure 6.3: *Residual norm (rn) of the Reynolds equation (left) and absolute value of the residual (r) of the force balance equation (right) on different grids as a function of the number of relaxations (s). $M = 20$, $L = 10$, incompressible lubricant, Barus viscosity-pressure equation.*

The left graph in Figure 6.3 shows the evolution of the residual norm of the Reynolds equation. The graph on the right shows the evolution of the absolute value of the residual of the force balance condition. The figure shows that the relaxation indeed converges but with decreasing mesh size the convergence speed decreases. This is exactly the behaviour one would expect and it is caused by the fact that the low frequency error components are slow to converge. Notice the ripple on the Reynolds residual curve. This ripple is caused by the changes applied to H_0 when relaxing the force balance equation. The residual of this equation diminishes in a similar way as the residual of the Reynolds equation. This implies that both equations have roughly the same converge speed. In the first *few* relaxations the decrease is not monotonously and the residual of the force balance equation may increase in the first number of iterations. This behaviour is caused by the choice of the Hertzian pressure distribution as an initial pressure distribution. This approximation exactly satisfies the force balance condition.

As an illustration the Figures 6.4 and 6.5 show the computed approximations to the dimensionless pressure P^h and film thickness H^h on a grid with $n_x = n_y = 32$. Although the resolution is very crude the solution contains all characteristic aspects of the solution

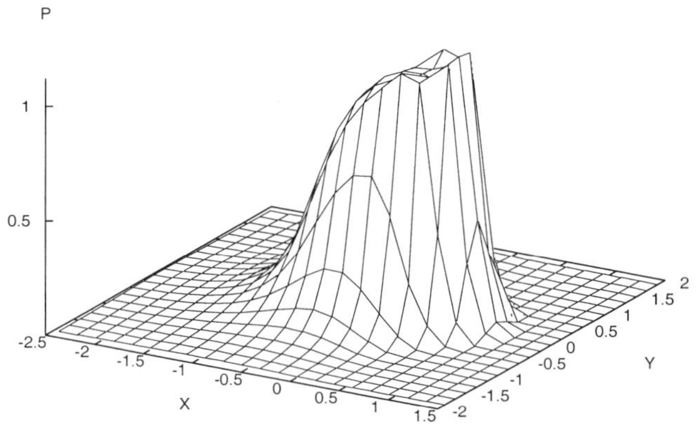

Figure 6.4: *Computed dimensionless pressure* P^h *as a function of* X *and* Y *on a grid with* $n_x = n_y = 32$. $M = 20$, $L = 10$, *incompressible lubricant, Barus viscosity-pressure equation.*

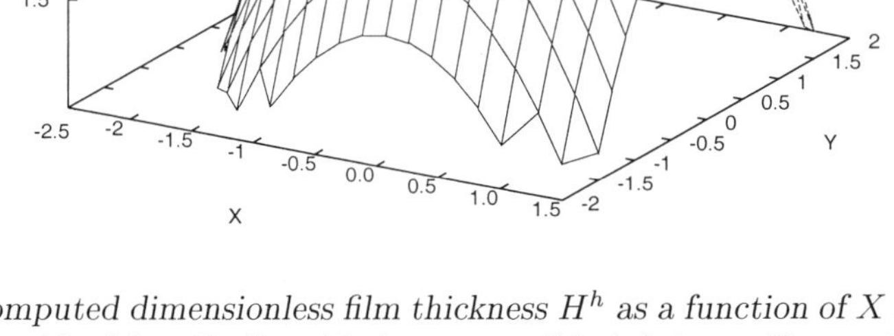

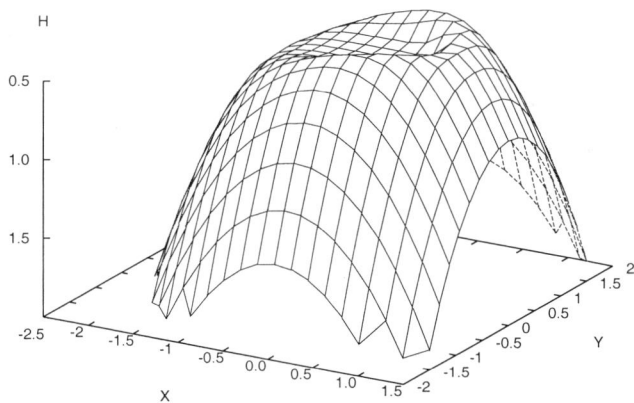

Figure 6.5: *Computed dimensionless film thickness* H^h *as a function of* X *and* Y *on a grid with* $n_x = n_y = 32$. $M = 20$, $L = 10$, *incompressible lubricant, Barus viscosity-pressure equation.*

of an EHL point contact problem. The pressure resembles the Hertzian semi-elliptical pressure but the singularity at $X^2 + Y^2 = 1$ is replaced by a gradual pressure increase in the inlet region and a rapid drop to the cavitation pressure in the outlet. Just before the cavitated region a 'pressure spike' occurs. This spike is due to the exponential viscosity-pressure equation: it does not occur for the isoviscous case $L = 0$. Furthermore, in solutions for small values of L it only appears as a local maximum in the pressure. As will be shown later, the spike actually forms a 'shield' wrapped around the solution in the outlet region. However, this can only be seen on a sufficiently fine grid. This 'spike region' is the two dimensional equivalent of the pressure spike first shown by Petrusevich for the line contact problem.

Figure 6.5 shows the computed film thickness. For convenience of interpretation the film thickness has been plotted upside down. In this way it can be seen as the shape of the ball for the case of a contact between a ball and a rigid flat surface. Due to the elastic deformation the surface is flattened in the central region. This effect decreases towards the outside of the contact as a result of which a horseshoe shape occurs. The overall minimum film thickness occurs in the 'side-lobes' of the horse-shoe. The flattening in the central region is in accordance with the asymptotic behaviour of the Reynolds equation. For small ξ this equation reduces to:

$$-\frac{\partial(\bar{\rho}H)}{\partial X} \approx 0 \tag{6.54}$$

with the solution $\bar{\rho}H = c(Y)$. This behaviour will be more clearly visible in a contour plot of the solution on a much finer grid, which will be discussed later.

The characteristic effects displayed in these figures will become stronger with increasing load as is shown by the results obtained for case 2. For this case the domain of computation is chosen smaller: $X \in [-2.5, 1.5]$ and $Y \in [-2, 2]$. Because of the higher load a smaller domain will be sufficient to accurately predict the film thickness values. Figure 6.6 shows the evolution of the residual of the Reynolds equation and the absolute value of the residual of the force balance equation as a function of the number of relaxations. The coarsest grid for which results are shown has $(32+1) \times (32+1)$ points. The results for the grid with $(16+1) \times (16+1)$ points have been omitted as this grid does not yield a physically acceptable solution. The converged result yields a minimum film thickness which is smaller than zero. From a mathematical and numerical point of view this is not a problem. However, such a solution has no physically meaning. This bizarre result indicates that on the $(16+1) \times (16+1)$ grid, the discretization error is locally larger than the solution.

Figure 6.6 shows that also for the higher load case, where the distributive relaxation is certainly used in the contact centre, the relaxation converges in the same way as for the previous case. The residual of the force balance equation convergences along with the residual of the Reynolds equation except for the first *few* iterations. The Figures 6.7 and 6.8 show the approximate solutions to the pressure P and the film thickness H as a function of X and Y on a grid with $n_x = n_y = 32$. The effects shown in the Figures 6.4-6.5 have indeed become stronger. The pressure distribution more closely resembles the semi-elliptic Hertzian pressure profile with the singularity at $X^2 + Y^2 = 1$ replaced by a continuous pressure increase in a small boundary layer. The pressure spike appears to

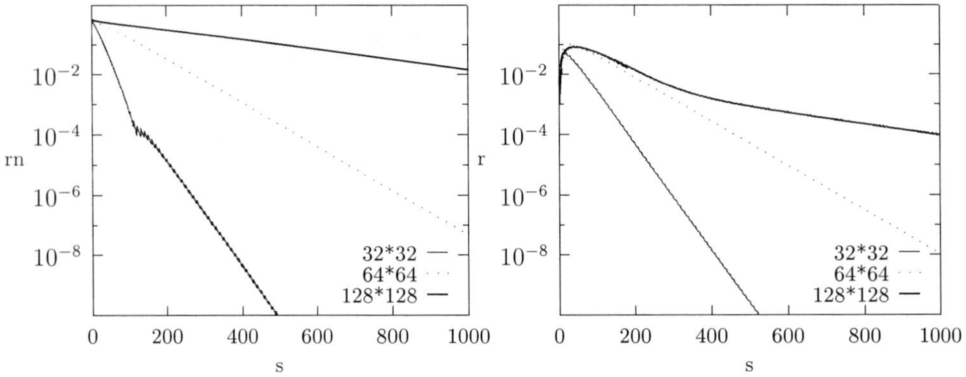

Figure 6.6: *Residual norm (rn)of the Reynolds equation (left) and absolute value of the residual (r) of the force balance equation (right) on different grids as a function of the number of relaxations (s). $M = 200$, $L = 10$, incompressible lubricant, Barus viscosity-pressure equation.*

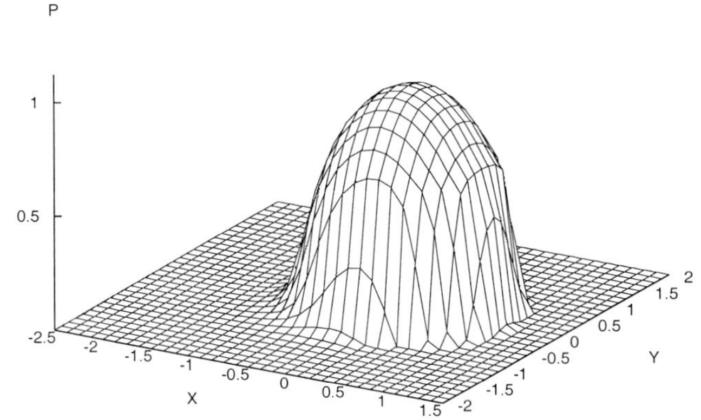

Figure 6.7: *Computed dimensionless pressure P^h as a function of X and Y on a grid with $n_x = n_y = 32$. $M = 200$, $L = 10$, incompressible lubricant, Barus viscosity-pressure equation.*

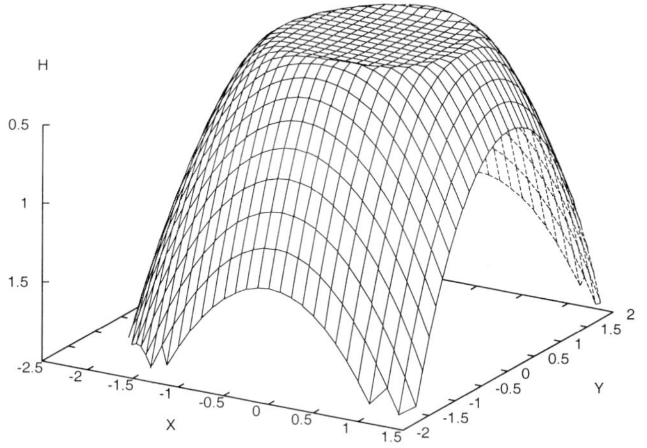

Figure 6.8: *Computed dimensionless film thickness H^h as a function of X and Y on a grid with $n_x = n_y = 32$. $M = 200$ $L = 10$, incompressible lubricant, Barus viscosity-pressure equation.*

be absent now, however, this is not true. With increasing load, the width of the pressure spike expressed in the dimensionless coordinate X decreases and therefore it will only be visible on a sufficiently fine grid. The film thickness resembles the dry contact result. In that case H would be zero in the unit disc $X^2 + Y^2 \leq 1$. For the EHL case the asymptotic solution satisfies Equation (6.54) which gives a small film thickness that is constant as a function of X but still varies as a function of Y. Finally, it is noted that the side lobes where the minimum film thickness occurs have decreased in (relative) size, and have moved to the outside of the contact disc $X^2 + Y^2 \leq 1$. Consequently, to resolve the minimum film thickness accurately for high loads, rather fine grids are needed.

Summarizing, it is concluded that a suitable relaxation process has been developed from the point of view of stability. However, its speed of convergence is far too low to efficiently obtain solutions on dense grids. This problem will be overcome by the use of Multigrid techniques to be described in the following sections.

6.8 Coarse Grid Correction Cycle

Having developed a stable relaxation process for the non-linear EHL problem, the next step is to introduce coarser grids and Multigrid techniques to overcome the slow convergence of fine grid relaxation.

The construction of a coarse grid correction cycle for the EHL problem can be done in the same way as for the dry contact problem, see Section 5.4. The FAS scheme should be applied to *all* equations as was done for the dry contact problem in the previous chapter, and once more, special attention should be given to the intergrid transfers in the cavitated region and the treatment of the force balance equation. To obtain good

asymptotic convergence the transfer of the residuals in the region near the cavitation boundary is very important. The best approach is to use full weighting and for all points $(i', j') \neq (i, j)$ appearing in the stencil define the residual to be zero if the point has zero pressure. This is a slightly more advanced approach than simply injecting the residual.

The developed coarse grid correction cycle proves to be an efficient tool to solve the EHL problem for a wide range of load conditions. However, its performance will depend on the choice of the relaxation factors and on the accuracy of the first approximation of H_0. The relaxation factors should be small enough to avoid instability. For most cases good performance can be obtained taking $\xi_{limit} = 0.3$, $0.2 \leq \omega_{ja} \leq 0.6$, $0.4 \leq \omega_{gs} \leq 0.8$, and $\omega_{H_0} \approx 0.1$. Also, in many cases W-cycles give better performance than V-cycles. The performance of the developed coarse grid correction cycle for the two load conditions considered in the previous section is outlined below.

W-cycle	level 3	level 4	level 5
	64×64	128×128	256×256
1	$6.43 \cdot 10^{-1}$	$7.30 \cdot 10^{-1}$	$9.02 \cdot 10^{-1}$
2	$1.39 \cdot 10^{-1}$	$2.18 \cdot 10^{-1}$	$3.90 \cdot 10^{-1}$
3	$2.70 \cdot 10^{-2}$	$7.01 \cdot 10^{-2}$	$2.13 \cdot 10^{-1}$
4	$4.03 \cdot 10^{-3}$	$2.24 \cdot 10^{-2}$	$1.79 \cdot 10^{-1}$
5	$7.44 \cdot 10^{-4}$	$6.21 \cdot 10^{-3}$	$1.06 \cdot 10^{-1}$
6	$1.48 \cdot 10^{-4}$	$8.87 \cdot 10^{-4}$	$7.88 \cdot 10^{-2}$
7	$3.17 \cdot 10^{-5}$	$1.87 \cdot 10^{-4}$	$3.28 \cdot 10^{-2}$
8	$7.52 \cdot 10^{-6}$	$4.08 \cdot 10^{-5}$	$1.15 \cdot 10^{-2}$
9	$1.99 \cdot 10^{-6}$	$9.68 \cdot 10^{-6}$	$2.36 \cdot 10^{-3}$
10	$5.61 \cdot 10^{-7}$	$2.34 \cdot 10^{-6}$	$4.97 \cdot 10^{-4}$
20	$6.37 \cdot 10^{-12}$	$6.22 \cdot 10^{-12}$	$6.62 \cdot 10^{-10}$

Table 6.4: *Residual norm of the Reynolds equation on three different grids as a function of the number of $W(2,1)$ cycles. $M = 20$, $L = 10$, incompressible lubricant, Barus viscosity-pressure equation. Relaxation parameters: $\omega_{ja} = 0.4$, $\omega_{gs} = 0.6$ and $\omega_{H_0} = 0.1$.*

The Tables 6.4 and 6.5 show the residual of the Reynolds equation and the force balance equation as a function of the number of $W(2,1)$ cycles for three grids for the case $M = 20$ and $L = 10$. The domain was chosen as $-4.5 \leq X \leq 1.5$ and $-3.0 \leq Y \leq 3.0$. and the coarsest grid (level 1) used in the cycle has $(16+1) \times (16+1)$ points. The Hertzian pressure profile was taken as an initial approximation for P^h. The first approximation for H^h was obtained by straightforward computation according to Equation (6.29) with an initial approximation to H_0 ($\tilde{H}_0 = -0.53$). Throughout the cycle Multilevel Multi-integration is used for the fast evaluation of the elastic deformation integrals. The algorithm as described in the previous chapter is generally used with 6th order transfers which ensures accurate evaluation for all cases presented in this chapter.

The results were further obtained using underrelaxation factors $\omega_{ja} = 0.4$ for the Jacobi changes and $\omega_{gs} = 0.6$ for the Gauss-Seidel changes. The relaxation factor used in the relaxation of the force balance condition was chosen $\omega_{H_0} = 0.1$. These values are suitable for low and moderately loaded cases. However, using W-cycles with increasing

W-cycle	level 3	level 4	level 5
	64×64	128×128	256×256
1	$2.62 \cdot 10^{-2}$	$2.76 \cdot 10^{-2}$	$6.91 \cdot 10^{-3}$
2	$3.03 \cdot 10^{-3}$	$7.29 \cdot 10^{-3}$	$1.02 \cdot 10^{-2}$
3	$2.54 \cdot 10^{-4}$	$1.89 \cdot 10^{-3}$	$7.82 \cdot 10^{-4}$
4	$1.78 \cdot 10^{-4}$	$2.27 \cdot 10^{-4}$	$1.65 \cdot 10^{-3}$
5	$9.11 \cdot 10^{-5}$	$3.12 \cdot 10^{-5}$	$2.95 \cdot 10^{-4}$
6	$2.67 \cdot 10^{-5}$	$1.09 \cdot 10^{-7}$	$1.33 \cdot 10^{-4}$
7	$8.58 \cdot 10^{-6}$	$3.83 \cdot 10^{-7}$	$8.13 \cdot 10^{-5}$
8	$2.83 \cdot 10^{-6}$	$1.31 \cdot 10^{-7}$	$2.07 \cdot 10^{-5}$
9	$9.30 \cdot 10^{-7}$	$1.39 \cdot 10^{-7}$	$1.25 \cdot 10^{-6}$
10	$2.89 \cdot 10^{-7}$	$3.96 \cdot 10^{-8}$	$8.22 \cdot 10^{-7}$
20	$1.23 \cdot 10^{-12}$	$7.87 \cdot 10^{-13}$	$2.39 \cdot 10^{-12}$

Table 6.5: *Residual norm of the force balance on three different grids as a function of the number of* $W(2,1)$ *cycles.* $M = 20$, $L = 10$, *incompressible lubricant, Barus viscosity-pressure equation. Relaxation parameters* $\omega_{ja} = 0.4$, $\omega_{gs} = 0.6$, $\omega_{H_0} = 0.1$.

number of gridlevels the number of visits to the coarser grids also increases. Consequently, on very fine grids, a smaller value of ω_{H_0} may give better results. This problem does not occur when V-cycles are used, but in that case the asymptotic performance is less good.

Table 6.4 and 6.5 show that the cycle converges and asymptotically the error reduction per cycle is roughly a factor of 3. This applies both to the residual of the Reynolds equation and to the residual of the force balance equation. Note that this is the asymptotic factor. Performance will often be better in the first few cycles and on coarser grids. Also larger underrelaxation factors can be used, which, on the coarser grids gives better performance. However, this is not necessarily true on very fine grids and for many cycles. If the underrelaxation factors are chosen too large this may lead to a situation where the cavitation boundary is moving back and forth which results in poor asymptotic convergence.

As the conditions of the problem vary significantly with varying parameter values, the relaxation parameters which give optimal performance vary as well. The general tendency is that with increasing load more under relaxation is needed and ω_{H_0} should be smaller. A safe choice for many cases is $\omega_{ja} = 0.2$ and $\omega_{gs} = 0.4$. The force balance constant should be taken $\omega_{H_0} < 0.05$ on very fine grids for W-cycles whereas it can be taken $O(0.1)$ for V-cycles. This conservative choice leads to a less than optimal performance on the coarser grids where a larger factor could be used. Nevertheless, as will be shown in the next section, this performance is sufficiently good to obtain a solution with an error that is below the discretization error in only a few cycles.

The results for load case number 2 with these relaxation factors are given in Tables 6.6 and 6.7. These tables show that both in terms of the residual of the Reynolds equation and in terms of the residual of the force balance equation these parameters give a stable convergence with an asymptotic error reduction of approximately 2 per cycle.

W-cycle	level 3	level 4	level 5
	64×64	128×128	256×256
1	$3.97 \cdot 10^{-1}$	$4.31 \cdot 10^{-1}$	$4.76 \cdot 10^{-1}$
2	$1.43 \cdot 10^{-1}$	$1.64 \cdot 10^{-1}$	$2.10 \cdot 10^{-1}$
3	$4.93 \cdot 10^{-2}$	$6.23 \cdot 10^{-2}$	$1.03 \cdot 10^{-1}$
4	$1.71 \cdot 10^{-3}$	$2.24 \cdot 10^{-2}$	$6.03 \cdot 10^{-2}$
5	$6.00 \cdot 10^{-3}$	$8.17 \cdot 10^{-3}$	$2.90 \cdot 10^{-2}$
6	$2.15 \cdot 10^{-3}$	$3.10 \cdot 10^{-3}$	$1.46 \cdot 10^{-2}$
7	$8.02 \cdot 10^{-4}$	$1.21 \cdot 10^{-3}$	$6.00 \cdot 10^{-3}$
8	$3.18 \cdot 10^{-4}$	$4.89 \cdot 10^{-4}$	$2.33 \cdot 10^{-3}$
9	$1.38 \cdot 10^{-4}$	$2.06 \cdot 10^{-4}$	$9.51 \cdot 10^{-4}$
10	$6.77 \cdot 10^{-5}$	$9.06 \cdot 10^{-5}$	$4.01 \cdot 10^{-4}$
20	$6.28 \cdot 10^{-7}$	$1.93 \cdot 10^{-7}$	$2.47 \cdot 10^{-7}$

Table 6.6: *Residual norm of the Reynolds equation on three different grids as a function of the number of $W(2,1)$ cycles. $M = 200$, $L = 10$, incompressible lubricant, Barus viscosity-pressure equation. Relaxation parameters $\omega_{ja} = 0.2$, $\omega_{gs} = 0.4$, and $\omega_{H_0} = 0.05$. $-2.5 \leq X \leq 1.5$, $-2 \leq Y \leq 2$.*

6.9 Full MultiGrid

The algorithm for the EHL problem is completed with the extension to a Full MultiGrid (FMG) algorithm. This is a very straightforward step. Instead of only using the coarser grids for convergence acceleration, they fulfill an additional task in generating an accurate first approximation on the grid. With respect to the solution of the problem for high load conditions (large M, large L) it is important to separate these two functions of the coarse grid. With increasing load the discretization error on a given grid increases. If the grid is rather coarse the solution of the discrete problem on this grid will be a poor approximation to the solution of the differential problem. Consequently, it will be a bad initial approximation to the solution on the next finer grid. Characteristic for situations where the discretization error is large is the occurrence of a significant negative film thickness values in the solution. In combination with the non-linearity and the cavitation condition these negative values of the film thickness can seriously frustrate the convergence of the relaxation process and even cause it to diverge. Hence, the first grid used in the FMG algorithm must have a sufficiently small mesh size. In very extreme cases the coarsest grid in the cycle should also be finer, but this is not often the case. In the correction cycle the coarser grids serve to generate a smooth correction to the fine grid problem, based on residual information from the more accurate fine grids.

Summarizing, for high loads the FMG process should not be started with simple relaxations on the coarsest grid. Rather one should start with a few correction cycles on a sufficiently fine grid. Subsequently, the FMG process can be continued as usual.

In Chapters 4 and 5 it was shown that using a FMG algorithm with one or two cycles per level yielded a solution with an error small compared to the discretization error. Below it will be shown that this also applies to the EHL problem.

Table 6.8 shows the residual of the Reynolds equation for four grids as a function of

W-cycle	level 3	level 4	level 5
	64×64	128×128	256×256
1	$1.05 \cdot 10^{-2}$	$2.29 \cdot 10^{-2}$	$1.06 \cdot 10^{-2}$
2	$8.28 \cdot 10^{-3}$	$2.76 \cdot 10^{-3}$	$1.30 \cdot 10^{-3}$
3	$5.61 \cdot 10^{-3}$	$4.42 \cdot 10^{-4}$	$1.54 \cdot 10^{-3}$
4	$3.17 \cdot 10^{-3}$	$3.59 \cdot 10^{-5}$	$4.72 \cdot 10^{-4}$
5	$1.68 \cdot 10^{-3}$	$5.00 \cdot 10^{-5}$	$1.25 \cdot 10^{-4}$
6	$8.69 \cdot 10^{-4}$	$2.59 \cdot 10^{-5}$	$4.13 \cdot 10^{-5}$
7	$4.38 \cdot 10^{-4}$	$8.59 \cdot 10^{-6}$	$1.55 \cdot 10^{-5}$
8	$2.19 \cdot 10^{-4}$	$1.69 \cdot 10^{-6}$	$6.04 \cdot 10^{-6}$
9	$1.08 \cdot 10^{-4}$	$5.82 \cdot 10^{-7}$	$1.90 \cdot 10^{-6}$
10	$5.26 \cdot 10^{-5}$	$1.07 \cdot 10^{-6}$	$4.21 \cdot 10^{-7}$
20	$3.15 \cdot 10^{-8}$	$1.54 \cdot 10^{-8}$	$6.79 \cdot 10^{-9}$

Table 6.7: *Residual norm of the force balance equation on three different grids as a function of the number of $W(2,1)$ cycles. $M = 200$, $L = 10$, incompressible lubricant, Barus viscosity-pressure equation. Relaxation parameters: $\omega_{ja} = 0.2$, $\omega_{gs} = 0.4$, and $\omega_{H_0} = 0.05$, $-2.5 \leq X \leq 1.5$, $-2 \leq Y \leq 2$.*

the number of $W(2,1)$ cycles carried out in a Full Multigrid algorithm for load case 1. From this table it can be seen that when using a FMG process with a single $W(2,1)$ cycle per level, the norm of the residual of the Reynolds equation for the final solution on level 5 is already smaller than the residual norm for the solution that was obtained applying 6 cycles starting with the Hertzian approximation as an initial solution, see Table 6.4. The work invested to obtain the FMG solution is only slightly larger than the work of a single cycle. Consequently, using the converged solution of a coarser grid as the initial solution on a finer grid, really pays off.

To determine how many cycles are needed to obtain a solution with an error of the magnitude of the discretization error Tables 6.9-6.10 can be used. These tables list the central and the minimum film thickness for the solutions presented in Table 6.8. The

W-cycle	level 3	level 4	level 5	level 6
	64×64	128×128	256×256	512×512
1	$2.39 \cdot 10^{-2}$	$1.45 \cdot 10^{-2}$	$7.45 \cdot 10^{-3}$	$5.31 \cdot 10^{-3}$
2	$5.31 \cdot 10^{-3}$	$1.48 \cdot 10^{-3}$	$7.68 \cdot 10^{-4}$	$1.02 \cdot 10^{-3}$
3	$1.32 \cdot 10^{-3}$	$2.82 \cdot 10^{-4}$	$2.53 \cdot 10^{-4}$	$6.37 \cdot 10^{-4}$
4	$3.67 \cdot 10^{-4}$	$5.89 \cdot 10^{-5}$	$7.38 \cdot 10^{-5}$	$3.14 \cdot 10^{-4}$
5	$1.04 \cdot 10^{-4}$	$1.28 \cdot 10^{-5}$	$2.50 \cdot 10^{-5}$	$2.28 \cdot 10^{-4}$
10	$1.44 \cdot 10^{-7}$	$1.62 \cdot 10^{-8}$	$1.52 \cdot 10^{-7}$	$4.16 \cdot 10^{-5}$

Table 6.8: *Residual norm of the Reynolds equation on four different grids as a function of the number of $W(2,1)$ cycles used in a FMG algorithm. $M = 20$, $L = 10$, incompressible lubricant, Barus viscosity-pressure equation. Relaxation parameters: $\omega_{ja} = 0.4$, $\omega_{gs} = 0.6$, and $\omega_{H_0} = 0.1$.*

W-cycle	level 3	level 4	level 5	level 6
	64×64	128×128	256×256	512×512
1	$4.7940 \cdot 10^{-1}$	$4.9336 \cdot 10^{-1}$	$4.9644 \cdot 10^{-1}$	$4.9739 \cdot 10^{-1}$
2	$4.8185 \cdot 10^{-1}$	$4.9358 \cdot 10^{-1}$	$4.9655 \cdot 10^{-1}$	$4.9724 \cdot 10^{-1}$
3	$4.8190 \cdot 10^{-1}$	$4.9357 \cdot 10^{-1}$	$4.9652 \cdot 10^{-1}$	$4.9729 \cdot 10^{-1}$
4	$4.8188 \cdot 10^{-1}$	$4.9357 \cdot 10^{-1}$	$4.9653 \cdot 10^{-1}$	$4.9727 \cdot 10^{-1}$
5	$4.8187 \cdot 10^{-1}$	$4.9357 \cdot 10^{-1}$	$4.9653 \cdot 10^{-1}$	$4.9728 \cdot 10^{-1}$
10	$4.8187 \cdot 10^{-1}$	$4.9357 \cdot 10^{-1}$	$4.9653 \cdot 10^{-1}$	$4.9727 \cdot 10^{-1}$

Table 6.9: *Central film thickness defined as H^h at $(X, Y) = (0, 0)$ on four different grids as a function of the number of $W(2, 1)$ cycles in a FMG algorithm. $M = 20$, $L = 10$, incompressible lubricant, Barus viscosity-pressure equation.*

W-cycle	level 3	level 4	level 5	level 6
	64×64	128×128	256×256	512×512
1	$2.9531 \cdot 10^{-1}$	$3.0399 \cdot 10^{-1}$	$3.0605 \cdot 10^{-1}$	$3.0661 \cdot 10^{-1}$
2	$2.9826 \cdot 10^{-1}$	$3.0509 \cdot 10^{-1}$	$3.0631 \cdot 10^{-1}$	$3.0652 \cdot 10^{-1}$
3	$2.9863 \cdot 10^{-1}$	$3.0515 \cdot 10^{-1}$	$3.0629 \cdot 10^{-1}$	$3.0657 \cdot 10^{-1}$
4	$2.9874 \cdot 10^{-1}$	$3.0516 \cdot 10^{-1}$	$3.0630 \cdot 10^{-1}$	$3.0655 \cdot 10^{-1}$
5	$2.9877 \cdot 10^{-1}$	$3.0516 \cdot 10^{-1}$	$3.0630 \cdot 10^{-1}$	$3.0655 \cdot 10^{-1}$
10	$2.9878 \cdot 10^{-1}$	$3.0516 \cdot 10^{-1}$	$3.0630 \cdot 10^{-1}$	$3.0655 \cdot 10^{-1}$

Table 6.10: *Minimum film thickness on three different levels as a function of the number of $W(2, 1)$ cycles in a FMG algorithm. $M = 20$, $L = 10$, incompressible lubricant, Barus viscosity-pressure equation.*

results show that already using two $W(2, 1)$ cycles per gridlevel an approximation to the minimum and central film thickness is obtained with an error that is very small compared to the discretization error. This can be seen from a comparison of the difference between the value after 2 cycles and 10 cycles with the difference between the values on the different grids. Finally, the order of the discretization error itself can be checked comparing the values obtained on the different grids. Each time the mesh size is refined the change in the central film thickness becomes approximately 4 times smaller which is consistent with the fact that a second order accurate discrization has been used.

To conclude the results for this load case, Figures 6.9 and 6.10 show the dimensionless pressure and film thickness for the level 6 solution as a function of X and Y. Comparing Figure 6.9 with Figure 6.4 shows that the higher resolution results in a much more pronounced pressure spike. Furthermore, this spike forms a shield of high pressure values preceding the drop to the cavitation pressure $P^h = 0$. Figure 6.11 (left) shows a contour plot of the film thickness. This figure clearly shows the horse-shoe shape of the film thickness and the flat region in the centre. In accordance with Equation (6.54) for constant ρ the contour lines tend to be straight lines.

An alternative way to present the film thickness results is by means of a pseudo-interferometry graph. Instead of drawing contour lines, an intensity $\mathcal{I}$ is defined according to:

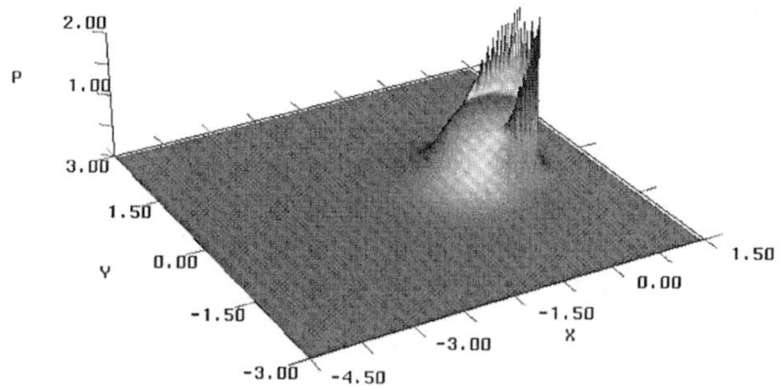

Figure 6.9: *Computed dimensionless pressure* P^h *on grid level 6* ($n_x = n_y = 512$) *as a function of* X *and* Y. $M = 20$, $L = 10$, *incompressible lubricant, Barus viscosity-pressure equation.*

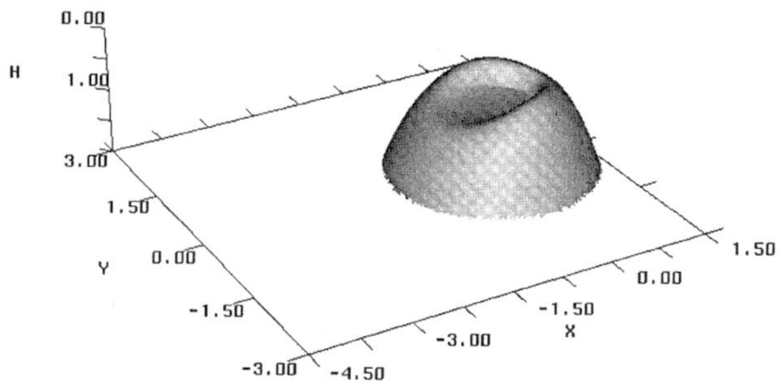

Figure 6.10: *Computed dimensionless film thickness* H^h *on grid level 6* ($n_x = n_y = 512$) *as a function of* X *and* Y. $M = 20$, $L = 10$, *incompressible lubricant, Barus viscosity-pressure equation.*

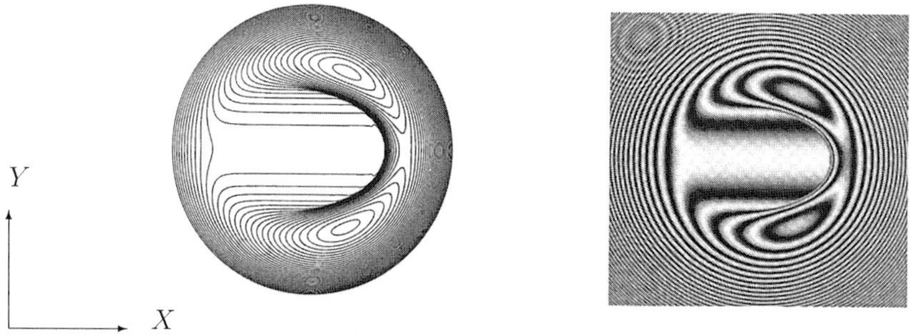

Figure 6.11: *Contour plot (left,* $\Delta H = 0.01$*) and pseudo-interferometry plot (right,* $\Delta H = 0.05$*) of the computed dimensionless film thickness* H^h *as a function of* X *and* Y. $M = 20$, $L = 10$, *incompressible lubricant, Barus viscosity-pressure equation.*

$$\mathcal{I}(X,Y) = \mathcal{I}_0 + \mathcal{I}_0 \cos\left(\frac{2\pi H(X,Y)}{\Delta H}\right) \tag{6.55}$$

where $\mathcal{I} = 1$ gives a square $h_x \times h_y$ of white color, $\mathcal{I} = 0$ gives a black square, and squares with grey shades are obtained for $0 < \mathcal{I} < 1$. Figure 6.11 (right) shows such a pseudo-interferometry graph of the film thickness for load case 1. This type of graph is interesting since it allows a direct comparison with the film thickness maps obtained with a popular measurement technique referred to as 'Optical Interferometry'. This technique was developed in the 1960's at Imperial College, London, by Foord et al. [44] and has presently matured to a level where the film thickness maps can be obtained for transient problems, and also very thin films, e.g. see Cann [29] and Kaneta [67, 68]. In an optical interferometry rig, a steel ball is run against a glass disk that is covered with a semi reflective layer (and a spacer layer for the thin film version). Light is shone into the contact. Part of it is reflected by the layer, and part of it goes through the film and is reflected later by the ball. The result is an interference pattern representing a film thickness map of the contact.

Tables 6.11-6.13 show the results obtained using a FMG algorithm for load case 2. In terms of the residual norm of the Reynolds equation, starting with the converged solution of a coarser grid on level 5 saves the work of about 5 cycles. Furthermore, the values of the central and the minimum film thickness given in Table 6.12 and 6.13 can be compared. A comparison of the values after many cycles with those obtained after 2 cycles shows that already after 2 cycles the error in the central and minimum film thickness is small compared to the discretization error. Finally, these values converge in a second order manner with decreasing mesh size. The difference between the values of the central and minimum film thickness obtained on two consequent grids decreases with a factor four, when decreasing the mesh size.

Finally, the Figures 6.12 and 6.13 show the computed dimensionless pressure and film thickness for the level 6 solution as a function of X and Y. Comparing Figure 6.12 with Figure 6.7 shows that because of the higher resolution the pressure spike is now clearly visible in the solution. Figure 6.14 shows a contourplot and a pseudo-interferometry graph of the dimensionless film thickness. Comparing Figure 6.14 with Figure 6.11 shows that for the higher load case the almost flat region in the center with straight contour lines covers almost the entire Hertzian dry contact region. Consequently the side-lobes where the minimum film thickness occurs have become very small.

The Full Multigrid algorithm completes a fast and efficient solver for the EHL circular contact problem. The computing time required by the solver is proportional to $O(N \ln(N))$ if N is the number of nodes on the grid. With the algorithm presented, the circular contact problem can be solved efficiently for a wide range of load conditions. The program in its standard form as given in Appendix H will run many load cases without any problem. However, as the EHL problem is a non-linear problem in which different effects play a stronger or weaker role with variation of the parameter values, it is very difficult to give one set of relaxation parameters and domain sizes that ensures good convergence for all cases. The program can be an efficient solver for many cases if it is used with some care. This *care* applies to the choice of the size of the domain which must be sufficiently large on the inlet side, in particular for cases with small M values. On the

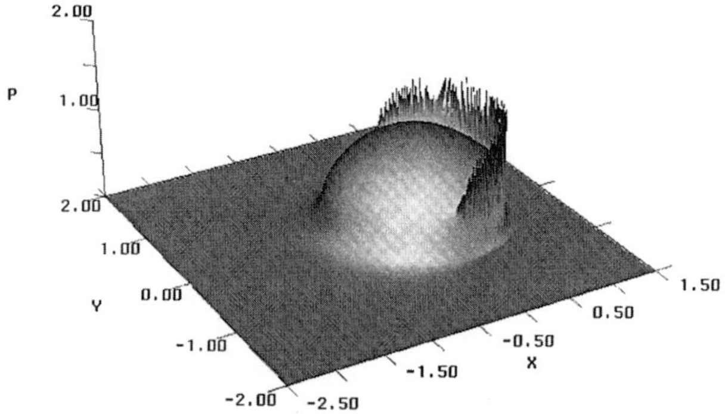

Figure 6.12: *Computed dimensionless pressure P^h on level 6 ($n_x = n_y = 512$) as a function of X and Y. $M = 200$, $L = 10$, incompressible lubricant, Barus viscosity-pressure equation.*

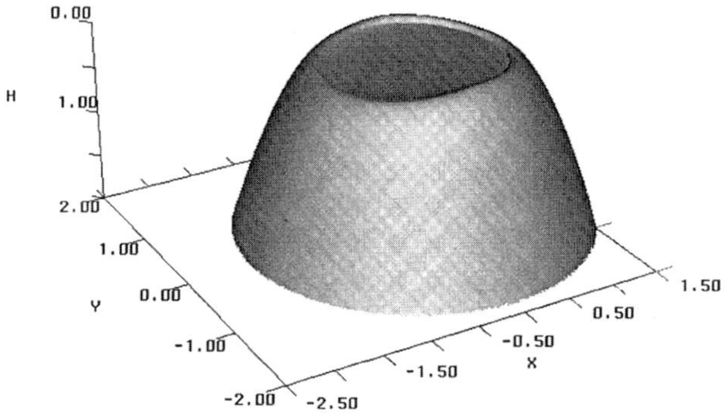

Figure 6.13: *Computed dimensionless film thickness H^h on level 6 as a function of X and Y. $M = 200$, $L = 10$, incompressible lubricant, Barus viscosity-pressure equation.*

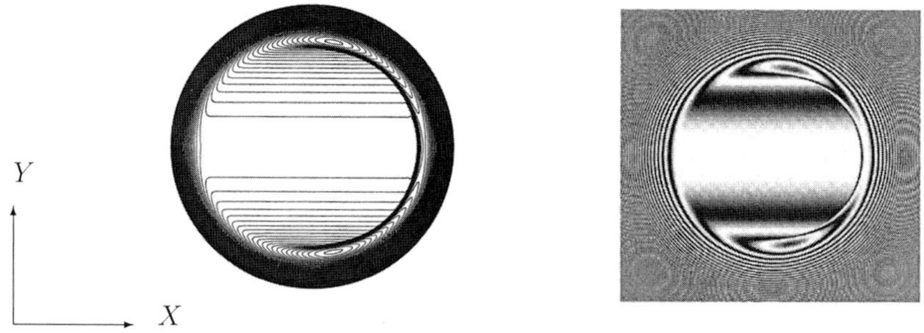

Figure 6.14: *Contour plot (left, $\Delta H = 2.5 \cdot 10^{-3}$) and pseudo-interferometry plot (right, $\Delta H = 2.5 \cdot 10^{-2}$) of the computed dimensionless film thickness H^h as a function of X and Y. $M = 200$, $L = 10$, incompressible lubricant, Barus viscosity-pressure equation.*

W-cycle	level 3	level 4	level 5	level 6
	64 × 64	128 × 128	256 × 256	512 × 512
1	$9.72 \cdot 10^{-2}$	$5.89 \cdot 10^{-2}$	$3.63 \cdot 10^{-2}$	$3.30 \cdot 10^{-2}$
2	$2.18 \cdot 10^{-2}$	$1.07 \cdot 10^{-2}$	$3.43 \cdot 10^{-3}$	$1.35 \cdot 10^{-3}$
3	$9.78 \cdot 10^{-3}$	$3.23 \cdot 10^{-3}$	$9.89 \cdot 10^{-4}$	$5.48 \cdot 10^{-4}$
4	$4.90 \cdot 10^{-3}$	$1.24 \cdot 10^{-3}$	$4.03 \cdot 10^{-4}$	$2.14 \cdot 10^{-4}$
5	$2.59 \cdot 10^{-3}$	$5.13 \cdot 10^{-4}$	$1.86 \cdot 10^{-4}$	$1.08 \cdot 10^{-4}$
10	$1.22 \cdot 10^{-4}$	$1.42 \cdot 10^{-5}$	$9.33 \cdot 10^{-6}$	$6.12 \cdot 10^{-6}$

Table 6.11: *Residual norm of the Reynolds on four different grids as a function of the number of $W(2,1)$ cycles in a FMG algorithm. $M = 200$, $L = 10$, incompressible lubricant, Barus viscosity-pressure equation.*

W-cycle	level 3	level 4	level 5	level 6
	64 × 64	128 × 128	256 × 256	512 × 512
1	$7.4704 \cdot 10^{-2}$	$9.4595 \cdot 10^{-2}$	$9.8932 \cdot 10^{-2}$	$9.9473 \cdot 10^{-2}$
2	$8.3089 \cdot 10^{-2}$	$9.6190 \cdot 10^{-2}$	$9.8879 \cdot 10^{-2}$	$9.9542 \cdot 10^{-2}$
3	$8.5665 \cdot 10^{-2}$	$9.6286 \cdot 10^{-2}$	$9.8895 \cdot 10^{-2}$	$9.9538 \cdot 10^{-2}$
4	$8.6321 \cdot 10^{-2}$	$9.6351 \cdot 10^{-2}$	$9.8901 \cdot 10^{-2}$	$9.9539 \cdot 10^{-2}$
5	$8.6512 \cdot 10^{-2}$	$9.6377 \cdot 10^{-2}$	$9.8902 \cdot 10^{-2}$	$9.9538 \cdot 10^{-2}$
10	$8.6682 \cdot 10^{-2}$	$9.6387 \cdot 10^{-2}$	$9.8902 \cdot 10^{-2}$	$9.9538 \cdot 10^{-2}$

Table 6.12: *Central film thickness defined as H^h at $(X, Y) = (0, 0)$ on four different grids as a function of the number of $W(2,1)$ cycles in a FMG algorithm. $M = 200$, $L = 10$, incompressible lubricant, Barus viscosity-pressure equation.*

other hand for large M values, i.e. $M > 500$ care should be taken that the FMG process is started on a sufficiently fine grid, see Figure 5.10.

Finally, with increasing value of the L parameter, the problem becomes more difficult to solve. In that case the piezoviscous effects become very strong and the film becomes very stiff. This implies that the process is very sensitive to the changes of H_0 and to the quality of the first approximation on a given grid. In fact it becomes very sensitive to any change. Changes in the position of the cavitation boundary condition, or interpolation errors in the coarse grid correction may render the algorithm unstable for the extreme cases of $L > 20$. The use of a less extreme viscosity-pressure relation may make these problems easier to solve.

Generally, if stability problems occur for a particular load case various measures can be taken. For example, one should ensure that the first approximation to H_0 is a good one. Secondly, the grid on which the FMG process is started should be sufficiently fine. Next, the coarsest grid in the cycle should be sufficiently fine. This can often be ensured by choosing a smaller domain for higher load cases (large M). And finally, the relaxation parameters can be reduced.

W-cycle	level 3	level 4	level 5	level 6
	64×64	128×128	256×256	512×512
1	$3.0780 \cdot 10^{-2}$	$3.8239 \cdot 10^{-2}$	$4.0236 \cdot 10^{-2}$	$4.1414 \cdot 10^{-2}$
2	$3.3918 \cdot 10^{-2}$	$3.9675 \cdot 10^{-2}$	$4.1386 \cdot 10^{-2}$	$4.1702 \cdot 10^{-2}$
3	$3.4848 \cdot 10^{-2}$	$4.0179 \cdot 10^{-2}$	$4.1440 \cdot 10^{-2}$	$4.1695 \cdot 10^{-2}$
4	$3.5373 \cdot 10^{-2}$	$4.0302 \cdot 10^{-2}$	$4.1448 \cdot 10^{-2}$	$4.1696 \cdot 10^{-2}$
5	$3.5579 \cdot 10^{-2}$	$4.0331 \cdot 10^{-2}$	$4.1450 \cdot 10^{-2}$	$4.1695 \cdot 10^{-2}$
10	$3.5701 \cdot 10^{-2}$	$4.0342 \cdot 10^{-2}$	$4.1452 \cdot 10^{-2}$	$4.1696 \cdot 10^{-2}$

Table 6.13: *Minimum film thickness on four different grids as a function of the number of $W(2,1)$ cycles in a FMG algorithm. $M = 200$, $L = 10$, incompressible lubricant, Barus viscosity-pressure equation.*

6.10 Design Graphs

The program as described can be used to compute the film thickness and pressure distribution as a function of the contact operating conditions. Consequently, the evolution of specific aspects of the solution (pressure or film thickness) can be analysed using parametric studies. Obviously, the film thickness itself is important, as, compared with an estimate of the roughness on the surfaces, it can yield an indication of the lubrication condition of the contact in real applications. The question arises what film thickness value should be used, the minimum film thickness because it is the smallest value, or the central film thickness because it extends over a much larger region?

As an example a design graph for the central film thickness is given in Figure 6.15. This figure shows the central film thickness defined as the film thickness at the point $(0, 0)$ as a function of M and L. Note that the dimensionless film thickness according to Equation (6.19) is used. In the chart the four asymptotic regimes mentioned in Table 6.1 are indicated. The markers indicate the results of numerical solutions. Also shown in the figure are the predictions given by two film thickness formulas. The dashed lines indicate the predictions of the formula for the central film thickness presented by Hamrock and Dowson [56]. This formula was obtained by curve-fitting using a set of numerical results as input. As can be seen from the figure, the Hamrock and Dowson equation yields accurate results for $M > 10$ and $2.5 \leq L \leq 25$. However, for small values of M and L the prediction yields values which are much too low.

An alternative way to construct a film thickness formula is to create a function fit using solutions for the asymptotic regimes as building blocks. This appraoch was introduced by Moes, e.g. see [86, 87, 99]. It has the advantage that it yields a formula that is applicable in the entire parameter range, including the asymptotic regimes. For the line contact case the solutions for the asymptotic regimes can be derived analytically. For the point contact problem approximate solutions for the asymptotic regimes can be used as building blocks, see [87]. A formula to predict the central film thickness for the circular contact can be obtained using the following approximate asymptotic solutions:

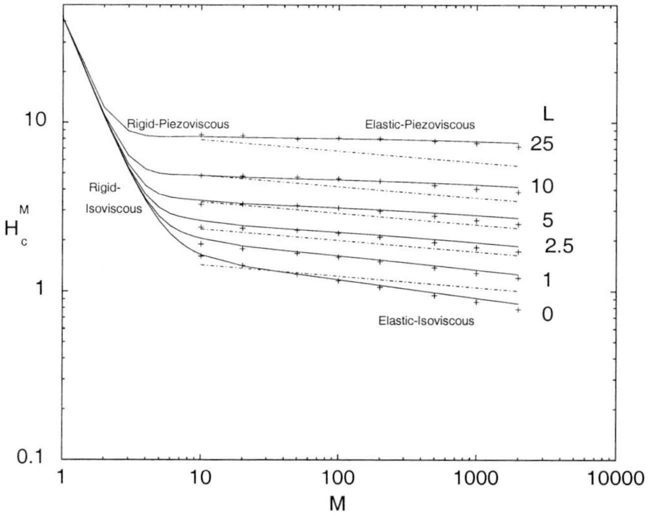

Figure 6.15: *Central film thickness chart H_c^M as a function of M and L. The drawn lines indicate the predictions of (6.61). The dashed lines indicate the predictions of Hamrock and Dowson [56].*

- *Rigid Isoviscous:*

$$H_{ri}^M = 41.4M^{-2} \tag{6.56}$$

- *Rigid Piezoviscous:*

$$H_{rp}^M = 0.91L^{2/3} \tag{6.57}$$

- *Elastic Isoviscous:*

$$H_{ei}^M = 2.42M^{-2/15} \tag{6.58}$$

A more accurate equation based on a fit through numerically obtained values for this case, see [99], is:

$$H_{ei}^M = 1.96M^{-1/9} \tag{6.59}$$

- *Elastic Piezoviscous:*

$$H_{ep}^M = 1.25M^{-1/12}L^{3/4} \tag{6.60}$$

Introducing a parameter s to ensure a smooth transition between the asymptotic regimes, the following film thickness formula is proposed:

$$H_c^M = \left\{ \left[(H_{ri}^M)^{3/2} + [(H_{ei}^M)^{-4} + 0.1]^{-3/8} \right]^{2s/3} + \left[(H_{rp}^M)^{-8} + (H_{ep}^M)^{-8} \right]^{-s/8} \right\}^{1/s} \tag{6.61}$$

where:

$$s = \frac{3}{2} \left[1 + exp\left(-\frac{6}{5} \frac{H_{ei}^M}{H_{ri}^M} \right) \right]$$

The drawn lines in Figure 6.15 have been obtained using Equation (6.61) with (6.56), (6.57), (6.59), and (6.60). As can be seen from this graph the formula accurately predicts the computed central film thickness over the entire range of parameter values.

6.11 Conclusion

In this chapter the various aspects of the previous solvers have been combined into an efficient solver for the circular EHL problem. These include the solvers for the Poisson problem, for the Hydrodynamic Lubrication problem, for the Dry Contact problem and in particular the Multilevel Multi-integration. As each step in the EHL algorithm requires at most $O(N \ln(N))$ operations, the total cpu time needed by the solver is also $O(N \ln(N))$. This enables the solution of the EHL problem on dense grids and thus makes the algorithm well suited for extensive parameter studies. In this chapter only a few pressure and film thickness results were presented. Once the pressure and film thickness are known, the pressure and film thickness distribution can be used as input to compute many other quantities. For example using the multilevel multi-integration the sub-surface stresses can be computed efficiently.

The algorithm can be extended in many ways to cater for more complex (realistic) EHL problems. For example different viscosity-pressure equations and density pressure equations can be used. The solver can be extended to elliptic contacts as described in [110]. Non-Newtonian lubricant behaviour can also be incorporated, by means of effective viscosities, involving only minor changes to the program. A related extension involves the 'slip-at-the-wall' as described by Ehret et al. [38, 39].

More involved are the addition of fractional film content as a variable to study the effects of starvation and variable lubricant supply, see Chevalier et al. [31, 32].

The algorithm has served as a basis for a solver of the time dependent problem of roughness moving through the contact, see Venner and Lubrecht [100]-[106]. Alternatively a transient analysis was used to obtain insight in the dynamics of the contact such as stiffness and damping, see Wijnant and Venner [109]-[112].

In general these efficient numerical tools allow one to generate huge amounts of data in relatively short times. As an example: the transient 2d EHL solver easily produces $2 \times 10^4 \times 512 \times 512 = O(10^{10})$ data points. As the calculations are performed in *double precision* just storing the pressure and film thickness distribution at each time step, would require $O(100)$ GBytes. It is clear that the old adagio "calculation is not about generating numbers but all about generating insight" is particularly valid for these solvers. In many cases the vast amount of numbers has been reduced to simple approximative equations that can be interpreted in physical terms and solved on a pocket calculator.

6.12 Advanced Topics

As mentioned above, the low complexity of the algorithm creates room for more complex EHL problems to be studied. Some of these problems are briefly described below.

6.12.1 Roelands Equation, Compressible Lubricant

The results presented in this chapter were obtained assuming an incompressible lubricant and using the Barus viscosity-pressure equation. This was done on purpose as it represents the simplest situation and hence the problem can be described by only two parameters. However, as was explained in Chapter 1 the Barus equation is valid only for a limited range of pressures. With increasing pressures the predicted viscosity is generally too high. An improved equation is the Roelands equation (6.3). In addition, due to the high pressures occurring in EHL contacts, it is no longer justified to assume the lubricant to be incompressible. The relation often used to model the increase of the density with pressure is the Dowson and Higginson equation (6.4).

The program given in Appendix H is fully written in terms of a variable density. It can be used to solve the pressure and film thickness distribution using either of the viscosity equations and/or assuming an incompressible or compressible lubricant. This section presents some results obtained for a compressible lubricant and using the Roelands viscosity-pressure equation. From a numerical point of view, adding these equations makes the problem easier to solve. First, because the Roelands equation predicts a viscosity that increases less rapidly with pressure than according to the Barus equation. Secondly, because the compressibility adds some flexibility to the system. As a result, the problem becomes less sensitive to changes in the value of H_0 when relaxing the force balance equation.

The results presented apply to the same load cases considered throughout this chapter. However, as mentioned in Section 6.4 using Roelands' equation and assuming a compressible lubricant the problem is no longer governed by two parameters. In addition to M and L, two other parameters have to be given. The results presented here have been obtained using $\alpha = 2.2 \cdot 10^{-8}$ $[Pa^{-1}]$ and $\eta_0 = 40 \cdot 10^{-3}$ $[Pa\,s]$. Using Equation (6.22) this gives $z = 0.67$.

Tables 6.14 and 6.15 give the dimensionless minimum and central film thickness as a function of the gridlevel obtained using a FMG algorithm with 3 W(2,1) cycles per level for the two load cases. The Figures 6.16 and 6.18 show the film thickness contour plots and pseudo-interferometry graphs. Finally, Figures 6.17 and 6.19 show the computed pressure at the centerline $Y = 0$ as a function of X and at the line $X = 0$ as a function of Y. For reference the results obtained using the Barus viscosity-pressure equation and assuming an incompressible lubricant are shown in the latter two figures.

From Tables 6.14 and 6.15 it can be concluded that the convergence behaviour of the

level	$n_x \times n_y$	H_c	H_m
2	32×32	$3.8174 \cdot 10^{-1}$	$2.6291 \cdot 10^{-1}$
3	64×64	$4.1904 \cdot 10^{-1}$	$2.8622 \cdot 10^{-1}$
4	128×128	$4.2872 \cdot 10^{-1}$	$2.9094 \cdot 10^{-1}$
5	256×256	$4.3116 \cdot 10^{-1}$	$2.9218 \cdot 10^{-1}$
6	512×512	$4.3177 \cdot 10^{-1}$	$2.9237 \cdot 10^{-1}$

Table 6.14: *Central film thickness defined as H^h at $(X,Y) = (0,0)$ and minimum film thickness as a function of the gridlevel computed with a FMG algorithm with 3 W(2,1) cycles. $M = 20$, $L = 10$, compressible lubricant, Roelands viscosity-pressure equation.*

level	$n_x \times n_y$	H_c	H_m
2	32×32	$4.0557 \cdot 10^{-2}$	$1.6921 \cdot 10^{-2}$
3	64×64	$7.0686 \cdot 10^{-2}$	$3.3080 \cdot 10^{-2}$
4	128×128	$7.8872 \cdot 10^{-2}$	$3.7120 \cdot 10^{-2}$
5	256×256	$8.0935 \cdot 10^{-2}$	$3.8480 \cdot 10^{-2}$
6	512×512	$8.1447 \cdot 10^{-2}$	$3.8760 \cdot 10^{-2}$

Table 6.15: *Central film thickness defined as H^h at $(X,Y) = (0,0)$ and minimum film thickness as a function of the gridlevel computed with a FMG algorithm with 3 W(2,1) cycles. $M = 200$, $L = 10$, compressible lubricant, Roelands viscosity-pressure equation.*

minimum and central film thickness with decreasing mesh size displays the second order discretization. Each time the mesh size is refined the change in the calculated values decreases by roughly a factor of 4. Comparing the values of the central and minimum film thickness given in Table 6.14 for load case 1 with those given in Tables 6.9 and 6.10 it can be concluded that the minimum film thickness has hardly changed whereas the central film thickness is now smaller by about 13%. The same applies to the second load case as can be seen from a comparison of the values given in Table 6.15 with those given in Tables 6.12 and 6.13. For this load case a reduction of 18% of the central film thickness is obtained. However, not only the value of the central film thickness has changed. As can be seen from a comparison of Figures 6.16 and 6.18 with the Figures 6.11 and 6.14 the shape of the film in the central region has also changed. This is more clearly observed in Figures 6.17 and 6.19.

The origin of the different film thickness shapes can be understood from the equations. As has been explained before, due to the high viscosity and small film thickness, the Reynolds equation in the central region of the contact reduces to:

$$\frac{\partial(\bar{\rho}H)}{\partial X} \approx 0 \tag{6.62}$$

with the solution for the incompressible case $H = c(Y)$. The function $c(Y)$ will be determined by the flow into the contact. In principle it will depend on the viscosity-pressure equation that is used. However, as in the inlet region the pressures are small, the

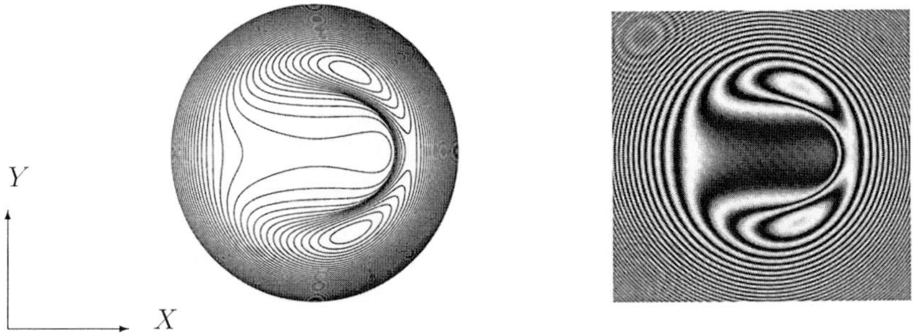

Figure 6.16: *Contour plot (left, $\Delta H = 0.01$) and pseudo-interferometry plot (right, $\Delta H = 0.05$) of the computed dimensionless film thickness H^h as a function of X and Y. $M = 20$, $L = 10$, compressible lubricant, Roelands viscosity-pressure equation.*

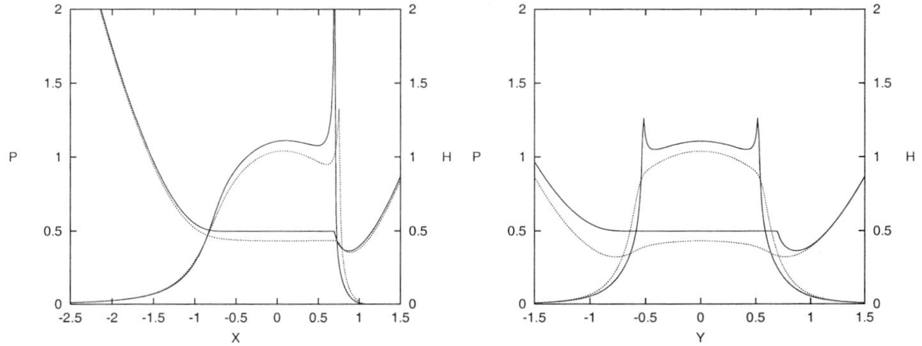

Figure 6.17: *Dimensionless pressure P and film thickness H as a function of X at the line $Y = 0$ (left) and as a function of Y at the line $X = 0$ (right) for $M = 20$, $L = 10$. Drawn line: incompressible lubricant, Barus viscosity-pressure equation. Dashed line: compressible lubricant, Roelands viscosity-pressure equation.*

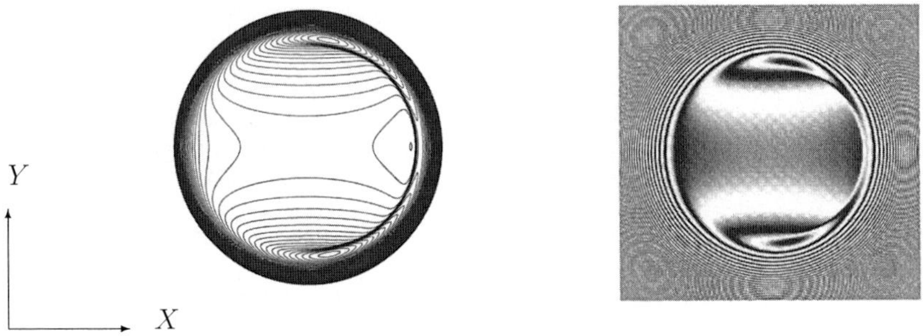

Figure 6.18: *Contour plot (left $\Delta H = 2.5 \cdot 10^{-3}$) and pseudo-interferometry plot (right $\Delta H = 2.5 \cdot 10^{-2}$) of the computed dimensionless film thickness H^h as a function of X and Y. $M = 200$, $L = 10$, compressible lubricant, Roelands viscosity-pressure equation.*

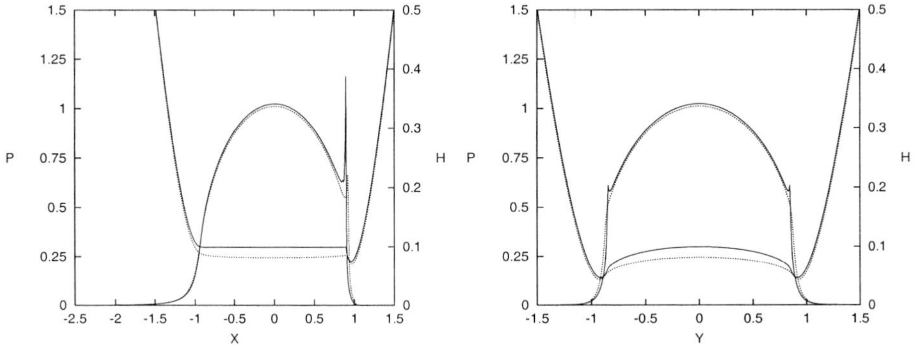

Figure 6.19: *Dimensionless pressure P and film thickness H as a function of X at the line $Y = 0$ (left) and as a function of Y at the line $X = 0$ (right) for $M = 200$, $L = 10$. Drawn line: Incompressible lubricant, Barus viscosity-pressure equation. Dashed line: compressible lubricant, Roelands viscosity-pressure equation.*

dimensionless viscosity will roughly be unity. After all, for small values of the pressure the difference in the viscosity predicted by these equations is small. It can therefore be concluded that $c(Y)$ is the same when using either the Roelands equation or the Barus equation. If $c(Y)$ is the same for both cases this implies that the level of the film thickness will be the same for both cases and thus neither the minimum nor the central film thickness will change much if the Roelands equation is used instead of the Barus equation. Consequently, the changes observed in the central film thickness, and the central region of the contact must be due to the compressibility. This is indeed the case.

For the compressible case Equation (6.62) imposes $\bar{\rho}H = \bar{c}(Y)$ with $\bar{c}(Y)$ also determined by the inlet region of the contact. However, as the pressures are low in the inlet region, compressibility effects only play a minor role as $\bar{\rho} \approx 1$. Hence $\bar{c}(Y) \approx c(Y)$. Thus the level of the product $\bar{\rho}H$ for the compressible case will roughly be the same as the level of H for the incompressible case. The consequences for the minimum and central film thickness can be derived straightforwardly. The minimum film thickness occurs close to the contact periphery $X^2 + Y^2 \approx 1$ where the pressures are small. Hence, $\bar{\rho} = 1$ and thus the minimum film thickness will roughly be the same as in the incompressible case.

For the central film thickness the situation is different. If $\bar{\rho}H$ is the same for an incompressible and compressible lubricant, H can no longer be the same in the central region. In this region the pressures are high and $\bar{\rho} > 1$. Along a line in the X direction, the relation $\bar{\rho}H = c(Y)$ dictates that the product of the density and the film thickness remains constant. Consequently, changes in the density have to be compensated by changes in the film thickness. With a semi-elliptical pressure variation along this line, this implies that the film thickness profile will no longer be a constant, but will display a curved line with a local minimum at $X = 0$. This behaviour will be similar for every line in the X direction, however, each line will have a different constant. Moreover, the density increase along each of these lines is different, as the pressure decreases going towards the periphery of the contact. These changes in the film thickness imply that the contour plot will no longer show straight lines in the center of the contact. This is illustrated by the Figures 6.16 and 6.18 which show the contour plots and pseudo-interferometry graphs of the film thickness for the solutions obtained assuming a compressible lubricant and using the Roelands pressure viscosity equation.

It should be noted that the change of the central film thickness due to the compressibility of the lubricant can be predicted quite accurately. Because the conditions in the inlet of the contact will be the same for an incompressible and compressible lubricant the function $c(Y)$ will also be the same. As a first order approximation one can use:

$$H_c^c = \left(1 - \frac{1}{\bar{\rho}(P_c)}\right) H_c^i \tag{6.63}$$

if H_c^c is the central film thickness obtained for the compressible case, H_c^i its value for the incompressible case, and P_c the pressure in the centre of the contact. For the present cases $P_c \approx 1$ which gives $\bar{\rho} = 1.15$ for load case 1 ($p_h = 0.45$ $[GPa]$) and $\bar{\rho} = 1.21$ for load case 2 ($p_h = 0.97$ $[GPa]$). Consequently, the central film thickness obtained for the compressible case should be 15% lower than for the incompressible case, for load case 1. For load case 2 it should be about 20% lower. These values are indeed close to the variations obtained numerically. This approach can be used to predict the effect of compressibility on the film thickness for any density pressure equation, e.g. see [103] where results obtained using the density pressure equation proposed by Jacobson and Vinet [62] are presented.

Finally, the pressure distribution changes too. For load case 1 the compressibility has a significant effect on the pressure profile. This results in lower pressures in the central region and in particular in a much lower pressure spike. However, as the compressibility predicted by the Dowson and Higginson equation is limited to 34%, the differences in the pressure profile decrease with increasing load. For load case 2, the only visible effect is the decrease of the pressure spike which becomes less pronounced.

6.12.2 Time Dependent Problems

The EHL problem is generally a transient problem. Only in the idealized situation of perfectly smooth surfaces and stationary operating conditions does the problem reduce to a steady state problem. If the effect of surface roughness or micro geometry are to be studied, the problem becomes inherently time dependent. Thus the pressure and film thickness have to be solved as a function of time with the micro geometry moving through the contact. The variable time adds another dimension to the problem and augments the time required for a numerical solution. The stationary solver presented in this chapter can be extended to include transient effects. For the time dependent problem the Reynolds equation is given by:

$$\frac{\partial}{\partial x}\left(\frac{\rho h^3}{12\eta}\frac{\partial p}{\partial x}\right) + \frac{\partial}{\partial y}\left(\frac{\rho h^3}{12\eta}\frac{\partial p}{\partial y}\right) - u_m\frac{\partial(\rho h)}{\partial x} - \frac{\partial(\rho h)}{\partial t} = 0 \tag{6.64}$$

with $p = 0$ on the boundary and the cavitation condition $p(x, y, t) > 0$. The film thickness equation for the general case including surface roughness can be written as:

$$h(x,y) = h_0(t) + \frac{x^2}{2R_x} + \frac{y^2}{2R_y} - r(x,y,t) + \frac{2}{\pi E'}\int_{-\infty}^{+\infty}\int_{-\infty}^{+\infty}\frac{p(x',y')\,dx'\,dy'}{\sqrt{(x-x')^2+(y-y')^2}} \tag{6.65}$$

where:

$$r(x, y, t) = r_1(x - u_1 t, y) + r_2(x - u_2 t, y) \tag{6.66}$$

stands for the undeformed roughness of the two surfaces, and u_1 and u_2 are the velocities of the two surfaces. The integration funtion $h_0(t)$ is in general determined by an equation of motion, see [110]. If the acceleration forces are neglected and a constant load is assumed, this equation reduces to the usual condition of force balance. When both load, speed and geometry vary as a function of time the same approach can be used, see Messé [82].

$$\mathrm{w}(t) = \int_{-\infty}^{+\infty}\int_{-\infty}^{+\infty} p(x', y', t)\,dx'\,dy' \tag{6.67}$$

Using the dimensionless variables defined by (6.10) and defining the dimensionless time as:

$$T = tu_m/a \tag{6.68}$$

results in the following dimensionless Reynolds equation:

$$\frac{\partial}{\partial X}\left(\xi\frac{\partial P}{\partial X}\right) + \frac{\partial}{\partial Y}\left(\xi\frac{\partial P}{\partial Y}\right) - \frac{\partial(\bar{\rho}H)}{\partial X} - \frac{\partial(\bar{\rho}H)}{\partial T} = 0 \tag{6.69}$$

where:

$$\xi = \frac{\bar{\rho}H^3}{\bar{\eta}\bar{\lambda}} \quad \text{with} \quad \bar{\lambda} = \frac{12u_m\eta_0R_x^2}{a^3p_h}$$

The boundary conditions are $P(X_a, Y) = P(X_b, Y) = P(X, -Y_a) = P(X, Y_a) = 0$. The cavitation condition imposes $P(X, Y, T) \geq 0$. The dimensionless film thickness equation is given by:

$$H(X,Y,T) = H_0(T) + \frac{X^2}{2} + \frac{Y^2}{2} - \mathcal{R}(X,Y,T) + \frac{2}{\pi^2}\int_{-\infty}^{+\infty}\int_{-\infty}^{+\infty}\frac{P(X',Y',T)\,dX'dY'}{\sqrt{(X-X')^2+(Y-Y')^2}} \tag{6.70}$$

with:

$$\mathcal{R}(X,Y,T) = \mathcal{R}_1(X - u_1/u_mT, Y) + \mathcal{R}_2(X - u_2/u_mT, Y) \tag{6.71}$$

and H_0 is determined by the dimensionless force balance condition:

$$\int_{-\infty}^{+\infty}\int_{-\infty}^{+\infty} P(X,Y,T)\,dX\,dY = \frac{2\pi}{3} \tag{6.72}$$

The introduction of the roughness moving through the contact adds one additional parameter to the problem, i.e. only the parameter u_1/u_m needs to be specified. If u_1/u_m is given then by definition u_2/u_m follows from $u_1 + u_2 = 2u_m$.

The objective is now to solve $P(X, Y, T)$ and $H(X, Y, T)$ as a function of time. In the same way as was done for the steady state problem, the nature of the transient problem can be analysed by looking at the limit for very high viscosities. For the transient problem one obtains:

$$-\frac{\partial(\bar{\rho}H)}{\partial X} - \frac{\partial(\bar{\rho}H)}{\partial T} = 0 \tag{6.73}$$

whilst for the steady state problem $\bar{\rho}H = c(Y)$ is obtained as explained before. For the case of surface roughness the stationary condition will imply that inside the high viscosity region all roughness will be flattened. For the transient problem this equation imposes $\bar{\rho}H = \bar{\rho}H(X - T)$. Thus, the product $\bar{\rho}H$ will tend to be a function of $X - T$ in the high viscosity region. This is a natural consequence of demanding mass conservation in a flow between flexible walls where the viscous terms can be neglected. Note that this will happen regardless of the dimensionless surface velocities u_1/u_m and u_2/u_m. This implies that any change in the film thickness induced by roughness upon entering a high viscosity region will be propagated through the high viscosity region at the dimensionless speed of unity. In physical terms this is the average velocity of the surfaces. This behaviour leads to very interesting and sometimes confusing aspects in transient solutions of the problem which at any time will consist of a combination of components. Only for pure rolling the situation reduces to a simple situation as in this case the velocity of the surfaces equals the average velocity. For more details regarding the transient aspects of the problem the reader is referred to publications such as [76, 77], [99]-[112] and the references therein.

The problem can be discretized in exactly the same way as the steady state problem. In addition to the grid in space a 'mesh' in time can be assumed with mesh size (timestep) h_T. Subsequently, the discrete Reynolds equation for each timestep can be derived as:

$$\frac{\xi^h_{i-1/2,j,k}P^h_{i-1,j,k} - (\xi^h_{i-1/2,j,k} + \xi^h_{i+1/2,j,k})P^h_{i,j,k} + \xi^h_{i+1/2,j,k}P^h_{i+1,j,k}}{h^2} +$$
$$\frac{\xi^h_{i,j-1/2,k}P^h_{i,j-1,k} - (\xi^h_{i,j-1/2,k} + \xi^h_{i,j+1/2,k})P^h_{i,j,k} + \xi^h_{i,j+1/2,k}P^h_{i,j+1,k}}{h^2} -$$
$$(\bar{\rho}H)^h_x - (\bar{\rho}H)^h_T \;=\; 0 \qquad (6.74)$$

with the coefficients defined by:

$$\begin{aligned} \xi^h_{i\pm1/2,j,k} &= (\xi^h_{i,j,k} + \xi^h_{i\pm1,j,k})/2 \\ \xi^h_{i,j\pm1/2,k} &= (\xi^h_{i,j,k} + \xi^h_{i,j\pm1,k})/2 \end{aligned} \qquad (6.75)$$

The discrete 'wedge' term can be taken as:

$$(\bar{\rho}H)^h_x \doteq \frac{1.5\bar{\rho}^h_{i,j,k}H^h_{i,j,k} - 2\bar{\rho}^h_{i-1,j,k}H^h_{i-1,j,k} + 0.5\bar{\rho}^h_{i-2,j,k}H^h_{i-2,j,k}}{h} \qquad (6.76)$$

and the discrete 'squeeze' term as:

$$(\bar{\rho}H)^h_T \doteq \frac{1.5\bar{\rho}^h_{i,j,k}H^h_{i,j,k} - 2\bar{\rho}^h_{i,j,k-1}H^h_{i,j,k-1} + 0.5\bar{\rho}^h_{i,j,k-2}H^h_{i,j,k-2}}{h_T} \qquad (6.77)$$

At this point it is noted that an accurate approximation of these terms is very important. The wedge and squeeze term together form a so-called *advection* operator, i.e. they describe a propagation mechanism in the solution along a characteristic $X = T$. If the discretization error is too large this will lead to artificial amplitude decay effects in the solution. One should at least choose a discrete approximation to the combined wedge and squeeze term such that it has a zero discretization error for the characteristic direction. If the mesh size and timestep are chosen equal the standard second order upstream discretization given above satisfies this requirement. A more advanced option is to use a combined discretization of the wedge and squeeze term as is common for advective terms in Computational Fluid Dynamics, see [107].

Having obtained the discrete system of equations the next step is to solve the equations. For this purpose the multigrid solver as presented in this chapter can be used. The extension involves incorporation of the discrete wedge term in the discrete Reynolds equation and in the definition of the systems of equations to be solved for each line in the line relaxation. Subsequently, the coarse grid correction cycle can be used for each timestep, using the solution of a previous timestep as a first approximation. Alternatively the more advanced F-cycle explained at the end of Chapter 4 can be used.

6.12.3 Starved Lubrication

In the previous chapters on lubrication it was always assumed that pressure is generated whenever the gap between the two bodies narrows in the direction of motion. This supposes that the gap between the two bodies is completely filled with oil, and thus that sufficient amounts of oil are present in the inlet. Whenever the quantity of oil is insufficient, the generated oil film thickness will be smaller than predicted by the fully flooded theory. In order to correctly describe this problem the Reynolds equation (6.1) has to be expanded to include the parameter θ which describes the ratio of the oil film thickness to the gap height. As a consequence θ is dimensionless, and $0 \leq \theta \leq 1$. An extended Reynolds equation was proposed by Elrod et al. [40, 41] describing the flow in the complete and incomplete zone:

$$\frac{\partial}{\partial x}\left(\frac{\rho h^3}{12\eta}\frac{\partial p}{\partial x}\right) + \frac{\partial}{\partial y}\left(\frac{\rho h^3}{12\eta}\frac{\partial p}{\partial y}\right) - u_m\frac{\partial(\rho\theta h)}{\partial x} = 0 \tag{6.78}$$

Please note that this equation reduces to the classical equation in the complete zone $p \geq 0$ and $\theta = 1$. In the incomplete zone no pressure is generated: $p = 0$ and $0 \leq \theta \leq 1$. This complete equation is valid for $x_a < x < x_b$ and $-y_a < y < y_a$ with the boundary conditions $p(x_a, y) = p(x_b, y) = p(x, -y_a) = p(x, y_a) = 0$. These boundary conditions have to be extended to include a boundary condition for θ: $\theta(x_a, y) = \theta_0(y)$. When the amount of oil available in the inlet is reduced, the pressure build-up will be delayed, and the generated film thickness will be reduced. Chevalier et al. [31, 32] showed that it is advantageous to express the solution in terms of the amount of lubricant present on the surfaces in front of the contact, H_{oil}. Furthermore, the starved film thickness was expressed as a ratio, by dividing it by the fully flooded film thickness H_{cff}.

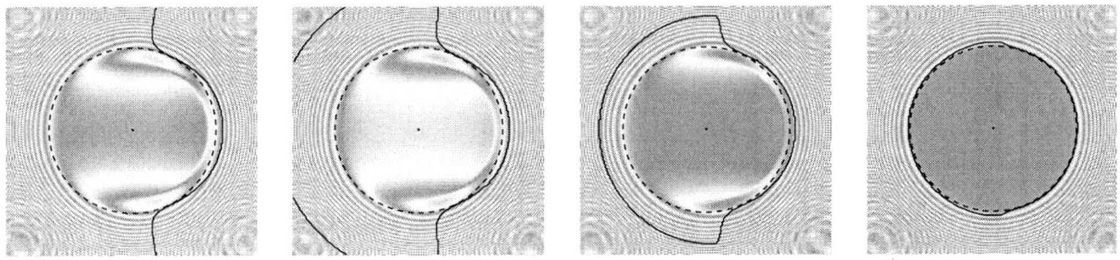

H_{oil}/H_{cff}=4 H_{oil}/H_{cff}=2 H_{oil}/H_{cff}=1 H_{oil}/H_{cff}=0.2

Figure 6.20: *Film thickness graphs using pseudo interferometric techniques showing the influence of the inlet oil film H_{oil}, $M = 100$, $L = 10$.*

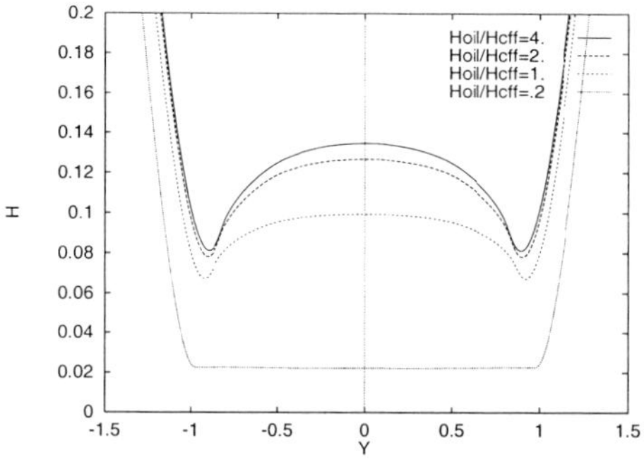

Figure 6.21: *Film thickness graphs $H(X=0,Y)$ showing the influence of the inlet oil film H_{oil},* $M=100$, $L=10$.

Figure 6.20 shows the film thickness evolution as a function of the inlet parameter H_{oil}. The drawn line represents the inlet and outlet meniscus. When the amount of available lubricant is reduced, the film thickness diminishes, but also the film thickness variation reduces. The film thickness profile becomes more and more Hertzian. This can be observed even better in Figure 6.21 which shows a cross section of the film thickness distribution for $x=0$.

In order to reduce the complexity, one can study the evolution of the central film thickness H_c. The central film thickness depends on three parameters: M, L and H_{oil}, the amount of oil present on the surfaces. However, it was found that the ratio H_c/H_{cff} is virtually independent of M and L. Consequently, H_c/H_{cff} can be expressed as a function of H_{oil}/H_{cff} only.

Figure 6.22 shows this relation for various operating conditions, indeed showing only a small dependence on M and L. The following equation accurately describes the film thickness ratio $\mathcal{R}=H_c/H_{cff}$, as a function of the dimensionless amount of oil available on the surfaces $r=H_{oil}/H_{cff}$:

$$\mathcal{R}=\frac{r}{\sqrt[\gamma]{1+r^{\gamma}}} \tag{6.79}$$

The value of γ for EHL contacts lies between 2 and 5. In this equation and in Figure 6.22, two asymptotes can be observed: $r\rightarrow 0$ and $r\rightarrow\infty$ which give $\mathcal{R}\rightarrow r$ and $\mathcal{R}\rightarrow 1$ respectively. Of these two, the asymptote for thin films is the most interesting, as it shows that for very thin oil films, the EHL film thickness is the same as the film in front of the contact. In other words: the contact becomes very efficient in building up an oil film when very little oil is available in the inlet. Thus for $r\rightarrow 0$ and $H_c\rightarrow H_{oil}$ or in dimensional terms: when $h_{oil}/h_{cff}\rightarrow 0$ then $h_c\rightarrow h_{oil}$.

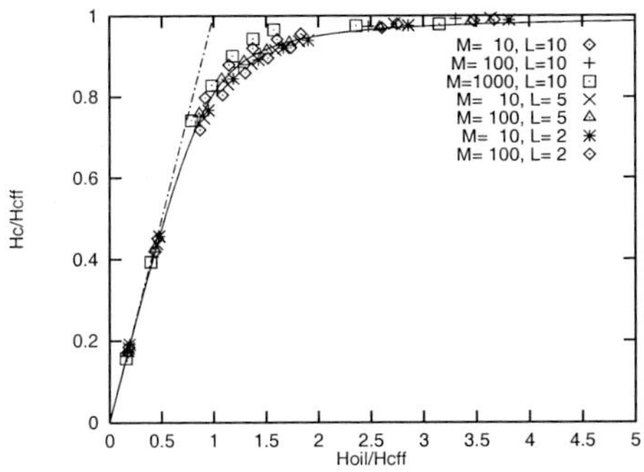

Figure 6.22: *Numerical calculations of the central film thickness* H_c *as a function of* H_{oil} *for various sets of operating conditions, and a circular contact.*

Bibliography

[1] **Alcouffe, R.E., Brandt, A., Dendy, J.E. and Painter, J.W.,** 1981, "The Multigrid Method for the Diffusion Equation with Strongly Discontinuous Coefficients," *SIAM J. of Sci. Stat. Comp.*, **2**, pp. 430-454.

[2] **Alpert, B., Beylkin, G., Coifman, R. and Rohklin, V.,** 1993, "Wavelet-like Bases for fast Solution of Second-Kind Integral Equations," *SIAM J. of Sci. Comp.,* **14**, pp. 159-184.

[3] **Appel, A.W.,** 1985, "An Efficient Program for Many-Body Simulation," *SIAM J. of Sci. Statist. Comp.,* **6**, pp. 85-103.

[4] **Bai, D. and Brandt, A.,** 1987, "Local Mesh Refinement MultiLevel Techniques," *SIAM J. of Sci. Stat. Comp.*, **8**, pp. 109-134.

[5] **Barus, C.,** 1893, "Isothermals, Isopiestics and Isometrics relative to Viscosity," *Am. J. of Science*, **45**, 87-96.

[6] **Barnes, J. and Hut, P.,** 1986, "A Hierarchical $O(n \ln n)$ Force Calculation Algorithm," *Nature,* **324**, pp. 446-449.

[7] **Beylkin, G., Coifman, R. and Rokhlin, V.,** 1991, "Fast Wavelet Transforms and Numerical Algorithms," *Comm. Pure Appl. Math.,* **44**, pp. 141-183.

[8] **Blok, H.**, 1957, discussion on paper 71 by Burke and Neal, pp. 118, *Proc. Conf. Lubr. and Wear, IMechE*, London, pp. 745-747.

[9] **Bos, J. and Moes, H.,** 1994, "Frictional Heating of Elliptic Contacts," Proceedings of the 20th Leeds-Lyon Symposium on Tribology, *Elseviers Tribology Series, Ed. D. Dowson et al.,* pp. 491-500.

[10] **Bos, J.,** 1995, "Frictional Heating of Tribological Contacts," *PhD Thesis, University of Twente, Enschede, The Netherlands,* ISBN 90-9008920-9.

[11] **Bos, J. and Moes, H.,** 1995, "Frictional Heating of Tribological Contacts," *ASME J. of Tribology,* **117**, pp. 171-177.

[12] **Brandt, A.,** 1977, "Multi-Level Adaptive Technique (MLAT) for fast Numerical Solutions to Boundary Value Problems," *Proc. 3rd Int. Conf. Numerical methods in Fluid Mechanics (Paris, 1972); Lecture Notes in Physics*, **18**, Springer-Verlag, Berlin, pp. 82-89.

[13] **Brandt, A.**, 1977, "Multi-Level Adaptive Solutons to Boundary Value Problems," *Mathematics of Computation*, **31**, pp. 333-390.

[14] **Brandt, A. and Cryer, C.W.**, 1983, "Multigrid Algorithms for the Solution of Linear Complementarity Problems arising from Free Boundary Problems," *SIAM J. of Sci. Stat. Comp.*, **4**, 4, pp. 655-684.

[15] **Brandt, A.**, 1984, "MultiGrid Techniques: 1984 Guide with Applications to Fluid Dynamics," GMD Bonn, ISBN 3884570811.

[16] **Brandt, A.**, 1987, "Multi-Level Approaches to Large Scale Problems," *Proc. of ICM86, Berkeley California*, pp. 1319-1334.

[17] **Brandt, A.**, 1990, "Rigorous Local Mode Analysis of Multigrid," *in Prelim. Proc. 4th Copper Mountain Conference on MultiGrid Methods*, Copper Mountain, Colorado.

[18] **Brandt, A. and Lubrecht, A.A.**, 1990, "Multilevel Matrix Multiplication and Fast Solution of Integral Equations", *J. of Comp. Phys.*, **90**, 2, pp. 348-370.

[19] **Brandt, A.**, 1991, "Multilevel Computations of Integral Transforms and Particle Interactions with Oscillatory Kernels," *Computer Physiscs Communications,* **65**, pp. 24-38.

[20] **Brandt, A. and Greenwald, J.** 1991, "Parabolic Multigrid Revisited," *International Series of Numerical Mathematics,* **98**, Birkhauser Verlag, Basel, pp. 143-153.

[21] **Brandt, A. and Yavneh, I.**, 1992, "On Multigrid Solution of High-Reynolds Incompressible Entering Flows," *J. of Comp. Phys.*, **101**, 1, pp. 151-164.

[22] **Brandt, A. and Livshits, I.**, 1997, "Wave-Ray Multigrid Method for Standing Wave Equations," *Electronic Transactions on Numerical Analysis*, **6**, pp. 162-181.

[23] **Brandt, A. and Venner, C.H.**, 1998, "Multilevel Evaluation of Integral Transforms with Asymptotically Smooth Kernels," *SIAM J. of Sci. Comp.*, **19**, 2, pp. 468-492.

[24] **Brandt, A. and Venner, C.H.**, 1999, "Fast Evaluation of Integral Transforms on Adaptive Grids," *Multigrid Methods V: Lecture Notes in Computational Science and Engineering* **3**, Ed. W. Hackbusch and G. Wittum, Springer Verlag, pp. 20-44.

[25] **Briggs, W.L.**, 1987, "A Multigrid Tutorial," *SIAM*, ISBN 0-89871-221-1.

[26] **Brigham, E.O.**, 1974, "The Fast Fourier Transform," Englewood Cliffs, NJ, Prentice Hall.

[27] **Cameron, A. and Wood, W.L.**, 1949, "The Full Journal Bearing," *Proc. IMechE*, **26**, pp. 59-64.

[28] **Cameron, A.,** 1966, "The Principles of Lubrication," Longmans Green and co, ltd, London, UK.

[29] **Cann, P.M., Spikes, H.A. and Hutchinson, J.,** 1996, "The Development of a Spacer Layer Imaging Method for Mapping Elastohydrodynamic Contacts," *STLE Tribology Transactions*, **36**, 4, pp. 915-921.

[30] **Carrier, J., Greengard, L. and Rokhlin, V.,** 1988, "A fast Adaptive Multipole Algorithm for Particle Simulations," *SIAM J. of Sci. Stat. Comp.*, **9**, pp. 669-686.

[31] **Chevalier, F.,** 1996, "Modélisation des Conditions d'Alimentation en Lubrifiant dans les Contacts Elastohydrodynamiques Ponctuels," PhD thesis, INSA de Lyon, France.

[32] **Chevalier, F., Lubrecht A.A., Cann, P.M.E., Colin, F. and Dalmaz, G.,** 1998, "Film Thickness in Starved EHL Point Contacts," *ASME J. of. Tribology.*, **120**, pp. 126-133.

[33] **Dahlquist, G. and Björck, A.,** 1974, "Numerical Methods," *Prentice-Hall Series in Automatic Computation*, Ed. G. Forsythe, Prentice-Hall, Inc. Englewood Cliffs, New Jersey, US, ISBN 0-13-627315-7.

[34] **Dowson, D. and Higginson, G.R.,** 1959, "A Numerical Solution to the Elastohydrodynamic Problem," *J. of Mech. Eng. Sci.*, **1**, pp. 6.

[35] **Dowson, D. and Higginson, G.R.,** 1966, "Elastohydrodynamic Lubrication, The Fundamentals of Roller and Gear Lubrication," Pergamon Press, Oxford, Great Britain.

[36] **Dumont, M.-L.,** 1997, "Etude des Endommagements de Surface induits par Fatigue de Roulement dans les Contacts Elastohydrodynamiques pour des Aciers M50 et 100Cr6," PhD thesis, INSA de Lyon, France.

[37] **Dowson, D.,** 1998, "History of Tribology," 2nd Edition, Professional Engineering Publishing Ltd, London, Burry St Edmunds, UK, ISBN 1-86058-070-X.

[38] **Ehret, P., Dowson, D. and Taylor C.M.,** 1998, "On Lubricant Transport Conditions in ElastoHydroDynamic Conjunctions," *Proc. Roy. Soc. London*, pp. 763-787.

[39] **Ehret, P., Dowson, D. and Taylor C.M.,** 1999, "Transient EHL Solutions with Interfacial Slip," *ASME J. of Tribology*, **121**, pp. 703-710.

[40] **Elrod, H.G. and Adams, M.L.,** 1974, "A Computer Program for Cavitation and Starvation Problems," Proceedings of the 1st Leeds-Lyon Symposium on Tribology, pp. 37-41.

[41] **Elrod, H.G.,** 1981, "A Cavitation Algorithm," *ASME J. of Lub. Tech*, **103**, pp. 350-354.

[42] **Ertel, A.M.,** 1939, "Hydrodynamic Lubrication based on New Principles," *Akad. Nauk SSSR Prikadnaya Mathematica i Mekhanika,* 3, 2, pp. 41-52.

[43] **Fedorenko, R.P.,** 1964, "On the Speed of Convergence of an Iteration Process," *Ž. Vyčisl. Mat. i Mat. Fiz.*, **4**, pp. 559-564.

[44] **Foord, C.A., Wedeven, L.D., Westlake, F.J. and Cameron, A.,** 1969, "Optical Elastohydrodynamics," *Proc. ImechE,* **184**, 1, 28, pp. 487-505.

[45] **Gohar, R.,** 1988, "Elastohydrodynamics," *Ellis Horwood series in Mechanical Engineering,* Ed. J.M. Alexander, Ellis Horwood Limited, Chichester, UK, ISBN 0-85312-820-0

[46] **Golub, G.H. and van Loan, C.F.,** 1989, "Matrix Computations," second edition, The John Hopkins University Press, Baltimore London.

[47] **Greengard, L. and Rokhlin, V.,** 1987, "A fast Algorithm for Particle Simulations," *J. of Comput. Phys.*, **73**, pp. 325-348.

[48] **Greengard, L.,** 1994, "Fast Algorithms for Classical Physics," *Science,* **265**, pp. 909-914.

[49] **Greenwood, J.A. and Johnson, K.L.,** 1992, "The Behaviour of Transverse Roughness in Sliding Elastohydrodynamically Lubricated Contacts," *WEAR*, **153**, pp. 107-117.

[50] **Grubin, A.N.,** 1949, "Fundamentals of the Hydrodynamic Theory of Lubrication of Heavily Loaded Cylindrical Surfaces," Central Scientific Research Institute for Technology and Mechanical Engineering, Book no 30, Moscow, D.S.I.R. translations pp. 115-166.

[51] **Gümbel, L.,** 1916, "Über Geschmierte Arbeitsräder," *Z. ges. Turbinenwesen*, **13**, p. 357.

[52] **Hackbusch, W.,** 1984, "Local defect Correction Method and Domain Decomposition Techniques," in: Defect Correction Methods: Theory and Applications, K. Böhmer and H.J. Stetter, eds. Computations Supplementation, **5**, Springer Verlag, Wien pp. 89-113.

[53] **Hackbusch, W.,** 1992, "Elliptic Differential Equations, Theory and Numerical Treatment," Springer Series in Computational Mathematics, **18**, Springer Verlag Berlin, ISBN 3-540-54822-X.

[54] **Hackbusch, W.,** 1994, "Iterative Solution of Large Sparse Systems of Equations," Springer Verlag, ISBN 0-387-94064-2.

[55] **Hamrock, B.J. and Dowson D.,** 1976, "Isothermal Elastohydrodynamic Lubrication of Point Contacts, part I, Theoretical Formulation," *ASME J. of Lub. Tech.*, **98**, pp. 223-229.

[56] **Hamrock, B.J. and Dowson D.**, 1977, "Isothermal Elastohydrodynamic Lubrication of Point Contacts, part III, Fully Flooded Results," *ASME J. of Lub. Tech.*, **99**, pp. 264-276.

[57] **Hamrock, B.J. and Brewe D.**, 1983, "Simplified Solution for Stresses and Deformations," *ASME J. of Lub. Tech.*, **105**, pp. 171-177.

[58] **Hemker, P.W.**, 1980, "On the Structure of an Adaptive Multi-Level Algorithm," *BIT*, **20**, pp. 289-301.

[59] **Hemker, P.W.**, 1990, "On the Order of Prolongations and Restrictions in Multigrid Procedures," *J. Comp. and Applied Math.*, **32**, pp. 423-429.

[60] **Hertz, H.**, 1881, "Über die Berührung Fester Elastischer Körper," *J. reine und angew. Math.*, **92**, pp. 156-171.

[61] **Ioannides, E. and Harris, T.A.**, 1985, "A New Fatigue Life Model for Rolling Bearings," *ASME J. Lub. Tech.*, **107**, pp. 367-378.

[62] **Jacobson, B.O. and Vinet, P.**, 1987, "A Model for the Influence of Pressure on the Bulk Modulus and the Influence of the Temperature on the Solidification Pressure of Liquid Lubricants," *ASME J. of Tribology*, **109**, pp. 709-713.

[63] **Jacobson, B.O.**, 1991, "Rheology and ElastoHydrodynamic Lubrication," *Elseviers Tribology Series, Ed. D. Dowson*, ISBN 0 444 88146 8.

[64] **Kalker, J.J.**, 1986, "Numerical Calculation of the Elastic Field in a Half-Space," *Comm. Appl. Num. Methods*, **2**, pp. 401-410.

[65] **Kalker, J.J.**, 1990, "Three-Dimensional Elastic Bodies in Rolling Contact," Kluwer Academic Publishers, ISBN 0-7923-0712-7.

[66] **Johnson, K.L.**, 1985, "Contact Mechanics," Cambridge University Press, ISBN 0 521 34796.

[67] **Kaneta, M.**, 1992, "Effects of Surface Roughness in Elastohydrodynamic Lubrication," *JSME*, III, **35**, 4, pp. 535-546.

[68] **Kaneta, M., Sakai, T. and Nishikawa, H.**, 1992, "Optical Interferometric Observations of the Effects of a Bump on Point Contact EHL," *ASME J. of Tribology*, **114**, pp. 779-784.

[69] **Kevorkian, J.**, 1993, "Partial Differential Equations: Analytical Solution Techniques," Chapman & Hall Mathematics Series, Chapman & Hall, ISBN 0-412-05131-1.

[70] **Kernighan, B.W. and Ritchie D.M.**, 1988, "The C Programming Language," Second edition, Prentice Hall, ISBN 0-13-110362-8.

[71] **Lubrecht, A.A.,** 1987, "Numerical Solution of the EHL Line and Point Contact Problem Using Multigrid Techniques," Ph.D. Thesis, University of Twente, Enschede, The Netherlands, ISBN 90-9001583-3.

[72] **Lubrecht, A.A., Breukink, G.A.C., Moes, H., ten Napel, W.E. and Bosma, R.,** 1987, "Solving Reynolds' Equation for EHL Line Contacts by Application of a Multigrid Method," Proc. 13th Leeds Lyon Symposium on Tribology, *Elsevier Tribology Series*, **11**, *Ed. D. Dowson et al.*, pp. 175-182.

[73] **Lubrecht, A.A., ten Napel, W.E. and Bosma, R.,** 1987, "Multigrid, an Alternative Method of Solution for Two-Dimensional Elastohydrodynamically Lubricated Point Contact Calculations," *ASME J. of Tribology*, **109**, pp. 437-443.

[74] **Lubrecht, A.A. and Ioannides, E.,** 1991, "A Fast Solution to the Dry Contact Problem and the Associated Sub-surface Stress Field, Using Multilevel Techniques," *ASME J. of Tribology*, **113**, pp. 128-133.

[75] **Lubrecht, A.A.,** 1997, "Influence of Local and Global Features in EHL Contacts," Proceedings of the 23rd Leeds-Lyon Symposium on Tribology, *Elseviers Tribology Series,* **32**, pp. 17-25.

[76] **Lubrecht, A.A., Graille, D., Venner C.H. and Greenwood, J.A.,** 1998, "Waviness Deformation in EHL Line Contacts, under Rolling/Sliding," *ASME J. of Tribology*, **120**, pp. 705-709.

[77] **Lubrecht, A.A. and Venner, C.H.,** 1999, "Elastohydrodynamic Lubrication of Rough Surfaces," *Proc. ImechE, part J, J. of Eng. Tribology*, **213**, J5, pp. 397-404.

[78] **Lundberg, G. and Palmgren A.,** 1947, "Dynamic Capacity of Rolling Bearings," *Acta Polytech. Mech. Eng.*, Ser. 1. R.S.A.E.E. No. 3.

[79] **Lundberg, G. and Palmgren A.,** 1952, "Dynamic Capacity of Rolling Bearings," *Acta Polytech. Mech. Eng.*, Ser. 2. R.S.A.E.E. No. 4.

[80] **Martin, H.M.,** 1916, "Lubrication of Gear Teeth," *Engineering*, **102**, pp. 119-121.

[81] **McCormick, S.F.,** 1984 "Fast Adaptive Composite Grid (FAC) Methods: Theory for the Variational Case," in: Defect Correction Methods: Theory and Applications, K. Böhmer and H.J. Stetter, eds. Computations Supplementation, **5**, Springer Verlag, Wien pp. 89-113.

[82] **Messé, S. and Lubrecht A.A.,** 2000, "Transient Elastohydrodynamic Analysis of an OHC Cam/Tappet Contact," submitted to *IMechE J. of Eng. Tribology.*

[83] **Moes, H.,** 1969, Discussion on a contribution by K. Jakobsen and H. Christensen, Tribology Convention, Gothenburg, *Proc. IMechE.*, **183**, pp. 205-206.

[84] **Moes H. and Bosma, R.,** 1971, "Design Charts for Optimum Bearing Configuration, I The Full Journal Bearing," *ASME J. of Tribology*, **93**, pp. 302-306.

[85] **Moes, H. and Bosma, R.,** 1973 "Film Thickness and Traction in EHL at Point Contact," *Proceedings 1972 Symposium on Elastohydrodynamic Lubrication, IMechE,* pp. 149-152.

[86] **Moes, H.,** 1992, "Optimum Similarity Analysis, an Application to Elastohydrodynamic lubrication," *WEAR*, V 159, pp. 57-66.

[87] **Nijenbanning, G., Venner, C.H., and Moes, H.**, 1994, "Film Thickness in Elastohydrodynamically Lubricated Elliptic Contacts," *WEAR*, **176**, pp. 217-229.

[88] **Nowak, Z.P. and Hackbusch, W.,** 1986, "On the Complexity of the Panel Method," *International Conference on Modern Problems in Numerical Analysis*, Moscow, 1986.

[89] **Press, W.H., Teukolsky, S.A., Vetterling, W.T. and Flannery, B.,** 1995, "Numerical Recipes in C: the Art of Scientific Computing," Cambridge University Press, Cambridge, UK, ISBN 0-521-43108-5.

[90] **Pertusevich, A.I.,** 1951, "Fundamental Conclusions from the Contact Hydrodynamic Theory of Lubrication," *Izv. Akad. Nauk SSR (OTN),* **3**, pp. 209-223.

[91] **Reichel, L.,** 1986, "A fast Method for Solving certain Integral Equations of the First Kind with Application to Conformal Mapping," *J. of Comp. Appl. Math.*, **14**, pp. 125-142.

[92] **Reynolds, O.,** 1886, "On the Theory of Lubrication and its Application to Mr. Beauchamps Tower's Experiments, including an Experimental Determination of the Viscosity of Olive Oil," *Phil. Trans.*, **177**, pp. 157-234.

[93] **Roelands, C.J.A.,** 1966, "Correlational Aspects of the Viscosity-Temperature-Pressure Relationship of Lubricating Oils," Ph.D. Thesis, Technical University Delft, Delft, The Netherlands, (V.R.B., Groningen, The Netherlands).

[94] **Rokhlin, V.,** 1983, "Rapid Solution of Integral Equations of Classical Potential Theory," *J. of Comp. Phys.*, **60**, pp. 187-207.

[95] **Sassenfeld, H. and Walther, A.,** 1954, "Gleitlagerberechnungen," VDI-Forschungsheft 441, Deutscher Ingenieur-Verlag GMBH, Dušseldorf Germany.

[96] **Sommerfeld, A.,** 1904, "Zur Hydrodynamischen Theory der Schmiermittelreibung," *Zeitschrift für Mathematic und Physik*, **50**, pp. 97-155.

[97] **van der Stegen, R.H.M.,** 1997, "Numerical Modelling of Self-Acting Gas Lubricated Bearings with Experimental Verification," Ph.D. Thesis, University of Twente, Enschede, The Netherlands. ISBN 90-3651033-3.

[98] **Stuben, K. and Trottenberg, U.,** 1984, "Multigrid Methods: Fundamental Algorithms, Model Problems Analysis, and Applications," GMD Studien 96, GMD, Skt. Augustin, Germany.

[99] **Venner, C.H.,** 1991, "Multilevel Solution of the EHL Line and Point Contact Problems," Ph.D. Thesis, University of Twente, Enschede, The Netherlands. ISBN 90-9003974-0.

[100] **Venner, C.H. and Lubrecht, A.A.,** 1994, "Transient Analysis of Surface Features in an EHL Line Contact in the case of Sliding," *ASME J. of Tribology*, **116**, pp. 186-193.

[101] **Venner, C.H. and Lubrecht, A.A.,** 1994, "Numerical Simulation of a Transverse Ridge in a Circular EHL Contact, under Rolling/Sliding," *ASME J. of Tribology*, **116**, pp. 751-761.

[102] **Venner, C.H. and Lubrecht, A.A.,** 1995, "Numerical Simulation of Waviness in a Circular EHL Contact, under Rolling/Sliding," Proc. 22nd Leeds-Lyon Symposium on Tribology, *Elsevier Tribology Series*, **30**, *Ed. D. Dowson et al.*, pp. 259-272.

[103] **Venner, C.H., and Bos, J.,** 1994, "Effects of Lubricant Compressibility on the Film Thickness in EHL Line and Circular Contacts," *WEAR*, **173**, pp. 151-165.

[104] **Venner, C.H. and Lubrecht, A.A.,** 1996, "Numerical Analysis of the influence of Waviness on the Film Thickness of a Circular EHL Contact," *ASME J. of Tribology*, **118**, pp. 153-161.

[105] **Venner, C.H., Couhier, F., Lubrecht, A.A. and Greenwood, J.,** 1997 "Amplitude Reduction of Waviness in Transient EHL Line Contacts," Proc. 23rd Leeds-Lyon Symposium on Tribology, *Elsevier Tribology Series*, **32**, *Ed. D. Dowson et al.*, pp. 103-112.

[106] **Venner, C.H. and Lubrecht, A.A.,** 1999. "Amplitude Reduction of Anisotropic Harmonic Surface Patterns in EHL Circular Contacts under Pure Rolling," Proceedings of the 25th Leeds-Lyon Symposium on Tribology, *Elseviers Tribology Series,* **36**, *Ed. D. Dowson et al.,* pp. 151-162.

[107] **Venner, C.H. and Morales Espejel, G.E.,** 1999, "Amplitude reduction of small amplitude waviness in transient EHL line contacts," *Proc. IMechE, J. of Engineering Tribology,* **213**, pp. 487-504.

[108] **Wesseling, P.**, 1991, "An Introduction to Multigrid Methods," John Wiley & Sons, New York, ISBN 0-471-93083-0

[109] **Wijnant, Y.H. and Venner, C.H.,** 1997, "Analysis of an EHL circular contact incorporating Rolling Element Vibration," Proceedings of the 23rd Leeds-Lyon Symposium on Tribology, *Elseviers Tribology Series*, **36**, *Ed. D. Dowson et al.,* pp. 445-456.

[110] **Wijnant, Y.H.,** 1998, "Contact Dynamics in the field of Elastohydrodynamic Lubrication," Ph.D. Thesis, University of Twente, Enschede, The Netherlands. ISBN 90-36512239.

[111] **Wijnant, Y.H. and Venner, C.H.,** 1999, "Contact Dynamics of Starved EHL Contacts," Proceedings of the 25th Leeds-Lyon Conference, *Elseviers Tribology Series*, **36**, *Ed. D. Dowson et al.,* pp. 705-716.

[112] **Wijnant, Y.H., Venner, C.H. and Larsson, R.,** 1999, "Effects of Structural Vibrations on the Film Thickness in an EHL circular contact," *ASME J. of Tribology*, **121**, pp. 259-264.

Appendix A

MultiLevel Routines

A.1 1d Poisson Multilevel Building Blocks

This section outlines very briefly the datastructure, the functions and the subroutines performing the different tasks in the **MG1d** program which solves the one dimensional Poisson problem. The overall structure and the components will be extended to two dimensions for **MG2d**, but will undergo only few fundamental changes.

In order to organize all the grid-dependent data a *structure* called *Level* is introduced. The total MultiGrid solver employs different *Levels* that are organized into a *Stack* together with all the variables that are level-independent. A word, such as *structure*, is written in italics whenever the language C concept is meant. The star $*$ in front of a variable denotes a pointer to this variable. These pointers are used to denote vectors: $*a$ denotes a pointer to a vector a with components $a[i]$. The variables include integers $\{int\}$, double precision $\{double\}$ and double precision vectors $\{*double\}$. Variables inside a *structure* are designated using an arrow $->$: for instance the variable hx inside the *Level structure* L is designated by $L->hx$, and the vector u inside the *Level structure* L is designated by $L->u$.

Level A *structure* containing variables describing a single level (all information needed on a certain grid). The variables include integers $\{int\}$, double precision $\{double\}$ and double precision vectors $\{*double\}$.

hx	grid size in x direction (h_x)	$\{double\}$
ii	number of grid intervals in x direction	$\{int\}$
$*u$	solution vector, element u_i is denoted $u[i]$	$\{*double\}$
$*f$	right hand side vector, element f_i is denoted $f[i]$	$\{*double\}$
$*uconv$	converged solution vector	$\{*double\}$
$*uold$	initial coarse grid solution vector	$\{*double\}$

Stack A *structure* containing a number of level-independent variables and all the individual levels.

$nx0$	number of grid intervals on coarsest level in x direction	$\{int\}$
$maxlevel$	number of *Levels* in the *Stack*	$\{int\}$
xa	x coordinate of the start of the computational domain	$\{double\}$
xb	x coordinate of the end of the computational domain	$\{double\}$
wu	work unit counter	$\{double\}$
$*Lk$	array of *Levels*	$\{*Level\}$

The entire program uses dynamic memory allocation, which means that during the execution of the code, when the user inputs the number of levels, the program will allocate the memory necessary for the execution. This is done by means of the library function *calloc*(). For example the statement:

$$u = (double*)calloc(nx + 1, sizeof(double))$$

allocates storage space for a vector of nx+1 $\{double\}$'s. The first element of the vector is $u[0]$, the last element $u[nx]$.

Datastructure routines:

initialize() Initializes the level-independent variables of the *Stack* such as xa etc. Subsequently it allocates space for the number of *Level*'s needed. Finally it loops over all *Levels*, filling the level-dependent variables such as hx etc. and allocates memory space for the different vectors (like u inside a *Level*). Input are the number of nodes on the coarsest grid n_0 the number of gridlevels and the values of the xa and xb. $\{void\}$.

finalize() Frees the memory space allocated by *initialize*, $\{void\}$.

Functions of a single variable. They are used to fill the vectors u and f of a certain level or to provide comparisons for the error calculations:

U_a(x) Provides the analytical solution of u in the point (x), $\{double\}$.

U_i(x) Provides the initial solution as a function of the coordinate (x), $\{double\}$.

F_i(x) Provides the right hand side function f as a function of the coordinate (x), $\{double\}$.

Lu() Calculates the value of the discrete operator L^h working on the current approximation to the solution u in the discrete point (i), (2.17), $\{double\}$.

The following routines perform operations on the points of a single level k:

init_uf() Subroutine fills the u vector with the initial solution and the boundary conditions on level k, and also fills the f vector with the right hand side function on that level, $\{void\}$.

relax() Subroutine that updates the u vector point by point on level k, using lexicographic Gauss-Seidel relaxation, to make the local residual zero. Computes the average absolute residual for the new approximation and prints it to the screen together with the number of the level at which the relaxation was performed and the total amount of work in Work Units (WU) invested so far, $\{void\}$.

conver_a() Computes en (3.78), the L_1 norm (mean absolute difference) between the current approximation to the solution u on level k and the analytical solution U_a on that level, $\{double\}$.

The following routines perform operations involving more than one single level. In general these intergrid routines interpolate from a coarser grid $k-1$ to a finer grid k, or restrict from a finer grid k to a coarser grid $k-1$:

coarsen_u() Computes the initial approximation of u on level $k-1$ after coarsening from level k, (3.73). Because FAS is used this first approximation is the restriction of the current fine grid solution to the coarse grid. The restriction used is full weighting. The routine also stores the result in the vector *uold* for later use in the computation of the corrections according to the FAS formulation, $\{void\}$.

coarsen_f() Computes the level $k-1$ right hand side function f coarsening from level k, (3.74). Because FAS is used the right hand side consists of two parts: the restriction of the fine grid residuals, and the discrete operator on the coarse grid acting on the restricted fine grid solution, $\{void\}$.

refine() Correction of u on level k based on the interpolation of the difference between u and *uold* computed on level $k-1$, (3.75). The routine uses linear interpolation, $\{void\}$.

fmg_interpolate() High order interpolation of the converged solution u on level $k-1$ to provide an accurate initial solution for u on level k. (3.28,3.29). The routine uses cubic interpolation. This routine also stores the converged coarse grid solution u on level $k-1$, in the vector *uconv* for later use in the computation of the approximate error norm *aen*, $\{void\}$.

conver() Calculates aen (3.81), the L_1 norm of the converged solution on level $k-1$ with respect to the converged solution on level k. This routine approximates *conver_a()* when the analytical solution U_a is unknown, $\{double\}$.

The next routines are the driving routines that perform cycles involving all levels:

cycle() Performs a coarse grid correction cycle on level k. Input parameters are the level k the number of pre-relaxations ν_1, post relaxations ν_2, coarsest grid relaxations ν_0 and the cycle index γ. For $\gamma = 1$ a V cycle is performed and for $\gamma = 2$ a W cycle. Notice the recursive way it is programmed. For $\gamma = 1$ it simply decends until level 1 and then ascends until it reaches level k again, $\{void\}$.

fmg() Performs a Full MultiGrid cycle. Uses the previous routine *cycle*. Input parameters are the target level k (the finest grid) the cycle parameters ν_0, ν_1, ν_2, γ and ncy which is the number of cycles performed per gridlevel. Notice the recursion. It starts with relaxations on the coarsest grid and subsequently works its way up to attain finally the required maximum level, $\{void\}$.

Main program:

main() Reads parameters such as the maximum level $maxlev$, and the number of cycles to be performed per level ncy. Initializes the datastructure. Subsequently calls the routine *fmg* which carries out the solution process. Finally using the obtained converged solutions on each of the grids during the FMG process the error norms en and aen are computed, $\{void\}$.

A.2 2d Poisson Multilevel Building Blocks

This section outlines very briefly the datastructure, the functions and the subroutines performing the different tasks in the **MG2d** program which solves the two dimensional Poisson problem. The overall structure and the components are the same as described previously for the **MG1d** program. Straightforward extensions to two dimensions are used troughout the program. One of these extensions is the definition of a matrix. For the one dimensional program MG1d pointers were used to denote vectors. In the program MG2d a matrix is constructed simply as a vector of vectors. Thus if $*a$ denotes a vector a with elements $a[i]$ then $**b$ denotes a matrix b with elements $b[i][j]$. As before, variables inside a *structure* are designated using an arrow $->$: for instance the variable hx inside the *Level structure* L is designated by $L->hx$, and the matrix u inside the *Level structure* L is designated by $L->u$.

Level *Structure* containing variables describing a single level The variables include integers $\{int\}$, double precision $\{double\}$ and double precision matrices $\{**double\}$.

hx	grid size in x direction (h_x)	$\{double\}$
hy	grid size in y direction (h_y)	$\{double\}$
ii	number of grid intervals in x direction	$\{int\}$
jj	number of grid intervals in y direction	$\{int\}$
$**u$	solution matrix, element $u_{i,j}$ is denoted $u[i][j]$	$\{**double\}$
$**f$	right hand side matrix, element $f_{i,j}$ is denoted $f[i][j]$	$\{**double\}$
$**uconv$	converged solution matrix	$\{**double\}$
$**uold$	initial coarse grid solution matrix	$\{**double\}$

Stack *Structure* containing the level-independent variables and all the individual levels.

$nx0$	number of grid intervals on coarsest level in x direction	$\{int\}$
$ny0$	number of grid intervals on coarsest level in y direction	$\{int\}$
$maxlevel$	number of *Levels* in the *Stack*	$\{int\}$
xa	x coordinate of the start of the computational domain	$\{double\}$
xb	x coordinate of the end of the computational domain	$\{double\}$
ya	y coordinate of the start of the computational domain	$\{double\}$
yb	y coordinate of the end of the computational domain	$\{double\}$
wu	work unit counter	$\{double\}$
$*Lk$	array of *Levels*	$\{*Level\}$

The program uses dynamic memory allocation. The procedure is the same as in the MG1d program. However, to allocate the storage space for a matrix of $\{double\}$'s an auxillary subroutine $**matrix()$ is used. The statement:

$$u = matrix(nx, ny)$$

allocates storage space for a matrix of {*double*}'s having the size $(nx + 1) \times (ny + 1)$. The matrix has indices ranging from 0 to nx and from 0 to ny. The matrix elements are $u[i][j]$.

Datastructure routines:

initialize() Initializes the level-independent variables as xa etc. Then it allocates space for the *Level* strucure, then it loops over all *Levels*, filling the level-dependent variables as hx etc. and finally allocates memory space for the different matrices (like u inside a *Level*), {*void*}. Input are the number of nodes in x and y direction on the coarsest grid, the number of levels, and the values of xa, xb, ya and yb.

finalize() Frees the memory space allocated by *initialize*, {*void*}.

The following subroutines are functions of two variables. They are used to fill the matrices u and f either of a certain level or to provide comparisons for the error calculations:

U_a(x,y) Provides the analytical solution of u in the point (x, y), {*double*}.

U_b(x,y) Provides the analytical solution for the domain boundary $(x, y) \in \partial\Omega$, {*double*}.

U_i(x,y) Provides the initial solution as a function of the coordinates (x, y), {*double*}.

F_i(x,y) Provides the right hand side function f as a function of the coordinates (x, y), {*double*}.

Lu() Calculates the value of the discrete operator L^h working on the current approximation to the solution u in the discrete point (i, j), (2.20), {*double*}.

The following routines perform operations on the points of a single level k:

init_uf() Subroutine that fills the u matrix with the initial solution and the boundary conditions on level k, and also fills the f matrix with the right hand side function on that level, {*void*}.

relax() Subroutine that updates the u matrix point by point on level k, using lexicographic Gauss-Seidel relaxation, to make the local residual zero. Computes the average absolute residual for the new approximation and prints it to the screen together with the number of the level at which the relaxation was performed and the total amount of work in Work Units (WU) invested so far, {*void*}.

conver_a() Computes en, (3.88), the L_1 norm of the difference between the current approximation to the solution u on level k and the analytic solution U_a, {*double*}.

The following routines perform operations involving more than one single level. In general these intergrid routines interpolate from a coarser grid $k - 1$ to a finer grid k, or restrict from a finer grid k to a coarser grid $k - 1$:

coarsen_u() Computes the initial approximation of u on level $k-1$ after coarsening from level k, (3.73). Uses full weighting. The routine also stores the current coarse grid $k-1$ solution u in the matrix *uold* to facilitate the correction in *refine()*, $\{void\}$.

coarsen_f() Computes the level $k-1$ matrix f when coarsening from level k. FAS is used with full weighting for the restriction of the fine grid residuals.

refine() Correction of u on level k based on the interpolation of the difference between u and *uold* computed on level $k-1$. Uses bi-linear interpolation, $\{void\}$.

fmg_interpolate() This routine first stores the converged coarse grid solution u on level $k-1$, in the matrix *uconv* for later use when computing the approximate error norm *aen*. Next the solution is interpolated to level k to serve as a first approximation. (3.28,3.29). Bi-cubic interpolation is programmed one dimension at the time.

conver() Calculates *aen* (3.89) the L_1 norm of the converged solution on level $k-1$ with respect to the converged solution on level k. This routine approximates *conver_a()* when an analytical solution *U_a* is unknown, $\{void\}$.

The next routines are the driving routines that perform cycles involving all levels:

cycle() Performs a coarse grid correction cycle on level k. Input parameters are the level k the number of pre-relaxations ν_1, post relaxations ν_2, coarsest grid relaxations ν_0 and the cycle index γ, $\{void\}$.

fmg() Performs a Full MultiGrid cycle, using the previous routine *cycle*. Input parameters are the target level k (the finest grid) the cycle parameters ν_0, ν_1, ν_2, γ and *ncy* which is the number of cycles performed per gridlevel. It starts from level 1 to attain finally the required maximum level, $\{void\}$.

Main program:

main() Reads input parameters such as maximum level *maxlev* and the number of cycles to be performed per level *ncy*. Initializes datastructure. Calls the routine *fmg* which carries out the solution process. The obtained converged solutions on each grid during the FMG process are used to compute the error norms *en* and *aen*, $\{void\}$.

A.3 2d Hydrodynamic Multilevel Extensions

This section outlines very briefly the additions to the datastructure, functions and subroutines in the Hydrodynamic Lubrication code **HL2d**.

Level A matrix containing the film thickness values h denoted by $**h$ is added, the element $h_{i,j}$ is denoted $h[i][j]$, $\{**double\}$. The matrices $**u$, $**uold$ and $**uconv$ are replaced by $**p$, $**pold$ and $**pconv$ to clearly indicate the physical origin of the variable (pressure).

Global Variables

Three global variables have been added, *epsilon* denoting the dimensionless eccentricity, $\{double\}$, *LoR* denoting the ratio $k = L/R$, $\{double\}$, and RoL denoting the ratio $R/L = 1/k$ as it appears in the discrete equation 4.30, $\{double\}$. In the program *LoR* is set at its definition and *RoL* is computed as $1/LoR$ in the routine *initialize*. Presently *LoR* is set to 1, but the program can be used for other values of *LoR* too. However, the reader is reminded that to efficiently run the program for $LoR \neq 1$ the number of nodes in θ and Y direction should be adjusted as is explained in Section 4.8. This can be done by changing the values in the call of the routine *initialize()*.

Functions used to fill the matrices p and h and f at a given level:

P_b(x,y) Boundary pressure at point (x, y), $\{double\}$.

P_i(x,y) Initial pressure at point (x, y), $\{double\}$.

H_i(x,y) Film thickness at point (x, y), $\{double\}$.

F_i() Right hand side at point (x, y), $\{double\}$.

Lu() Calculates the value of the discrete Reynolds operator (4.30) in the point (i, j), using a long central derivative for the $\partial h/\partial x$ term, $\{double\}$.

Single level routines:

init_uf() Initializes pressure, film thickness and right hand side, $\{void\}$.

prerelax1() Performs a local relaxation around the cavitation line, where the residuals are likely to be reduced less, due to their non-smooth character. The domain is simply choosen to cover the θ parameter from 1/2 the domain to 3/4 of the domain, $\{void\}$.

prerelax2() Performs a local relaxation around the cavitation line, where the residuals are likely to be reduced less, due to their non-smooth character. The domain is choosen to cover 5 points in the θ direction around the cavitation boundary, $\{void\}$.

relax() Performs a point by point Gauss-Seidel relaxation sweep over the entire domain, imposing the cavitation condition $p \geq 0$. Computes the average absolute value of the residual after the sweep and prints it to screen together with the level index k and the current value of the Work Unit counter, $\{void\}$.

Routines acting between a coarse level $k-1$ and a fine grid k:

coarsen_p() Performs a restriction of p from the fine grid k to the coarse grid $k-1$, (3.74). The restriction uses simple injection (3.18) in order not to mix values from the pressurised domain Ω_1 and the cavitated domain Ω_2, $\{void\}$.

coarsen_f() Computes the FAS right hand side function f on the coarse grid $k-1$ (4.44). The transfer uses simple injection (3.18) in order not to mix residuals from the pressurised domain Ω_1 and the cavitated domain Ω_2, $\{void\}$.

refine() Corrects the level k solution with the level $k-1$ result. Uses bi-linear interpolation. To minimize a possible shift of the cavitation boundary only non-zero points are corrected. The corrected approximation is subjected to the cavitation condition $p \geq 0$, $\{void\}$.

fmg_interpolate() Stores level $k-1$ solution for later use in convergence check. Interpolates the level $k-1$ solution to level k to serve as accurate first approximation. Uses bi-cubic interpolation programmed one dimension at the time, $\{void\}$.

Output routine:

output() Used at the end of the program. Computes dimensionless load (4.32), Sommerfeld number S_0^h and attitude angle Ψ^h. Writes datafile with the centerline pressure profile and film thickness, $\{void\}$.

Main program:

main() Solves the 2d Hydrodynamic Lubrication problem for a given value of ϵ (and L/R) on level *maxlev* using *ncy* cycles and FMG, $\{void\}$.

A.4 2d Dry Contact Multilevel Extensions

The organisation of the grid dependent and grid independent data is once again carried out by two *structures* called *Level* and *Stack*. The two *structures* are described anew since they have undergone some important changes. The **DRY2d** program contains two different multilevel solvers, one classical to accelerate the convergence of the contact problem, the second one to accelerate the calculation of the deformation integrals. The two algorithms can use the same variables and datastructure if it is extended in the following way. The fast calculation of the integral uses interpolation in one direction at the time. This requires the use of intermediate semi-refined or semi-coarsened grids. These intermediate grids (levels) are created between the "normal" uniform grids that are used by the solver. Consequently, to solve the problem using $maxlev$ levels, *internally* in the program $2*maxlev-1$ levels are created where the levels with the odd indices $1,3,5,\ldots,2*maxlev-1$ are the classical grids (with equal mesh size in both directions) used by the *solution* algorithm and the levels with even indices $2,4,\ldots$ are the semi-coarsened grids used by the *fast evaluation* only. The result is that procedures acting between grids in the solution algorithm jump from level $k-2$ to level k when refining and from level k to level $k-2$ when coarsening. The intermediate level $k-1$ is only used when the elastic deformation is computed on level k. However, when output is given to the user by routines in the solution algorithm, e.g. in the relaxation routine, or in the output routine, the level index printed on screen refers to the "normal" grids.

Level A *structure* containing variables describing a single level (all information needed on a certain grid). The variables include integers $\{int\}$, double precision $\{double\}$ and double precision matrices $\{**double\}$.

hx	grid size in x direction (h_x)	$\{double\}$
hy	grid size in y direction (h_y)	$\{double\}$
ii	number of grid intervals in x direction	$\{int\}$
jj	number of grid intervals in y direction	$\{int\}$
$**p$	matrix containing pressure distribution	$\{**double\}$
$**w$	matrix containing elastic deformation integral	$\{**double\}$
$**f$	right hand side matrix of contact equation	$\{**double\}$
$**K$	matrix containing influence coefficients of the deformation	$\{**double\}$
$**K1$	correction matrix needed for fast integral calculation	$\{**double\}$
$**pconv$	converged solution matrix	$\{**double\}$
$**pold$	initial coarse grid solution matrix	$\{**double\}$
$**hfi$	matrix containing 'film thickness' or more exactly the gap height	$\{**double\}$
$**hrhs$	matrix containing the 'film thickness' right hand side function, $\{**double\}$	
rg	right hand side force balance	$\{double\}$

Stack A *structure* containing a number of level-independent variables and all the individual levels.

$nx0$	number of grid intervals on coarsest level in x direction	{*int*}
$ny0$	number of grid intervals on coarsest level in y direction	{*int*}
$m1$	number of correction points in the direction of interpolation	{*int*}
$m2$	number of correction points perpendicular to the direction of interpolation	{*int*}
od	number of ghost points	{*int*}
$maxlevel$	number of *Levels* in the *Stack*	{*int*}
$deep$	number of *Levels* the fast integration uses below the present level	{*int*}
xa	x coordinate of the start of the computational domain	{*double*}
xb	x coordinate of the end of the computational domain	{*double*}
ya	y coordinate of the start of the computational domain	{*double*}
yb	y coordinate of the end of the computational domain	{*double*}
$h0$	constant: (solid body approach) in film thickness equation	{*double*}
wu	work unit counter	{*double*}
$*Lk$	array of *Levels*	{**Level*}

The entire program uses dynamic memory allocation, this means that during the execution of the code, when the user inputs the number of levels, the program will allocate the memory necessary for the execution.

The subroutine $**matrix(*U, nx, ny)$ allocates storage space for a matrix of {*doubles*} having the size $(nx + 1 + 4od) \times (ny + 1 + 4od)$, this matrix has indices ranging from 0 to $nx + 4od$ and from 0 to $ny + 4od$. The value of od is taken from the from *stack* U given as input: $U- > od$.

The subroutine $**smatri(*U, nx, ny)$ allocates storage space for a shifted matrix (including the ghost points) of {*doubles*} having the size $(nx + 1 + 2od) \times (ny + 1 + 2od)$, this matrix has indices ranging from $-od$ to $nx + od$ and from $-od$ to $ny + od$. The value of od is taken from the stack U given as input: $U- > od$. The advantage of the use of this shifted matrix is that the ghost points can be incorporated without changing the fact that a coarse grid point with index I coincides with a fine grid point $2I$.

All matrix variables depending on position as pressure p and film thickness h are generated using the shifted matrix *smatri()*. The matrix variables relating to distances K and $K1$ use the *matrix* allocation.

Datastructure routines:

initialize() Initializes the level-independent variables as xa etc. Then it allocates space for the *Level* strucure, then it loops over all *Levels*, filling the level-dependent variables as hx etc. and finally allocates memory space for the different matrices (like p and h) inside a *Level*, {*void*}.

finalize() Frees the memory space allocated by *initialize*, {*void*}.

Routines acting on a single level k:

init_f() On the level k, fills the right hand side matrix f with the right hand side function, $\{void\}$.

init_p() On the level k, fills the p matrix with the initial solution and the boundary conditions, $\{void\}$.

resnorm() Calculates the residual norm of the contact equation, in all points where $p(i,j) > 0$ the gap height should be 0, thus the residual is the value of the gap. The norm is taken as the sum of the absolute values of the gap in the points of positive pressure, $\{void\}$.

relax() Updates the p matrix point by point on level k, using Jacobi distributive relaxation. Naturally only used on the solution levels. Computes the average absolute residual and prints it to screen together with the level index (in terms of "solution grids") and the current value of the Work Unit counter. In order to assure stable convergence under all conditions, an underrelaxation factor ur is introduced. For the smooth surface problem values of 0.4 work very well. For the rough surface contact problem, values of $ur = 0.2$ work well. Note that the relaxation uses the five point distributed relaxation scheme, and imposes the cavitation condition in every point after the changes are implemented, $\{void\}$.

relaxh0() Updates the $h0$ value in order to satisfy the force balance equation. The value of $h0$ is only changed on the coarsest level, $\{void\}$.

The following routines perform operations involving more than one single level. In general these intergrid routines interpolate from a coarser grid $k-2$ to a finer grid k, or restrict from a finer grid k to a coarser grid $k-2$, as part of the multigrid *solution* process. As a reminder, the solution process only uses the levels with odd indices. The levels with even indices are auxillary grids used in the fast evaluation of the elastic deformation.

coarsen_p() Computes the initial approximation of p on level $l-2$ after coarsening from level l, (3.73), using injection. The routine also stores the current coarse grid $l-2$ solution p in the matrix *pold* to facilitate the correction in *refine()*, $\{void\}$.

coarsen_f() Computes the level $l-2$ (FAS) right hand side function f of the contact equation and of the force balance equation after coarsening from level l, (5.42). Injection is used for the restriction of the residuals of the contact equation, $\{void\}$.

refine() Correction of p on level k based on the interpolation of the difference between p and *pold* computed on level $k-2$, (5.43). Uses bi-linear interpolation. To minimize possible shifts of the location of the free boundary only points with $P > 0$ are corrected. After the correction the complementarity condition $P \geq 0$ is imposed, $\{void\}$.

fmg_interpolate() Interpolation of the converged solution p on level $k-2$ to provide an accurate initial solution for p on level k, (3.28,3.29). Uses bi-linear interpolation. This routine also stores the converged coarse grid solution p on level $k-2$, in the matrix *pconv*, $\{void\}$.

conver() Calculates *aen* (3.81), the L_1 norm of the converged solution on level $k-2$ with respect to the converged solution on level k. This routine approximates *conver_a()* when an analytical solution *U_a* is unknown, {*void*}.

The next routines are the driving routines that perform cycles involving all levels:

cycle() Performs a coarse grid correction cycle on level k. The solution process only uses the grids with equal mesh size in both directions (solution grids) and thus in coarsening and refining it always skips an intermediate semi-refined or semi-coarsened grid. Hence, it goes from k to $k-2$ when coarsening and from $k-2$ to k when refining. Modified such that at the coarsest grid in the cycle, level 1, a relaxation of the force balance equation is carried out using the routine *relaxh0()*, {*void*}

fmg() Performs a Full MultiGrid cycle, using the individual cycles from *cycle()*. It starts from level 1 to attain finally the required maximum level k. Only uses the grids with equal mesh size in both directions (solution grids). Hence, when refining it skips a semi-refined grid that is only used by the Multilevel Multi-Integration and goes from $k-2$ to k. Modified such that at level 1 it includes relaxation of the force balance equation using the routine *relaxh0()*, {*void*}.

Functions and routines for the calculation of the gap height using Multilevel Multi-Integration for the calculation of the elastic deformation:

ah(a,b) Function that calculates the arcsinh(a/b), needed for the computation of the kernel coefficients, {*double*}.

init_log() Computes the kernel coefficients for the computation of the elastic deformation w. Stored as a function of the distance.

kval() Function that calculates the fine grid kernel value transfered to the current level in the current point, {*double*}.

fillk() Subroutine that fills the matrix K^{HH} on all coarser grid by transfers from the finest grid K^{hh}, {*void*}.

calcku() Subroutine that calculates the deformation w on a given level through direct multiplication and addition, {*void*}.

sto6k1() Subroutine that calculates and stores the sixth order correction kernel $K1$ on level l. If the level l is even, the interpolation is performed in the x direction, if l is odd the interpolation is performed in the y direction, {*void*}.

coarsenp6x() Subroutine that calculates the coarse grid pressure distribution p on level $l-1$ using a the sixth order restriction in the x direction, of the pressure distribution on level l, (5.63), {*void*}.

coarsenp6y() Subroutine that calculates the coarse grid pressure distribution p on level $l-1$ using a the sixth order restriction in the y direction, of the pressure distribution on level l, (5.63), $\{void\}$.

refine6x() Subroutine that carries out the interpolation and correction of the integrals going from level $l-1$ to level l, using a the sixth order correction in the x direction. First the even points are corrected and trivially interpolated. Than the odd points are interpolated using sixth order interpolation and corrected, (5.74), $\{void\}$.

refine6y() Subroutine that carries out the interpolation and correction of the integrals going from level $l-1$ to level l, using a the sixth order correction in the y direction. First the even points are corrected and trivially interpolated. Than the odd points are interpolated using sixth order interpolation and corrected, (5.77), $\{void\}$.

calchi() Subroutine that calculates the film thickness on a certain level, starting with the calculation of the deformations, either through the classical integration *calcku()* or through the fast integration. Finally it adds the undeformed geometry, $\{void\}$.

Output routines:

outputP() Writes computed pressure profile on centerline $Y = 0$ as a function of X to file "px.dat", $\{void\}$.

outputH() Writes computed gap on centerline $Y = 0$ as a function of X to file "hx.dat", $\{void\}$.

Main program:

main() Solves the contact problem on level $maxl$ using the fast integration with $deep$ levels deep and interpolation order $order$, using ncy cycles in a cycling or FMG way, $\{void\}$.

A.5 2d EHL Contact Multilevel Extensions

In this section the changes that have been made in the process of changing the program DRY2d into EHL2d are given. It should be noted that the general structure remains the same. The MultiGrid solver uses a Stack of grids (*Level*). For the case of *maxlev* levels, inside the program $2 * maxlev-1$ levels are created. The *odd* levels, 1,3,5 etc are the ones by the MultiGrid *Solution* process. The *even* levels are the half-coarsened grids that are used in the *Multilevel Multi-Integration* routines. No changes have to be made to these latter routines as the task of evaluating the elastic deformation for a given approximation to the pressure is exactly the same in the EHL program as it is in the dry contact program.

Level

Additional variables in the structure *Level* are the Matrix and vectors for the line relaxation system $AX = Y$:

$**A$	Matrix for line relaxation	$\{**double\}$
$*X$	Vector with changes	$\{*double\}$
$*Y$	Vector with right hand side	$\{*double\}$

and some characteristic solution values:

Hm	minimum film thickness	$\{double\}$
Hc	central film thickness	$\{double\}$
Hcp	film thickness at $\partial P/\partial X = \partial P/\partial Y = 0$	$\{double\}$

global variables

Global variables are added for the load conditions, lubricant parameters, parameters of the numerical solution process, and to measure cpu time:

load parameters:

$MMoes$	Dimensionless load parameter M	$\{double\}$
$LMoes$	Dimensionless load parameter L	$\{double\}$
$rlambda$	Dimensionless parameter $1/\bar{\lambda}$	$\{double\}$
$alphabar$	Dimensionless paramter $\bar{\alpha}$	$\{double\}$

parameters Roelands equation:

$et0$	η_0	$\{double\}$
$p0r$	$p_0 = 1.96 \cdot 10^8$	$\{double\}$
zr	pressure viscosity index z	$\{double\}$
$alpha$	α	$\{double\}$

relaxation parameters:

xi_l	relaxation switch parameter ξ_{limit}	{*double*}
urja	underrelaxation jacobi part ω_{ja}	{*double*}
urgs	underrelaxation Gauss-Seidel part ω_{gs}	{*double*}
hfact	relaxation factor force balance ω_{H_0}	{*double*}

multigrid process parameters:

starl	start level FMG process	{*int*}
cfilev	current finest level, used for output control	{*int*}

auxillary:

cputimes[]	array to store cpu times for each level in FMG	{*unsigned long*}
t0	auxillary value used to read clocktime	{*unsigned long*}

initialize() Same as for the DRY2d program except for the additional allocation of the matrix A and the vectors X and Y. X and Y are vectors of $ii-1$ elements and A is a matrix of $ii-1$ lines and 11 columns. Only 5 of these columns serve for the storage of the band of the matrix (hexadiaginal). The other space is needed for filling in the Gaussian Elimination process.

finalize() Frees the allocated memory at the end of the program

Additional functions of a single variable:

reta() Function that calculates the value of $1/\bar{\eta}$ for a given value of P. The expressions for the case of the Barus and the Roelands viscosity pressure equation are given, {*double*}.

rho() Function that calculates the value of $\bar{\rho}$ for a given value of P. This value is unity for an incompressible lubricant. For a compressible lubricant it is computed using the Dowson and Higginson equation, {*double*}.

Lu() Calculates the value of the discrete operator working on the current approximation to the pressure at the point (i, j). Used for the computation of the residuals and the FAS coarse grid right hand side. For $i=1$ a first order upstream discretization of the wedge term is used. For $i>1$ a second order upstream discretization, {*double*}.

Modified routines acting on a single level:

resnorm() Calculates the residual norm of the Reynolds equation with the current approximation to the pressure and film thickness. The residual is only defined for points with $p(i, j) > 0$, {*double*}.

relax() Performs one relaxation sweep on level l. One relaxation consists of a film thickness computation followed by one sweep over the grid, line by line, carrying out the line relaxation. With the improved solution the residual of the force balance equation and the residual of the Reynolds equation are computed and printed to the screen, $\{void\}$.

Modified routines operating between grids. Restriction from fine grid k to coarse grid $k-2$ or interpolation from grid $k-2$ to grid k. As a reminder, the solution process only uses the levels with odd indices. The levels with even indices are auxillary grids used in the fast evaluation of the elastic deformation.

coarsen_p() Performs a restriction of p and hfi from the fine grid k to the coarse grid $k-2$. The restriction uses full weighting if in all nine points of the stencil $p(i,j) > 0$. Otherwise injection is used. For hfi injection is used everywhere. The restricted solution for p is stored in *pold* for later use when the fine grid solution is corrected using the coarse grid result, $\{void\}$.

coarsen_f() Computes the FAS right hand side function $prhs$ and $hfirhs$ for the Reynolds equation and film thickness equation on the coarse grid. Computes rg the right hand side of the force balance equation on the coarse grid. For the Reynolds equation the restriction of the residuals uses full weighting where in points of zero pressure the residual is defined to be 0. The film thickness equation on the fine grid has no residuals as the film thicknesses on the fine grid were just evaluated. Hence, the coarse grid right hand side of this equation only consists of the film thickness operator evaluated using the transferred pressure. In terms of the FAS formulation this is the $L^H(I_h^h\tilde{u}^h)$ term, $\{void\}$.

refine() Interpolation of pressure corrections from grid $k-2$ to grid k. The correction is done according to the FAS formulation. To compute the correction (error) the previously stored *pold* is used. The correction of the fine grid solution is done taking into account cavitation in the following way. Firstly, corrections are only applied to points that are not cavitated. This minimizes the disturbance caused to the cavitation boundary by the coarse grid. Secondly, the entire corrected fine grid result is subjected to the cavitation condition $p \geq 0$. Finally with the corrected values of p the new values of hfi are calculated, $\{void\}$.

fmg_interpolate() Fourth order interpolation of the (converged) solution of grid $k-2$ to grid k to serve as a first approximation. Used in the FMG process. Before interpolating first the level $k-2$ solution is stored in *pconv* for later use in convergence check, and some characteristic values of this solution are computed and stored for later use in output (Hm, Hc, and Hcp). The clock is read and the result stored for output of cpu time. The interpolation of p from the coarse grid to the fine grid is done with fourth order interpolation. The interpolation is implemented one dimension at the time. Central interpolation can be used everywhere, except near the boundaries. The interpolated pressure field is subjected to the cavitation condition $p \geq 0$. Finally the obtained first approximation to p on the fine grid is used to compute a first approximation to the film thickness hfi, $\{void\}$.

conver() Calculates the L_1 norm of the converged solution on level $k-2$ with respect to the converged solution on level k. For convergence checks, $\{double\}$.

Changes to the driving routines:

fmg() Modified to start at a given level indicated with ks, $\{void\}$.

Added input routines:

input_loadpar() Input of load and lubricant parameters. parameters, $MMoes$, $LMoes$, *rlambda* and *alphabar* standing for M, L, $1/\bar{\lambda}$ and $\bar{\alpha}$. Note that in the program *rlambda* $= 1.0/\bar{\lambda}$ is used. This saves divisions when computing the coefficients ξ^h. Gives values to the parameters appearing in the Roelands equation. Finally an initial value is given to the integration constant H_0, $\{void\}$.

input_solvepar() Input of parameters for the numerical solution process. These are the parameters of the FMG process: the number of levels $maxlev$, the starting level $starl$, the type of cycle and the number of cycles per level, and the parameters for the relaxation processes: relaxation switch limit *xi_l*, and relaxation factors $urja$, $urgs$ and $hfact$, $\{void\}$.

Modified output routine

output() Computes characteristic solution parameters Hm, Hc and Hcp for final solution on level k. Creates two datafiles *P.dat* and *H.dat* with the solution of P and H as a function of X and Y. Output is given to the screen of the values of Hm, Hc, and Hcp and of these parameters in terms of the Moes dimensionless film thickness definition on all grids that were used in the FMG process. Thus, the level index printed to screen in this routine refers to the "solution" level, i.e. the gridlevel counted not taking into account the semi-refined grids used for the fast evaluation of the deformation. Finally the cpu time used to obtain the converged solution on each of the levels is given, $\{void\}$.

Routines for fast evaluation of the film thickness: no changes

Auxillary routines for line relaxation:

init_line() Creates the system from which the changes are to be solved for line j as defined in appendix C. Only the band of the matrix is stored. Also, no changes have to be solved for $i=0$ and $i=ii$, so only a system of ii–1 equations needs to be solved. The coefficient $A^j_{i,i}$ is represented by $A[i-1][ml]$ and $A^j_{i,k}$ by $A[i-1][ml+k-i]$. Y is the right hand side vector and X the vector with changes to be solved, $\{void\}$.

pivot() Used by *solve_system* to determine the pivot element in Gaussian Elimination, $\{int\}$.

swap() Used by *solve_system* to swap rows of the matrix system $AX=Y$ if needed because of the pivoting, $\{void\}$.

bcksb() Used by *solve_system*. Constructs the solution vector Y using backsubstitution, $\{void\}$.

solve_system() Solves the system $AX = Y$ by means of Gaussian Elimination, $\{void\}$.

return_line() The changes computed in X are applied to the pressure at line j and neighbouring lines depending on the type of relaxation locally used, Jacobi distributive line or Gauss-Seidel line,{void}.

solve_line() Creates the system of equations for line j, solves it, and applies the changes, $\{void\}$.

The main program organizes the input, initialization of memory and variables, calls the FMG routine and at the end the output routine. Finally the allocated memory is freed.

Appendix B

Debugging Hints

The hints in this appendix are grouped per program, but assume that the reader has started by building (and debugging) the simpler programs as **MG1d.c** and **MG2d.c** before attempting the more complicated ones. The authors strongly suggest that even if the readers objective is to build only the **EHL2d** program, the reader should start to program and debug the **MG1d** and **MG2d** codes, until they perform as the examples given in chapter 3. Then the reader should use the **MG2d** code and modify it to obtain the **HL2d** code, debug it completely before moving on to the **DRY2d** code and finally to the **EHL2d** code. Even if the reader looses a little time doing the extra programming, he or she will eventually save a lot of time debugging. Needless to say that the gradual approach ensures a much better understanding, which will be helpful when extending the programs to solve more complicated problems.
The other basic rule which should be followed throughout the hints is that once dealing with a certain problem, the simplest possible configuration should be chosen, before trying the more complicated ones. Only if those simple configurations are fully debugged, the reader should move on and replace them one by one, by more complicated routines.

B.1 MG1d.c & MG2d.c

• *relax()* The first task is to ensure that the relaxation routine is correct, see Section 2.6, 3.9 and 3.10. Use many relaxation sweeps to solve the problem on level 1. Does the residual norm diminish correctly (Figure 3.14 and 3.16)?
If not check that the residual in the point which has just been relaxed is zero.
If not check if the residual is correctly computed, if necessary use the fact that the analytical solution is known. Does the solution obtained resemble the analytical solution?
If the relaxation on level 1 works correctly check the convergence of relaxation on level 2, then 3. Convergence becomes very slow on finer grids, and many relaxation sweeps are necessary (Figure 3.14).

If relaxation works correctly on all grids, it is time to start testing two level cycles. Start on level 2, relax twice, coarsen, relax ten times on level 1, refine, relax on level 2. Has this cycle reduced the residual norm by a factor of 10?

If not, study the residual norm during the cycle, it should be continuous when going from level 2 to 1. Is it reduced by a factor of at least 100 on grid 1? A useful test is to modify *coarsen_f()* slightly. Modify the coarse grid right hand side $\hat{f}$ by setting the fine grid residuals r_i^h to zero (3.74). The coarse grid solution u^H should give zero coarse grid residuals, and the fine grid correction should also be zero. This test effectively decouples the fine grid relaxation from the coarse grid correction cycle, and deviations from the described behaviour point directly at the faulty routine. This test also permits a (graphical) comparison of the approximate solutions u on fine and coarse grid. It should be pointed out that a transfer error in a single point is sometimes sufficient to hinder convergence.
In order to keep the transfer routines as simple as possible, start by using injection for u and r (Section 3.3). Once this works fine replace them by full weighting.
Another useful tool is to start the 2 level cycle from a solution which has already converged on level 2, either through the use of many relaxation sweeps on this grid, or through the use of the analytical solution.

If the single cycle, 2 level program works correctly, it is time to study the many cycle behaviour. Each V-cycle should reduce the residual norm by a factor of 10, until the machine precision is reached (10^{-15}). If the residual norm does not decrease as predicted, this can be caused by some rather subtle errors, sometimes in a single point. Reread the points outlined above.

If the 2 level many cycle behaviour is correct, it is time to study the 3 level one cycle behaviour, using the same techniques as described above for 2 levels. Then follows the level 3, many cycle test etc. When a program passes the level 2 many cycle test, the majority of the problems have been resolved.

The last step is to introduce the FMG cycle, FMG interpolation and the approximate error norm calculation. The final test are performed using FMG with 1, 2 and 3 V-cycles, checking the discretization error and the numerical error (Section 3.10.3).

B.2 HL2d.c

Debugging the **HL2d** program uses the same steps as debugging **MG2d**. First it is necessary to check the proper functioning of the relaxation as was described previously. In addition, the correct implementation on different levels of the function H has to be checked. When studying the V-cycle behaviour it is advised to use the problem without cavitation first, and to establish the correct convergence of this linear problem. The results should be compared with those obtained in Section 4.5. Because of the long and narrow domain, attention has to be paid on the coarsest grid with respect to the FMG high order interpolation. The fourth order interpolation requires at least four points in every direction.

When adding the cavitation condition, it is very important to maintain a boundary which does not move between the different grids. This excludes the full weighting operators for u and r (Section 3.3), at least in the vicinity of the cavitation boundary. Furthermore, the *prerelax()* routine can be suppressed initially and replaced by *relax()* itself. Once good convergence is established, a significant amount of work can be gained by using *prerelax()*, but for debugging the simplicity of using a single routine is important. Finally the cavitating results should be comparable to the ones of Section 4.5.2.

B.3 DRY2d.c

Debugging the **DRY2d** program requires a dual approach. The fast integration and the fast solver should be debugged separately. The order is not very important. To debug the fast integration start by checking the classical integration. When using a Hertzian pressure distribution, does one obtain the film thickness profile with nearly zero values for $X^2 + Y^2 < 1$? If so is this also the case on levels 2 and 3? Does the discretization error reduce as predicted?

The second step is the implementation of the fast integration itself, starting with the second order and correcting over the entire domain $m1 = m2 = 1000$. The transfer routines in x and y direction should be programmed only once, using 'cut and paste' to obtain the second routine. Calculating the integral on level 2 using a single level deep, should give exactly the same result as the classical integration (up to the level of the machine precision). Large errors in all points normally stem from programming errors in *fillK1()* or domain errors in the summations. Correct values in even points but large errors in odd points hint at errors in the interpolation and correction routines. Finally, fine grid integrals with many levels deep should be computed, and should again give the same result as the classical integrals up to the level of the machine precision. If all these tests are correct, the small values of $m1$ and $m2$ should be used, but this results rapidly in relatively large errors.

A next step involves programming the fourth order routines. This involves the use of the 'ghost' points and requires a careful analysis of the summation boundaries. The steps described above should again be followed, paying attention to errors that tend to creep in from the boundaries when using more levels deep. This type of error hints at incorrect treatment of the 'ghost' points or errors in the summation boundaries. Finally, the small values of $m1$ and $m2$ should be implemented, resulting in larger errors, but already an appreciable gain in computing speed is obtained.

The sixth order routines, and finally the eight order routines follow the same rules, where the 'cut and paste' technique really pays of, in limiting the programming and debugging effort.

When the fast integration work correctly, it is time to program and debug the fast solver, starting once again with the relaxation and the classical integration. Since this routine uses the distributed relaxation, it is important to test if the five point relaxation does indeed result in zero residuals. On finer levels the relaxation becomes unstable when starting from a zero initial solution. Thus the debugging should be carried out starting from the Hertzian pressure distribution. In a first time, the H_0 constant should be frozen at -1 and not be changed. When this results in a correct convergence, the H_0 value can be varied to account for the force balance equation, where overall convergence should only be slightly influenced by these changes in H_0.

When the fast solver works correctly, it can be combined with the fast integration, possibly using the large values of $m1 = m2 = 1000$ to obtain exactly the same results as with the classical integration. Small differences can thus be easily detected and traced back to the defective routine. Furthermore, this allows a direct comparison between the second, fourth, sixth and eight order fast integration. When all errors are removed, the small values of $m1$ and $m2$ should be introduced again, and the gain in computing time can be evaluated.

B.4 EHL2d.c

The main change from the DRY2d code to the EHL2d code is in the equations and the relaxation. In that case debugging the EHL2d.c code starts with debugging the relaxation. First construct the case where only Gauss-Seidel line relaxation is used. This leads to the simplest system of equations. For a low load case, the relaxation should converge to machine error residuals. If the relaxation diverges, check if the system of equations is solved exactly. If the residual stalls it may be due to an error in the system of equations near the cavitation boundary. When the Gauss-Seidel line relaxation on a single grid works correctly one can proceed to the mixed Gauss-Seidel/Jacobi relaxation. Again this mixed relaxation is debugged by checking single grid convergence. Apply the relaxation on a single grid to the model cases $M = 20$, $L = 10$ and $M = 200$, $L = 10$. Compare the performance and the result with the figures given in Chapter 6. Also values of the minimum and central film thickness can be compared with values given in Figure 6.15. Is the value roughly correct ? If the single grid performance is satisfactory one can proceed to the multigrid case. First check if a cycle converges for 2 levels, many cycles, then for 3 levels, many cycles, etc. Compare the results with the output given in Chapter 6 for the model cases. If the cycle does not converge check the coarse grid correction by not transferring residuals of the Reynolds equation in the routine *coarsen_f*. In that case the residuals on all coarser grids should be zero, and the solver should act as if it performs relaxations on a single grid. Naturally for this test H_0 should be fixed. If one has obtained a code that can reproduce to a good extend the results presented in Chapter 6 one can start to apply it to other cases.

Appendix C

Systems of Equations for Line Relaxation

For the model problems considered in Chapter 6 a line relaxation can be described as follows. Let $\tilde{P}^h_{i,j}$ denote the approximation to $P^h_{i,j}$ and $\tilde{H}^h_{i,j}$ the associated approximation to $H^h_{i,j}$. One relaxation sweep yielding an improved approximation $\bar{P}^h_{i,j}$ and $\bar{H}^h_{i,j}$ then consists of the following steps. The grid is scanned line by line, at each line applying changes to all the points of the line simultaneously, and, if distributive relaxation is used partly to the neighbouring lines. Subsequently, the new approximation $\bar{P}^h_{i,j}$ can be used to obtain $\bar{H}^h_{i,j}$. For line relaxation in X direction the changes $\delta^h_{i,j}$ to be applied at a given line j (and partly to the neighbouring lines) are solved from a system of equations of the form:

$$A^j \underline{\delta}^h_j = \underline{r}^h_j \tag{C.1}$$

where $\underline{\delta}^h_j$ is a vector of changes $\delta^h_{i,j}$ and $\underline{r}^h_j$ the vector with residuals $r^h_{i,j}$. Both are vectors of $n_x - 1$ elements. A^j is a $(n_x - 1) \times (n_x - 1)$ matrix with coefficients $A^j_{i,k}$. Their exact definition as well as the definition of $r^h_{i,j}$ depends on the type of relaxation applied. For the model problems discussed in Sections 6.6.2, 6.6.3, and the EHL problem discussed in Section 6.7, their definition is given below. For briefness of description it is assumed that the second order upstream discretization of the wedge term is used everywhere, including at the first line $i=1$ of the grid. The more general non-linear notation $L^h\langle \underline{P}^h \rangle$ representing the discrete operator working on $\underline{P}^h$ is used throughout this appendix to shorten the notation.

C.1 Model Problem Small ξ

For this problem the discrete equation at gridpoint (i,j) is given by:

$$\begin{aligned} L^h_{i,j}\langle \underline{P}^h \rangle &= \xi \frac{P^h_{i-1,j} - 2P^h_{i,j} + P^h_{i+1,j}}{h^2} + \xi \frac{P^h_{i,j-1} - 2P^h_{i,j} + P^h_{i,j+1}}{h^2} + \\ & \frac{1.5H^h_{i,j} - 2H^h_{i-1,j} + 0.5H^h_{i-2,j}}{h} = {}_P f^h_{i,j} \end{aligned} \tag{C.2}$$

where ${}_P f^h_{i,j}$ is the FAS right hand side of the 'Reynolds' equation and $H^h_{i,j}$ is defined by:

$$H^h_{i,j} - H_0 - \frac{X_i^2}{2} - \frac{Y_j^2}{2} + \sum_{i'}\sum_{j'} K^{hh}_{i,i',j,j'} P^h_{i',j'} = {}_H f^h_{i,j} \tag{C.3}$$

with ${}_H f^h_{i,j} = 0$ standing for the FAS right hand side of the film thickness equation. The coefficients $K^{hh}_{i,i',j,j'}$ are given by (6.31).

For this problem a distributive Jacobi line relaxation with a second order distribution in X and Y direction was proposed. In that case the residual to be used in the vector $\underline{r}^h_j$ of Equation (C.1) is defined as:

$$\begin{aligned} r^h_{i,j} &= {}_P f^h_{i,j} - \xi \frac{\tilde{P}^h_{i-1,j} - 2\tilde{P}^h_{i,j} + \tilde{P}^h_{i+1,j}}{h^2} - \\ &\quad \xi \frac{\tilde{P}^h_{i,j-1} - 2\tilde{P}^h_{i,j} + \tilde{P}^h_{i,j+1}}{h^2} + \frac{1.5\tilde{H}^h_{i,j} - 2\tilde{H}^h_{i-1,j} + 0.5\tilde{H}^h_{i-2,j}}{h} \end{aligned} \tag{C.4}$$

and the coefficients $A^j_{i,k}$ of the matrix are defined by:

$$A^j_{i,k} = \left\{ \frac{\partial L^h_{i,j}\langle \underline{P}^h \rangle}{\partial P_{k,j}} - \frac{1}{4}\left(\frac{\partial L^h_{i,j}\langle \underline{P}^h \rangle}{\partial P_{k+1,j}} + \frac{\partial L^h_{i,j}\langle \underline{P}^h \rangle}{\partial P_{k-1,j}} + \frac{\partial L^h_{i,j}\langle \underline{P}^h \rangle}{\partial P_{k,j+1}} + \frac{\partial L^h_{i,j}\langle \underline{P}^h \rangle}{\partial P_{k,j-1}} \right) \right\}_{\underline{P}^h = \underline{\tilde{P}}^h} \tag{C.5}$$

for $0<i<n_x$ and $0<k<n_x$. Defining:

$$\Delta K^{hh}_{k,l} = K^{hh}_{k,l} - \frac{1}{4}(K^{hh}_{k-1,l} + K^{hh}_{k+1,l} + K^{hh}_{k,l-1} + K^{hh}_{k,l+1}) \tag{C.6}$$

this gives for $|i-k|>2$:

$$A^j_{i,k} = -\frac{1.5\Delta K^{hh}_{|i-k|,0} - 2\Delta K^{hh}_{|i-k-1|,0} + 0.5\Delta K^{hh}_{|i-k-2|,0}}{h} \tag{C.7}$$

For $k = i$ one obtains:

$$A^j_{i,i} = -5\frac{\xi}{h^2} - \frac{1.5\Delta K^{hh}_{0,0} - 2\Delta K^{hh}_{1,0} + 0.5\Delta K^{hh}_{2,0}}{h} \tag{C.8}$$

Furthermore, for $i > 2$

$$A^j_{i,i-2} = -\frac{1}{4}\frac{\xi}{h^2} - \frac{1.5\Delta K^{hh}_{2,0} - 2\Delta K^{hh}_{1,0} + 0.5\Delta K^{hh}_{0,0}}{h} \tag{C.9}$$

For $i > 1$

$$A^j_{i,i-1} = 2\frac{\xi}{h^2} - \frac{1.5\Delta K^{hh}_{1,0} - 2\Delta K^{hh}_{0,0} + 0.5\Delta K^{hh}_{1,0}}{h} \tag{C.10}$$

For $i<n_x-1$

$$A^j_{i,i+1} = 2\frac{\xi}{h^2} - \frac{1.5\Delta K^{hh}_{1,0} - 2\Delta K^{hh}_{2,0} + 0.5\Delta K^{hh}_{3,0}}{h} \tag{C.11}$$

For $i<n_x-2$

$$A^j_{i,i+2} = -\frac{1}{4}\frac{\xi}{h^2} - \frac{1.5\Delta K^{hh}_{2,0} - 2\Delta K^{hh}_{3,0} + 0.5\Delta K^{hh}_{4,0}}{h} \tag{C.12}$$

Strictly speaking, Equation (C.5) should be modified for points near the boundary, ($i=1$, $i=n_x-1$, $j=1$, $j=n_y-1$). The equation should account for the fact that no changes will be applied to a neighbouring point if it is on the boundary. However, as relaxation does not solve the equation exactly anyway, the effect of using the expressions given above throughout the domain, has only a small influence on the the performance of the relaxation.

The system of equations defined above represents a full matrix and solving it by means of Gaussian elimination would require $O(n_x^3)$ operations. As was explained in Chapter 6 one full relaxation over the grid (creating the system of equations, solving it, and applying the changes to all lines of the grid) would then cost $O(N^2)$ operations leading to a very time-consuming solver. However, because ΔK decreases quickly with increasing distance $|i-k|$ it is justified to truncate the system of equations to a hexadiagonal system, taking $A^j_{i,k}=0$ for $k<i-3$ and $k>i+2$ ($0<k<n_x-1$). This hexadiagonal system of equations can be solved with Gaussian elimination in $O(n_x)$ operations and one relaxation sweep over the entire grid will cost $O(N)$ operations.

When the system for line j is solved, the changes $\delta^h_{i,j}$ multiplied with the underrelaxation factor ω_{ja} are applied to the current approximation to P^h at all points (i,j) of line j, and to its neighbours according to the distribution.

C.2 Model Problem Varying ξ

The discrete equation for gridpoint (i,j) is given by:

$$\begin{aligned} L^h_{i,j}\langle \underline{P}^h\rangle \;=\; & \frac{\xi^h_{i-1/2,j}P^h_{i-1,j} - (\xi^h_{i-1/2,j}+\xi^h_{i+1/2,j})P^h_{i,j} + \xi^h_{i+1/2,j}P^h_{i+1,j}}{h^2} + \\ & \frac{\xi^h_{i,j-1/2}P^h_{i,j-1} - (\xi^h_{i,j-1/2}+\xi^h_{i,j+1/2,j})P^h_{i,j} + \xi^h_{i,j+1/2}P^h_{i,j+1}}{h^2} - \\ & \frac{1.5H^h_{i,j} - 2H^h_{i-1,j} + 0.5H^h_{i-2,j}}{h} =_P f^h_{i,j} \end{aligned} \tag{C.13}$$

with

$$H^h_{i,j} - H_0 - \frac{X_i^2}{2} - \frac{Y_j^2}{2} + \sum_{i'}\sum_{j'} K^{hh}_{i,i',j,j'}P^h_{i',j'} =_H f^h_{i,j} \tag{C.14}$$

and

$$\begin{aligned} \xi^h_{i\pm1/2,j} &= (\xi^h_{i,j} + \xi^h_{i\pm1/2,j})/2 \\ \xi^h_{i,j\pm1/2} &= (\xi^h_{i,j} + \xi^h_{i,j\pm1/2})/2 \end{aligned} \tag{C.15}$$

with $\xi^h_{i,j} = \xi(x_i, y_j)$ a given function.

In this case two types of relaxation have to be combined. Gauss-Seidel line relaxation for large values of ξ and Jacobi distributive line relaxation for small ξ. This can be achieved by constructing the system (C.1) to be solved when relaxing a line j with the definition of the coefficients and the residual depending on the index i according to the *local* value of the coefficient ξ.

C.2.1 Gauss-Seidel Line Relaxation

A point i of the line with index j is defined as a Gauss-Seidel line relaxation point if $|\xi^h_{i+1/2,j}|/h^2$, $|\xi^h_{i-1/2,j}|/h^2$, $|\xi^h_{i,j+1/2}|/h^2$ and $|\xi^h_{i,j-1/2}|/h^2$ are larger than ξ^h_{limit}. For such a point the residual $r^h_{i,j}$ is defined by:

$$\begin{aligned} r^h_{i,j} &= {}_Pf^h_{i,j} - \\ &\frac{\xi^h_{i-1/2,j}\tilde{P}^h_{i-1,j} - (\xi^h_{i-1/2,j} + \xi^h_{i+1/2,j})\tilde{P}^h_{i,j} + \xi^h_{i+1/2,j}\tilde{P}^h_{i+1,j}}{h^2} - \\ &\frac{\xi^h_{i,j-1/2}\bar{P}^h_{i,j-1} - (\xi^h_{i,j-1/2} + \xi^h_{i,j+1/2,j})\tilde{P}^h_{i,j} + \xi^h_{i,j+1/2}\tilde{P}^h_{i,j+1}}{h^2} + \\ &\frac{1.5\tilde{H}^h_{i,j} - 2\tilde{H}^h_{i-1,j} + 0.5\tilde{H}^h_{i-2,j}}{h} \end{aligned} \tag{C.16}$$

and the coefficients $A^j_{i,k}$ for $0<i<n_x$ and $0<k<n_x$ by:

$$A^j_{i,k} = \left.\frac{\partial L^h_{i,j}\langle \underline{P}^h\rangle}{\partial P_{k,j}}\right|_{\underline{P}^h=\underline{\tilde{P}}^h} \tag{C.17}$$

for $|i-k| > 1$ this gives:

$$A^j_{i,k} = -\frac{1.5K^{hh}_{|i-k|,0} - 2K^{hh}_{|i-k-1|,0} + 0.5K^{hh}_{|i-k-2|,0}}{h} \tag{C.18}$$

For $i = k$ one obtains:

$$A^j_{i,i} = -\frac{\sum \xi^h}{h^2} - \frac{1.5K^{hh}_{0,0} - 2K^{hh}_{1,0} + 0.5K^{hh}_{2,0}}{h} \tag{C.19}$$

where

$$\sum \xi^h = \xi^h_{i+1/2,j} + \xi^h_{i-1/2,j} + \xi^h_{i,j+1/2} + \xi^h_{i,j-1/2} \tag{C.20}$$

For $i > 1$ one finds

$$A^j_{i,i-1} = \frac{\xi^h_{i-1/2,j}}{h^2} - \frac{1.5K^{hh}_{1,0} - 2K^{hh}_{0,0} + 0.5K^{hh}_{1,0}}{h} \tag{C.21}$$

and for $i<n_x-1$

$$A^j_{i,i+1} = \frac{\xi^h_{i+1/2,j}}{h^2} - \frac{1.5K^{hh}_{1,0} - 2K^{hh}_{2,0} + 0.5K^{hh}_{3,0}}{h} \tag{C.22}$$

C.2.2 Jacobi Distributive Line Relaxation

A point i is defined as a Jacobi distributive line relaxation point if $|\xi^h_{i+1/2,j}|/h^2$, $|\xi^h_{i-1/2,j}|/h^2$, $|\xi^h_{i,j+1/2}|/h^2$ *or* $|\xi^h_{i,j-1/2}|/h^2$ are smaller than or equal to ξ_{limit}. For such a point the residual $r^h_{i,j}$ is defined by:

$$\begin{aligned} r^h_{i,j} &= {}_Pf^h_{i,j} - \\ &\frac{\xi^h_{i-1/2,j}\tilde{P}^h_{i-1,j} - (\xi^h_{i-1/2,j} + \xi^h_{i+1/2,j})\tilde{P}^h_{i,j} + \xi^h_{i+1/2,j}\tilde{P}^h_{i+1,j}}{h^2} - \\ &\frac{\xi^h_{i,j-1/2}\tilde{P}^h_{i,j-1} - (\xi^h_{i,j-1/2} + \xi^h_{i,j+1/2,j})\tilde{P}^h_{i,j} + \xi^h_{i,j+1/2}\tilde{P}^h_{i,j+1}}{h^2} + \\ &\frac{1.5\tilde{H}^h_{i,j} - 2\tilde{H}^h_{i-1,j} + 0.5\tilde{H}^h_{i-2,j}}{h} \end{aligned} \tag{C.23}$$

and the coefficients $A^h_{i,k}$ are defined according to Equation (C.5). Introducing $\Delta^{hh}_{k,l}$ as defined by Equation (C.6) gives for $|i-k|>2$.

$$A^j_{i,k} = -\frac{1.5\Delta K^{hh}_{|i-k|,0} - 2\Delta K^{hh}_{|i-k-1|,0} + 0.5\Delta K^{hh}_{|i-k-2|,0}}{h} \tag{C.24}$$

For $i=k$ one obtains:

$$A^j_{i,i} - \frac{5}{4}\frac{\sum\xi^h}{h^2} - \frac{1.5\Delta K^{hh}_{0,0} - 2\Delta K^{hh}_{1,0} + 0.5\Delta K^{hh}_{2,0}}{h} \tag{C.25}$$

where

$$\sum\xi^h = \xi^h_{i+1/2,j} + \xi^h_{i-1/2,j} + \xi^h_{i,j+1/2} + \xi^h_{i,j-1/2} \tag{C.26}$$

For $i > 2$

$$A^j_{i,i-2} = -\frac{1}{4}\frac{\xi^h_{i-1/2,j}}{h^2} - \frac{1.5\Delta K^{hh}_{2,0} - 2\Delta K^{hh}_{1,0} + 0.5\Delta K^{hh}_{0,0}}{h} \tag{C.27}$$

For $i > 1$

$$A^j_{i,i-1} = \frac{\xi^h_{i-1/2,j}}{h^2} + \frac{1}{4}\frac{\sum\xi^h}{h^2} - \frac{1.5\Delta K^{hh}_{1,0} - 2\Delta K^{hh}_{0,0} + 0.5\Delta K^{hh}_{1,0}}{h} \tag{C.28}$$

For $i < n_x - 1$

$$A^j_{i,i+1} = \frac{\xi^h_{i+1/2,j}}{h^2} + \frac{1}{4}\frac{\sum\xi^h}{h^2} - \frac{1.5\Delta K^{hh}_{1,0} - 2\Delta K^{hh}_{2,0} + 0.5\Delta K^{hh}_{3,0}}{h} \tag{C.29}$$

For $i < n_x - 2$

$$A^j_{i,i+2} = -\frac{1}{4}\frac{\xi^h_{i+1/2,j}}{h^2} - \frac{1.5\Delta K^{hh}_{2,0} - 2\Delta K^{hh}_{3,0} + 0.5\Delta K^{hh}_{4,0}}{h} \tag{C.30}$$

Again it is not necessary to solve the full system of equations. It is sufficient to solve a hexadiagonal system. Notice that in practice this system is obtained directly using the above definitions for all $A^j_{i,k}$ with $A^j_{i,k} = 0$ for $k < i-3$ and $k > i+2$ $(0 < k < n_x - 1$ and $0 < i < n_x - 1)$. The resulting system is solved using Gaussian elimination. Subsequently, the changes are applied to all points i of the line j. If a point i is a Gauss-Seidel point the change $\omega_{gs}\delta^h_{i,j}$ is applied at point (i,j), where ω_{gs} is the underrelaxation factor. If the point i is a Jacobi distributive point the change $\omega_{ja}\delta^h_{i,j}$ is applied at the point (i,j) and a change $-\omega_{ja}\delta^h_{i,j}/4$ to its four neighbours.

C.3 EHL Problem

The discrete equation at gridpoint (i,j) is given by:

$$
\begin{aligned}
L^h_{i,j}\langle \underline{P}^h\rangle \; = \; & \frac{\xi^h_{i-1/2,j}P^h_{i-1,j} - (\xi^h_{i-1/2,j} + \xi^h_{i+1/2,j})P^h_{i,j} + \xi^h_{i+1/2,j}P^h_{i+1,j}}{h^2} + \\
& \frac{\xi^h_{i,j-1/2}P^h_{i,j-1} - (\xi^h_{i,j-1/2} + \xi^h_{i,j+1/2,j})P^h_{i,j} + \xi^h_{i,j+1/2}P^h_{i,j+1}}{h^2} - \\
& \frac{1.5\bar{\rho}^h_{i,j}H^h_{i,j} - 2\bar{\rho}^h_{i-1,j}H^h_{i-1,j} + 0.5\bar{\rho}^h_{i-2,j}H^h_{i-2,j}}{h} = {}_Pf^h_{i,j}
\end{aligned}
\tag{C.31}
$$

with $\bar{\rho}_{i,j} = \bar{\rho}(P^h_{i,j})$ and

$$
\begin{aligned}
\xi^h_{i\pm1/2,j} &= (\xi^h_{i,j} + \xi^h_{i\pm1/2,j})/2 \\
\xi^h_{i,j\pm1/2} &= (\xi^h_{i,j} + \xi^h_{i,j\pm1/2})/2
\end{aligned}
\tag{C.32}
$$

where

$$
\xi^h_{i,j} = \frac{\bar{\rho}(P^h_{i,j})(H^h_{i,j})^3}{\bar{\eta}(P^h_{i,j})\bar{\lambda}}
\tag{C.33}
$$

and

$$
H^h_{i,j} - H_0 - \frac{X_i^2}{2} - \frac{Y_j^2}{2} + \sum_{i'}\sum_{j'} K^{hh}_{i,i',j,j'}P^h_{i',j'} = {}_Hf^h_{i,j}
\tag{C.34}
$$

In the same way as described above, the system of equations (C.1), from which the changes associated with line j are to be solved, is constructed based on the local value of the coefficient ξ which now depends on $P^h_{i,j}$. Defining $\tilde{\xi}^h_{i,j} = \xi^h_{i,j}(\tilde{P}^h_{i,j})$ and $\tilde{\bar{\rho}}^h_{i,j} = \bar{\rho}(\tilde{P}^h_{i,j})$ the system of equations is constructed according to:

C.3.1 Gauss-Seidel Line Relaxation

A point i of the line with index j is defined as a Gauss-Seidel line relaxation point if $|\tilde{\xi}^h_{i+1/2,j}|/h^2$, $|\tilde{\xi}^h_{i-1/2,j}|/h^2$, $|\tilde{\xi}^h_{i,j+1/2}|/h^2$ and $|\tilde{\xi}^h_{i,j-1/2}|/h^2$ are larger than ξ^h_{limit}. For such a point the residual $r^h_{i,j}$ is defined by:

$$\begin{aligned} r^h_{i,j} &= {}_Pf^h_{i,j} - \\ &\frac{\tilde{\xi}^h_{i-1/2,j}\tilde{P}^h_{i-1,j} - (\tilde{\xi}^h_{i-1/2,j} + \tilde{\xi}^h_{i+1/2,j})\tilde{P}^h_{i,j} + \tilde{\xi}^h_{i+1/2,j}\tilde{P}^h_{i+1,j}}{h^2} - \\ &\frac{\tilde{\xi}^h_{i,j-1/2}\tilde{P}^h_{i,j-1} - (\tilde{\xi}^h_{i,j-1/2} + \tilde{\xi}^h_{i,j+1/2,j})\tilde{P}^h_{i,j} + \tilde{\xi}^h_{i,j+1/2}\tilde{P}^h_{i,j+1}}{h^2} + \\ &\frac{1.5\tilde{\bar{\rho}}^h_{i,j}\tilde{H}^h_{i,j} - 2\tilde{\bar{\rho}}^h_{i-1,j}\tilde{H}^h_{i-1,j} + 0.5\tilde{\bar{\rho}}^h_{i-2,j}\tilde{H}^h_{i-2,j}}{h} \end{aligned} \tag{C.35}$$

and the coefficients $A^j_{i,k}$ for $0<k<n_x$ by:

$$A^j_{i,k} = \left.\frac{\partial L^h_{i,j}\langle \underline{P}^h\rangle}{\partial P_{k,j}}\right|_{\underline{P}^h=\underline{\tilde{P}}^h} \tag{C.36}$$

Taking only the principal terms in the derivatives this gives for $|i-k|>1$:

$$A^j_{i,k} = -\frac{1.5\tilde{\bar{\rho}}^h_{i,j}K^{hh}_{|i-k|,0} - 2\tilde{\bar{\rho}}^h_{i-1,j}K^{hh}_{|i-k-1|,0} + 0.5\tilde{\bar{\rho}}^h_{i-2,j}K^{hh}_{|i-k-2|,0}}{h} \tag{C.37}$$

and for $i = k$

$$A^j_{i,i} = -\frac{\sum\tilde{\xi}^h}{h^2} - \frac{1.5\tilde{\bar{\rho}}^h_{i,j}K^{hh}_{0,0} - 2\tilde{\bar{\rho}}^h_{i-1,j}K^{hh}_{1,0} + 0.5\tilde{\bar{\rho}}^h_{i-2,j}K^{hh}_{2,0}}{h} \tag{C.38}$$

where

$$\sum\tilde{\xi}^h = \tilde{\xi}^h_{i+1/2,j} + \tilde{\xi}^h_{i-1/2,j} + \tilde{\xi}^h_{i,j+1/2} + \tilde{\xi}^h_{i,j-1/2} \tag{C.39}$$

Furthermore, for $i > 1$

$$A^j_{i,i-1} = \frac{\tilde{\xi}^h_{i-1/2,j}}{h^2} - \frac{1.5\tilde{\bar{\rho}}^h_{i,j}K^{hh}_{1,0} - 2\tilde{\bar{\rho}}^h_{i-1,j}K^{hh}_{0,0} + 0.5\tilde{\bar{\rho}}^h_{i-2,j}K^{hh}_{1,0}}{h} \tag{C.40}$$

and for $i< n_x-1$

$$A^j_{i,i+1} = \frac{\tilde{\xi}^h_{i+1/2,j}}{h^2} - \frac{1.5\tilde{\bar{\rho}}^h_{i,j}K^{hh}_{1,0} - 2\tilde{\bar{\rho}}^h_{i-1,j}K^{hh}_{2,0} + 0.5\tilde{\bar{\rho}}^h_{i-2,j}K^{hh}_{3,0}}{h} \tag{C.41}$$

C.3.2 Jacobi Distributive Line Relaxation

A point i is defined as a Jacobi distributive line relaxation point if $|\tilde{\xi}^h_{i+1/2,j}|/h^2$, $|\tilde{\xi}^h_{i-1/2,j}|/h^2$, $|\tilde{\xi}^h_{i,j+1/2}|/h^2$ *or* $|\tilde{\xi}^h_{i,j-1/2}|/h^2$ are smaller than or equal to ξ_{limit}. For such a point the residual $r^h_{i,j}$ is defined by:

$$\begin{aligned} r^h_{i,j} &= {}_P f^h_{i,j} - \\ &\frac{\tilde{\xi}^h_{i-1/2,j}\tilde{P}^h_{i-1,j} - (\tilde{\xi}^h_{i-1/2,j} + \tilde{\xi}^h_{i+1/2,j})\tilde{P}^h_{i,j} + \tilde{\xi}^h_{i+1/2,j}\tilde{P}^h_{i+1,j}}{h^2} - \\ &\frac{\tilde{\xi}^h_{i,j-1/2}\tilde{P}^h_{i,j-1} - (\tilde{\xi}^h_{i,j-1/2} + \tilde{\xi}^h_{i,j+1/2,j})\tilde{P}^h_{i,j} + \tilde{\xi}^h_{i,j+1/2}\tilde{P}^h_{i,j+1}}{h^2} + \\ &\frac{1.5\tilde{\bar{\rho}}^h_{i,j}\tilde{H}^h_{i,j} - 2\tilde{\bar{\rho}}^h_{i-1,j}\tilde{H}^h_{i-1,j} + 0.5\tilde{\bar{\rho}}^h_{i-2,j}\tilde{H}^h_{i-2,j}}{h} \end{aligned} \tag{C.42}$$

and the coefficients $A^h_{i,k}$ are defined according to Equation (C.5):

$$A^j_{i,k} = \left\{ \frac{\partial L^h_{i,j}\langle \underline{P}^h \rangle}{\partial P_{k,j}} - \frac{1}{4}\left(\frac{\partial L^h_{i,j}\langle \underline{P}^h \rangle}{\partial P_{k+1,j}} + \frac{\partial L^h_{i,j}\langle \underline{P}^h \rangle}{\partial P_{k-1,j}} + \frac{\partial L^h_{i,j}\langle \underline{P}^h \rangle}{\partial P_{k,j+1}} + \frac{\partial L^h_{i,j}\langle \underline{P}^h \rangle}{\partial P_{k,j-1}} \right) \right\} \Bigg|_{\underline{P}^h = \underline{\tilde{P}}^h} \tag{C.43}$$

Taking only the principal terms in the derivatives, and introducing $\Delta K^{hh}_{k,l}$ defined by Equation (C.6), gives:

$$A^j_{i,k} = -\frac{1.5\tilde{\bar{\rho}}^h_{i,j}\Delta K^{hh}_{|i-k|,0} - 2\tilde{\bar{\rho}}^h_{i-1,j}\Delta K^{hh}_{|i-k-1|,0} + 0.5\tilde{\bar{\rho}}^h_{i-2,j}\Delta K^{hh}_{|i-k-2|,0}}{h} \tag{C.44}$$

for $|i-k| > 2$. For $k = i$ one obtains:

$$A^j_{i,i} - \frac{5}{4}\frac{\sum\tilde{\xi}^h}{h^2} - \frac{1.5\tilde{\bar{\rho}}^h_{i,j}\Delta K^{hh}_{0,0} - 2\tilde{\bar{\rho}}^h_{i-1,j}\Delta K^{hh}_{1,0} + 0.5\tilde{\bar{\rho}}^h_{i-2,j}\Delta K^{hh}_{2,0}}{h} \tag{C.45}$$

where

$$\sum\tilde{\xi}^h = \tilde{\xi}^h_{i+1/2,j} + \tilde{\xi}^h_{i-1/2,j} + \tilde{\xi}^h_{i,j+1/2} + \tilde{\xi}^h_{i,j-1/2} \tag{C.46}$$

Furthermore, for $i > 2$

$$A^j_{i,i-2} = -\frac{1}{4}\frac{\tilde{\xi}^h_{i-1/2,j}}{h^2} - \frac{1.5\tilde{\bar{\rho}}^h_{i,j}\Delta K^{hh}_{2,0} - 2\tilde{\bar{\rho}}^h_{i-1,j}\Delta K^{hh}_{1,0} + 0.5\tilde{\bar{\rho}}^h_{i-2,j}\Delta K^{hh}_{0,0}}{h} \tag{C.47}$$

For $i > 1$

$$A^j_{i,i-1} = \frac{\tilde{\xi}^h_{i-1/2,j}}{h^2} + \frac{1}{4}\frac{\sum\tilde{\xi}^h}{h^2} - \frac{1.5\tilde{\bar{\rho}}^h_{i,j}\Delta K^{hh}_{1,0} - 2\tilde{\bar{\rho}}^h_{i-1,j}\Delta K^{hh}_{0,0} + 0.5\tilde{\bar{\rho}}^h_{i-2,j}\Delta K^{hh}_{1,0}}{h} \tag{C.48}$$

For $i < n_x - 1$

$$A^j_{i,i+1} = \frac{\tilde{\xi}^h_{i+1/2,j}}{h^2} + \frac{1}{4}\frac{\sum\tilde{\xi}^h}{h^2} - \frac{1.5\tilde{\bar{\rho}}^h_{i,j}\Delta K^{hh}_{1,0} - 2\tilde{\bar{\rho}}^h_{i-1,j}\Delta K^{hh}_{2,0} + 0.5\tilde{\bar{\rho}}^h_{i-2,j}\Delta K^{hh}_{3,0}}{h} \tag{C.49}$$

For $i < n_x - 2$

$$A^j_{i,i+2} = -\frac{1}{4}\frac{\tilde{\xi}^h_{i+1/2,j}}{h^2} - \frac{1.5\tilde{\bar{\rho}}^h_{i,j}\Delta K^{hh}_{2,0} - 2\tilde{\bar{\rho}}^h_{i-1,j}\Delta K^{hh}_{3,0} + 0.5\tilde{\bar{\rho}}^h_{i-2,j}\Delta K^{hh}_{4,0}}{h} \tag{C.50}$$

Unlike the model problems, the solution of the EHL problem is subject to a special condition, the cavitation condition $P^h_{i,j} \geq 0$. This condition should already be taken into account when constructing the system of equations to be solved in the relaxation. In Chapter 6 it has already been mentioned that solving the system and only imposing the condition $P^h_{i,j} \geq 0$ after the changes to the line are applied will lead to a situation where convergence of the relaxation may stall because a group of points of a given line is switched back and forth between cavitated and non-cavitated. This is caused by the fact that the changes of the line are solved simultaneously which implies that the change for point $i+1$ is solved as if a point i is not cavitated at all. This is fundamentally different from a Gauss-Seidel one point relaxation. In that case when relaxing the next point of a line, cavitation of the previous point is already taken into account. Summarizing, even when a line relaxation is used, it is desirable for the relaxation to behave as much as a point relaxation as possible in the vicinity of the cavitation boundary. This is achieved by setting $A^j_{i,k} = 0$ for $k \neq i$ if the current pressure is zero at one of the points $(i-2,j)$, $(i-1,j)$, (i,j), $(i+1,j)$, or $(i+2,j)$.

As for the previous cases only a hexadiagonal system is constructed. The resulting system of equations for all $1 \leq i \leq n_x - 1$ is solved using Gaussian elimination and subsequently the changes $\delta^h_{i,j}$ are applied to the line j. Also when applying the changes, the cavitation condition must be taken into account. The procedure to apply the changes solved for line j is the following. If a point (i,j) is a Gauss-Seidel point the change $\omega_{gs}\delta^h_{i,j}$ is applied where ω_{gs} is an underrelaxation factor. If the resulting pressure is smaller than zero it is set to zero. If the point is a Jacobi distributive point the change $\omega_{ja}\delta^h_{i,j}$ is applied to the point (i,j) where ω_{ja} is an underrelaxation factor. If the resulting pressure is smaller than zero it is set to zero. The net change, i.e. corrected for the cavitation condition, is then distributed to the neighbouring points according to the distribution. Finally, the change to the central point is always applied, however, the changes to the neighbouring points are only applied if the current pressure at these points is larger than zero and also, after these changes have been applied the cavitation condition $P \geq 0$ is imposed in these points.

Appendix D

Program Listing: MG1d.c

```
#include <stdio.h>
#include <stdlib.h>
#include <math.h>

/* FMG solver for the 1d poisson model problem   */

#define pi  3.1415926535897931


typedef struct
{
double hx;              /* mesh size                            */
int    ii;              /* number of nodes                      */
double *u, *f;          /* unknown and right hand side values   */
                        /* For convenience:                     */
double *uconv;          /* extra u for convergence check in FMG */
double *uold;           /* extra u for use in FAS coarsening    */
} Level;


typedef struct
{
int    nx0;               /* number of nodes coarsest grid      */
int    maxlevel;          /* number of grid-levels              */
double xa,xb;             /* begin,end of computational domain */
double wu;                /* work unit counter                  */
Level  *Lk ;              /* array of grid levels               */
} Stack;


/**********ROUTINES FOR DATASTRUCTURE**********/


void initialize(Stack *U, int nx0, int maxlevel, double xa, double xb)
{
/* initialize values in datastructure */

double  hx;
Level  *L;
```

```
int     i,ii;

U->xa=xa;
U->xb=xb;
U->maxlevel=maxlevel;
U->wu=0.0;

U->Lk=(Level *)calloc(maxlevel+1,sizeof(Level));

hx=(xb-xa)/nx0;
ii=nx0;
for (i=1;i<=maxlevel;i++)
  {
  L=U->Lk+i;
  L->hx=hx;
  L->ii=ii;
  L->f    =(double *)calloc(ii+1,sizeof(double));
  L->u    =(double *)calloc(ii+1,sizeof(double));
  L->uconv=(double *)calloc(ii+1,sizeof(double));
  L->uold =(double *)calloc(ii+1,sizeof(double));
  hx*=0.5;
  ii*=2;
  }
}


void finalize(Stack *U, int maxlevel)
{
/* frees memory at the end of the program */

Level  *L;
int     i;

for (i=1;i<=maxlevel;i++)
  {
  L=U->Lk+i;
  free(L->f);
  free(L->u);
  free(L->uconv);
  free(L->uold);
  }
free(U->Lk);
}


/*****************FUNCTIONS*****************/


double U_a(double x)
{
/* analytical solution */

return(sin(2*pi*x));
}
```

```
double U_i(double x)
{
/* initial approximation */

return(0.0);
}


double F_i(double x)
{
/* right hand side function */

return(-4.0*pi*pi*sin(2*pi*x));
}


double Lu(double u[], double rh2, int i)
{
/* Computes operator in point i */

return(rh2*(u[i-1]-2*u[i]+u[i+1]));
}


/**********SINGLE GRID ROUTINES**********/


void init_uf(Stack *U, double (*u0)(double x), double (*f0)(double x), int k)
{
/* initializes u and equation right hand side values on grid level k */

int    i;
Level  *L;
double x;

L =U->Lk+k;

for (i=0;i<=L->ii;i++)
    {
    x=U->xa+i*L->hx;
    L->u[i]=u0(x);
    L->f[i]=f0(x);
    }
}


void relax(Stack *U, int k)
{
/* perform Gauss-Seidel relaxation on gridlevel k */

Level  *L;
double hx,rh2,rn,*u,*f;
```

```
int    i,ii;

L =U->Lk+k;
hx=L->hx;
ii=L->ii;
u =L->u;
f =L->f;

rh2=1.0/(hx*hx);

for (i=1;i<=ii-1;i++)
    u[i]+=(f[i]-Lu(u,rh2,i))/(-2*rh2);

rn=0.0;
for (i=1;i<=ii-1;i++)
    rn+=fabs(f[i]-Lu(u,rh2,i));

U->wu += exp((U->maxlevel-k)*log(0.5));
printf("\nLevel %d    Residual Norm  %8.5e, WU %7.2e",k,rn/(ii-1),U->wu);
}


double conver_a(Stack *U, int k)
{
/* computes L1 norm of difference between U and analytic solution */
/* on level k                                                      */

Level  *L;
double x,en,*u;
int    i;

L =U->Lk+k;
u =L->u;

en=0.0;

for (i=1;i<=L->ii-1;i++)
  {
  x=U->xa+i*L->hx;
  en+=fabs(u[i]-U_a(x));
  }
return(en/(L->ii-1));
}


/****************INTER GRID ROUTINES********************/


void coarsen_u(Stack *U, int k)
{
/* compute initial approximation on level k-1 */
/* in coarsening step from level k            */

int    ic,iic;
```

```
Level  *Lc,*L;
double *uc,*u;

L =U->Lk+k;
u =L->u;

Lc =U->Lk+k-1;
iic=Lc->ii;
uc =Lc->u;

for (ic=1;ic<=iic-1;ic++)
  uc[ic]=0.25*(u[2*ic-1]+2.*u[2*ic]+u[2*ic+1]);

for (ic=0;ic<=iic;ic++)
  Lc->uold[ic]=uc[ic];
}


void coarsen_f(Stack *U, int k)
{
/* compute coarse grid right hand side on level k-1 */
/* in coarsening step from level k                  */

int    ic,iic;
Level  *Lc,*L;
double *uc,*u,*fc,*f,hc,hx, rhc2, rh2;
double rm,rc,rp;

L =U->Lk+k;
hx=L->hx;
u =L->u;
f =L->f;
rh2=1.0/(hx*hx);

Lc =U->Lk+k-1;
hc =Lc->hx;
iic=Lc->ii;
uc =Lc->u;
fc =Lc->f;
rhc2=1.0/(hc*hc);

for (ic=1;ic<=iic-1;ic++)
  {
  rm=f[2*ic-1]-Lu(u,rh2,2*ic-1);
  rc=f[2*ic  ]-Lu(u,rh2,2*ic  );
  rp=f[2*ic+1]-Lu(u,rh2,2*ic+1);

  fc[ic]=Lu(uc,rhc2,ic)+0.25*(rm+2*rc+rp);
  }
}


void refine(Stack *U, int k)
{
```

```
/* interpolation and addition of coarse grid correction from grid k-1*/
/* to grid k                                                          */

int    ic,iic;
Level  *Lc,*L;
double *uc,*u;

L =U->Lk+k;
u =L->u;

Lc =U->Lk+k-1;
iic=Lc->ii;
uc =Lc->u;

for (ic=1;ic<=iic-1;ic++)
  {
  u[2*ic-1]+=0.5*(uc[ic]-Lc->uold[ic]);
  u[2*ic  ]+=     uc[ic]-Lc->uold[ic] ;
  u[2*ic+1]+=0.5*(uc[ic]-Lc->uold[ic]);
  }
}


void fmg_interpolate(Stack *U, int k)
{
/* interpolation of coarse grid k-1 solution to fine grid k */
/* to serve as first approximation. Cubic interpolation     */

int     ic,i,iic,ii;
Level   *Lc,*L;
double  *uc,*u,*uconv;
double  den;

L =U->Lk+k;
ii=L->ii;
u =L->u;

Lc   =U->Lk+k-1;
iic  =Lc->ii;
uc   =Lc->u;
uconv=Lc->uconv;

/* store grid k-1 solution for later use in convergence check */

for (ic=0;ic<=iic;ic++) uconv[ic]=uc[ic];

den=1.0/16.0;

/* injection to points coinciding with coarse grid */

for (ic=0;ic<=iic;ic++) u[2*ic]=uc[ic];

/* non-central interpolation near boundary */
```

```
u[1]=(5.0*u[0]+15*u[2]-5*u[4]+u[6])*den;
u[ii-1]=(5.0*u[ii]+15*u[ii-2]-5*u[ii-4]+u[ii-6])*den;

/* central interpolation for all interior points */

for (i=3;i<=ii-3;i+=2)
  u[i]=(-u[i-3]+9.0*u[i-1]+9.0*u[i+1]-u[i+3])*den;

}


double conver(Stack *U, int k)
{
/* convergence check using converged solutions on level k and */
/* on next coarser grid k-1                                    */

int    ic;
Level  *Lc,*L;
double *uc,*u;
double aen;

L =U->Lk+k;

if (k==U->maxlevel) u=L->u;
  else u=L->uconv;

Lc =U->Lk+k-1;
uc =Lc->uconv;

aen=0.0;
for (ic=1;ic<=Lc->ii-1;ic++)
    aen+=fabs(uc[ic]-u[2*ic]);

return(aen/(L->ii-1));
}


/****************MULTIGRID DRIVING ROUTINES*************/


void cycle(Stack *U, int k, int nu0, int nu1, int nu2, int gamma)

/* performs coarse grid correction cycle starting on level k */
/* nu1 pre-relaxations, nu2 postrelaxations, nu0 relaxations */
/* on the coarsest grid, cycleindex gamma=1 for Vcycle,      */
/* gamma=2 for Wcycle                                         */

{
int i,j;

if (k==1)
  for (i=1;i<=nu0;i++) relax(U,k);
else
  {
```

```
  for (i=1;i<=nu1;i++) relax(U,k);
  coarsen_u(U,k);
  coarsen_f(U,k);
  for (j=1;j<=gamma;j++) cycle(U,k-1,nu0,nu1,nu2,gamma);
  refine(U,k);
  for (i=1;i<=nu2;i++) relax(U,k);
  }
}


void fmg(Stack *U, int k, int nu0, int nu1, int nu2, int gamma, int ncy)
{
/* performs FMG with k levels and ncy cycles per level */

int i,j;

if (k==1) for (i=1;i<=nu0;i++) relax(U,k);
else
  {
  fmg(U,k-1,nu0,nu1,nu2,gamma,ncy);
  fmg_interpolate(U,k);
  for (j=1;j<=ncy;j++)
    {
    cycle(U,k,nu0,nu1,nu2,gamma);
    printf("\n");
    }
  }
}


/**********MAIN PROGRAM**********/


void main()
{
Stack  U;
int j,maxlev,ncy;

printf("\ngive maxlevel");
scanf("%d",&maxlev);
printf("\ngive ncy      ");
scanf("%d",&ncy);

initialize(&U,4,maxlev,0.0,1.0);

for (j=maxlev;j>=1;j--) init_uf(&U,U_i,F_i,j);
fmg(&U,maxlev,10,2,1,1,ncy);

printf("\ner:%10.8e\n\n",conver_a(&U,maxlev));

if (maxlev>1)
   for (j=2;j<=maxlev;j++) printf("aen(%2d,%2d)=%8.5e\n",j,j-1,conver(&U,j));
finalize(&U, maxlev);
}
```

Appendix E

Program Listing: MG2d.c

```
#include <stdio.h>
#include <stdlib.h>
#include <math.h>

/* FMG solver for the 2d poisson model problem         */

#define pi  3.1415926535897931

typedef struct
{
double hx;              /* mesh size x                        */
double hy;              /* mesh size y                        */
int    ii;              /* number of nodes x                  */
int    jj;              /* number of nodes y                  */
double **u, **f;        /* unknown and right hand side values */
                        /* For convenience:                   */
double **uconv;         /* extra u for convergence check in FMG */
double **uold;          /* extra u for use in FAS coarsening  */
} Level;


typedef struct
{
int    nx0;              /* number of nodes coarsest grid     */
int    ny0;              /* number of nodes coarsest grid     */
int    maxlevel;         /* number of grid-levels             */
double xa,xb;            /* begin,end of computational domain */
double ya,yb;            /* begin,end of computational domain */
double wu;               /* work unit counter                 */
Level  *Lk ;             /* array of grid levels              */
} Stack;


/**********ROUTINES FOR DATASTRUCTURE**********/


double **matrix(int nx,int ny)
{
 int i;
```

```
 double **m=(double **)calloc(nx+1,sizeof(double*));

 for (i=0;i<=nx;i++) m[i]=(double *)calloc(ny+1,sizeof(double));
 return m;
}


void initialize(Stack *U, int nx0, int ny0, int maxlevel,
                double xa, double xb, double ya, double yb)
{
/* initialize values in datastructure */

double  hx,hy;
Level  *L;
int     i,ii,jj;

U->xa=xa;
U->xb=xb;
U->ya=ya;
U->yb=yb;
U->maxlevel=maxlevel;
U->wu=0.0;

U->Lk=(Level *)calloc(maxlevel+1,sizeof(Level));

hx=(xb-xa)/nx0;
hy=(yb-ya)/ny0;
ii=nx0;
jj=ny0;
for (i=1;i<=maxlevel;i++)
  {
  L=U->Lk+i;
  L->hx=hx;
  L->hy=hy;
  L->ii=ii;
  L->jj=jj;
  L->f    =matrix(ii,jj);
  L->u    =matrix(ii,jj);
  L->uold =matrix(ii,jj);
  L->uconv=matrix(ii,jj);
  if ((L->f==NULL)||(L->u==NULL)||(L->uold==NULL)||(L->uconv==NULL))
    {
    printf("\nproblem allocating memory\n");
    exit(-1);
    }
  hx*=0.5;
  hy*=0.5;
  ii*=2;
  jj*=2;
  }
}
```

```
void finalize(Stack *U, int maxlevel)
{
/* free memory at end of program */

Level  *L;
int     i;
for (i=1;i<=maxlevel;i++)
  {
  L=U->Lk+i;
  free(L->f);
  free(L->u);
  free(L->uold);
  free(L->uconv);
  }
free(U->Lk);
}


/**********FUNCTIONS**********/


double U_a(double x, double y)
{
/* analytical solution */

return(sin(2*pi*x)*sin(2*pi*y));
}


double U_b(double x, double y)
{
/* take analytical solution as boundary condition */

return(sin(2*pi*x)*sin(2*pi*y));
}


double U_i(double x, double y)
{
/* initial approximation */

return(0.0);
}


double F_i(double x, double y)
{
/* right hand side function */

return(-8.0*pi*pi*sin(2*pi*x)*sin(2*pi*y));
}


double Lu(double **u, double rhx2, double rhy2, int i, int j)
```

```
{
/* Computes operator in point i,j */

return(rhx2*(u[i-1][j  ]-2*u[i][j]+u[i+1][j  ])+
       rhy2*(u[i  ][j-1]-2*u[i][j]+u[i  ][j+1]));
}


/**********SINGLE GRID ROUTINES**********/


void init_uf(Stack *U, double (*u0)(double x, double y),
                       double (*ub)(double x, double y),
                       double (*f0)(double x, double y), int k)

/* initializes u, sets boundary condition for u, and  */
/* initializes right hand side values on grid level k */
{
int    i,j,ii,jj;
Level *L;
double x,y;

L =U->Lk+k;
ii=L->ii;
jj=L->jj;

for (i=0;i<=ii;i++)
  {
  x=U->xa+i*L->hx;
  for (j=0;j<=jj;j++)
    {
    y=U->ya+j*L->hy;
    if ((i==0)||(j==0)||(i==ii)||(j==jj))
      L->u[i][j]=ub(x,y);
    else
      L->u[i][j]=u0(x,y);
    L->f[i][j]=f0(x,y);
    }
  }
}


void relax(Stack *U, int k)
{
/* perform Gauss-Seidel relaxation on gridlevel k */

Level *L;
double hx, hy, rhx2, rhy2, err, **u, **f;
int    i,j,ii,jj;

L =U->Lk+k;
hx=L->hx;
hy=L->hy;
ii=L->ii;
```

```
jj=L->jj;
u =L->u;
f =L->f;

rhx2=1.0/(hx*hx);
rhy2=1.0/(hy*hy);

for (i=1;i<=ii-1;i++)
  for (j=1;j<=jj-1;j++)
    u[i][j]+=(f[i][j]-Lu(u,rhx2,rhy2,i,j))/(-2*rhx2-2*rhy2);

err=0.0;
for (i=1;i<=ii-1;i++)
  for (j=1;j<=jj-1;j++)
    err+=fabs(f[i][j]-Lu(u,rhx2,rhy2,i,j) );

U->wu += exp((U->maxlevel-k)*log(0.25));
printf("\nLevel %d Residual %8.5e Wu %8.5e",k,err/((ii-1)*(jj-1)),U->wu);
}


double conver_a(Stack *U, int k)
{
/* computes L1 norm of difference between U and analytic solution */
/* on level k                                                      */

Level  *L;
double x,y,err,**u;
int    i,j;

L =U->Lk+k;
u =L->u;

err=0.0;

for (i=1;i<=L->ii-1;i++)
  {
  x=U->xa+i*L->hx;
  for (j=1;j<=L->jj-1;j++)
    {
    y=U->ya+j*L->hy;
    err+=fabs(u[i][j]-U_a(x,y));
    }
  }
return(err/((L->ii-1)*(L->jj-1)));
}


/**********INTER GRID ROUTINES**********/


void coarsen_u(Stack *U, int k)
{
/* compute initial approximation on level k-1 */
```

```
/* in coarsening step from level k            */

int     ic,jc,iic,jjc;
Level  *Lc, *L;
double **uc, **uco, **u ;

L =U->Lk+k;
u =L->u;

Lc =U->Lk+k-1;
iic=Lc->ii; jjc=Lc->jj;
uc =Lc->u;
uco=Lc->uold;

for (ic=1;ic<=iic-1;ic++)
  for (jc=1;jc<=jjc-1;jc++)
    uc[ic][jc]=0.0625*(     u[2*ic-1][2*jc-1]+u[2*ic-1][2*jc+1]+
                            u[2*ic+1][2*jc-1]+u[2*ic+1][2*jc+1]+
                       2.0*(u[2*ic-1][2*jc  ]+u[2*ic+1][2*jc  ]+
                            u[2*ic  ][2*jc-1]+u[2*ic  ][2*jc+1])+

                        4.0*u[2*ic][2*jc]);

/* store coarse grid solution in uc0 array */
for (ic=0;ic<=iic;ic++)
  for (jc=0;jc<=jjc;jc++)
    uco[ic][jc]=uc[ic][jc];
}


void coarsen_f(Stack *U, int k)
{
/* compute coarse grid right hand side on level k-1 */
/* in coarsening step from level k                  */

int     ic,jc,iic,jjc;
Level  *Lc,*L;
double **u,**uc,**f,**fc ;
double hx,hy,rh2x,rh2y,hxc,hyc,rh2xc,rh2yc   ;
double rc,rn,re,rs,rw,rne,rse,rnw,rsw;

L  =U->Lk+k;
hx =L->hx; hy=L->hy;
u  =L->u;
f  =L->f;

Lc  =U->Lk+k-1;
iic =Lc->ii; jjc=Lc->jj;
hxc =Lc->hx; hyc=Lc->hy;
fc  =Lc->f;
uc  =Lc->u;

rh2xc=1.0/(hxc*hxc);
rh2yc=1.0/(hyc*hyc);
```

```
rh2x =1.0/(hx*hx);
rh2y =1.0/(hy*hy);

for (ic=1;ic<=iic-1;ic++)
  for (jc=1;jc<=jjc-1;jc++)
    {
    rc =(f[2*ic  ][2*jc  ] -Lu(u,rh2x,rh2y,2*ic  ,2*jc  ));
    rn =(f[2*ic  ][2*jc+1] -Lu(u,rh2x,rh2y,2*ic  ,2*jc+1));
    re =(f[2*ic+1][2*jc  ] -Lu(u,rh2x,rh2y,2*ic+1,2*jc  ));
    rs =(f[2*ic  ][2*jc-1] -Lu(u,rh2x,rh2y,2*ic  ,2*jc-1));
    rw =(f[2*ic-1][2*jc  ] -Lu(u,rh2x,rh2y,2*ic-1,2*jc  ));
    rne=(f[2*ic+1][2*jc+1] -Lu(u,rh2x,rh2y,2*ic+1,2*jc+1));
    rse=(f[2*ic+1][2*jc-1] -Lu(u,rh2x,rh2y,2*ic+1,2*jc-1));
    rsw=(f[2*ic-1][2*jc-1] -Lu(u,rh2x,rh2y,2*ic-1,2*jc-1));
    rnw=(f[2*ic-1][2*jc+1] -Lu(u,rh2x,rh2y,2*ic-1,2*jc+1));

    /* FAS coarse grid right hand side   */
    /* with full weighting of residuals  */

    fc[ic][jc]=Lu(uc,rh2xc,rh2yc,ic,jc)+
               0.0625*(rne+rse+rsw+rnw+2.0*(rn+re+rs+rw)+4.0*rc);
    }
}


void refine(Stack *U, int k)
{
/* Interpolation and addition of coarse grid correction from grid k-1 */
/* to grid k                                                           */

int     ic,jc,iic,jjc;
Level  *Lc,*L;
double **uc,**uco,**u;

L  =U->Lk+k;
u  =L->u  ;

Lc =U->Lk+k-1;
iic=Lc->ii; jjc=Lc->jj;
uc =Lc->u ;
uco=Lc->uold ;

for (ic=1;ic<=iic;ic++)
  for (jc=1;jc<=jjc;jc++)
    {
    if (ic<iic) u[2*ic  ][2*jc  ]+=(uc[ic][jc]-uco[ic][jc]);

    if (jc<jjc) u[2*ic-1][2*jc  ]+=(uc[ic  ][jc]-uco[ic  ][jc]+
                                    uc[ic-1][jc]-uco[ic-1][jc])*0.5;

    if (ic<iic) u[2*ic  ][2*jc-1]+=(uc[ic][jc  ]-uco[ic][jc  ]+
                                    uc[ic][jc-1]-uco[ic][jc-1])*0.5;

    u[2*ic-1][2*jc-1]+=(uc[ic  ][jc  ]-uco[ic  ][jc  ]+
```

```
                        uc[ic  ][jc-1]-uco[ic  ][jc-1]+
                        uc[ic-1][jc  ]-uco[ic-1][jc  ]+
                        uc[ic-1][jc-1]-uco[ic-1][jc-1])*0.25;
    }
}


void fmg_interpolate(Stack *U, int k)
{
/* interpolation of coarse grid k-1 solution to fine grid k */
/* to serve as first approximation. bi-cubic interpolation  */

int    ic,jc,iic,jjc,i,j,ii,jj;
Level  *Lc,*L;
double **uc,**u, **uconv;

L  =U->Lk+k;
ii =L ->ii; jj =L->jj;
u  =L->u  ;

Lc    =U->Lk+k-1;
iic   =Lc->ii; jjc=Lc->jj;
uc    =Lc->u ;
uconv =Lc->uconv;


/* store grid k-1 solution for later use in convergence check */

for (ic=1;ic<=iic-1;ic++)
  for (jc=1;jc<=jjc-1;jc++)
    uconv[ic][jc]=uc[ic][jc];

/* first inject to points coinciding with coarse grid points */

for (ic=1;ic<=iic-1;ic++)
  for (jc=1;jc<=jjc-1;jc++)
    u[2*ic][2*jc]=uc[ic][jc];

/* interpolate intermediate y direction */

for (i=2;i<=ii-2;i+=2)
  {
  u[i][1]=(5.0*u[i][0]+15.0*u[i][2]-5.0*u[i][4]+u[i][6])*0.0625;

  for (j=3;j<=jj-3;j+=2)
    u[i][j]=(-u[i][j-3]+9.0*u[i][j-1]+9.0*u[i][j+1]-u[i][j+3])*0.0625;

  u[i][jj-1]=(5.0*u[i][jj]+15.0*u[i][jj-2]-5.0*u[i][jj-4]+u[i][jj-6])*0.0625;
  }

/* interpolate in x direction */

for (j=1;j<=jj-1;j++)
  {
```

```
  u[1][j]=(5.0*u[0][j]+15.0*u[2][j]-5.0*u[4][j]+u[6][j])*0.0625;

  for (i=3;i<=ii-3;i+=2)
    u[i][j]=(-u[i-3][j]+9.0*u[i-1][j]+9.0*u[i+1][j]-u[i+3][j])*0.0625;

  u[ii-1][j]=(5.0*u[ii][j]+15.0*u[ii-2][j]-5.0*u[ii-4][j]+u[ii-6][j])*0.0625;
  }
}


double conver(Stack *U, int k)
{
/* convergence check using converged solutions on level k */
/* and on next coarser grid k-1                            */

int    ic,jc;
Level  *Lc,*L;
double **uc,**u;
double err;

L =U->Lk+k;

if (k==U->maxlevel) u=L->u;
  else u=L->uconv;

Lc =U->Lk+k-1;
uc =Lc->uconv;

err=0.0;
for (ic=1;ic<=Lc->ii-1;ic++)
  for (jc=1;jc<=Lc->jj-1;jc++)
  err+=fabs(uc[ic][jc]-u[2*ic][2*jc]);

return(err/((Lc->ii-1)*(Lc->jj-1)));
}


/**********MULTIGRID DRIVING ROUTINES**********/


void cycle(Stack *U, int k, int nu0, int nu1, int nu2, int gamma)

/* performs coarse grid correction cycle starting on level k */
/* nu1 pre-relaxations, nu2 postrelaxations, nu0 relaxations */
/* on the coarsest grid, cycleindex gamma=1 for Vcycle,      */
/* gamma=2 for Wcycle                                        */

{
int i,j;

if (k==1)
  for (i=1;i<=nu0;i++) relax(U,k);
else
  {
```

```
  for (i=1;i<=nu1;i++) relax(U,k);
  coarsen_u(U,k);
  coarsen_f(U,k);
  for (j=1;j<=gamma;j++) cycle(U,k-1,nu0,nu1,nu2,gamma);
  refine(U,k);
  for (i=1;i<=nu2;i++) relax(U,k);
  }
}


void fmg(Stack *U, int k, int nu0, int nu1, int nu2, int gamma, int ncy)
{
/* performs FMG with k levels and ncy cycles per level */

int i,j;

if (U->maxlevel==1)
  for (j=1;j< ncy;j++)
    for (i=1;i<=nu0;i++) relax(U,k);
if (k==1) for (i=1;i<=nu0;i++) relax(U,k);
else
  {
  fmg(U,k-1,nu0,nu1,nu2,gamma,ncy);
  fmg_interpolate(U,k);
  for (j=1;j<=ncy;j++)
    {
    cycle(U,k,nu0,nu1,nu2,gamma);
    printf("\n");
    }
  }
}


/**********MAIN PROGRAM**********/


void main()
{
Stack  U;
int j,maxlev,ncy;

printf("\ngive maxlevel"); scanf("%d",&maxlev);
printf("\ngive ncy      "); scanf("%d",&ncy);

initialize(&U,4,4,maxlev,0.0,1.0,0.0,1.0);

for (j=maxlev;j>=1;j--) init_uf(&U, U_i, U_b, F_i, j);

fmg(&U,maxlev,10,2,1,1,ncy);

printf("\n\nLevel %d: er=%10.8e\n\n",maxlev, conver_a(&U,maxlev));

if (maxlev>1)
  {
```

```
  printf("\n");
  for (j=2;j<=maxlev;j++) printf("\naen(%2d,%2d)=%8.5e",j,j-1,conver(&U,j));
  }

printf("\n");
finalize(&U,maxlev);
}
```

Appendix F

Program Listing: HL2d.c

```
#include <stdio.h>
#include <stdlib.h>
#include <math.h>

/* FMG solver for the 2d hydrodynamic lubrication problem          */
/* including the cavitation condition                              */

#define pi  3.1415926535897931
double epsilon, RoL, LoR = 1.0;

typedef struct
{
double hx;              /* mesh size x                                      */
double hy;              /* mesh size y                                      */
int    ii;              /* number of nodes x                                */
int    jj;              /* number of nodes y                                */
double **p, **h, **f;   /* pressure, film thickness and right hand side values */
                        /* For convenience:                                 */
double **pconv;         /* converged pressure for convergence check in FMG  */
double **pold;          /* 'old' pressure for use in FAS coarsening         */
} Level;


typedef struct
{
int    nx0;              /* number of nodes coarsest grid     */
int    ny0;              /* number of nodes coarsest grid     */
int    maxlevel;         /* number of grid-levels             */
double xa,xb;            /* begin,end of computational domain */
double ya,yb;            /* begin,end of computational domain */
double wu;               /* work unit counter                 */
Level  *Lk ;             /* array of grid levels              */
} Stack;


/**********ROUTINES FOR DATASTRUCTURE**********/


double **matrix(int nx,int ny)
```

```
{
 int i;

 double **m=(double **)calloc(nx+1,sizeof(double*));

 for (i=0;i<=nx;i++) m[i]=(double *)calloc(ny+1,sizeof(double));
 return m;
}


void initialize(Stack *U, int nx0, int ny0, int maxlevel,
                double xa, double xb, double ya, double yb)
{
/* initialize values in datastructure */

double  hx,hy;
Level  *L;
int     i,ii,jj;

U->xa=xa;
U->xb=xb;
U->ya=ya;
U->yb=yb;
U->maxlevel=maxlevel;
U->wu=0.0;

U->Lk=(Level *)calloc(maxlevel+1,sizeof(Level));

hx=(xb-ya)/nx0;
hy=(yb-ya)/ny0;
RoL = 1.0/LoR;
ii=nx0;
jj=ny0;
for (i=1;i<=maxlevel;i++)
  {
  L=U->Lk+i;
  L->hx=hx;
  L->hy=hy;
  L->ii=ii;
  L->jj=jj;
  L->p    =matrix(ii,jj);
  L->h    =matrix(ii,jj);
  L->f    =matrix(ii,jj);
  L->pold =matrix(ii,jj);
  L->pconv=matrix(ii,jj);
  if ((L->f==NULL)||(L->h==NULL)||(L->p==NULL)||
      (L->pold==NULL)||(L->pconv==NULL))
    {
    printf("\nproblem allocating memory\n");
    exit(-1);
    }
  hx*=0.5;
  hy*=0.5;
  ii*=2;
```

```
  jj*=2;
  }
}


void finalize(Stack *U, int maxlevel)
{
/* free memory at end of program */

Level  *L;
int     i;
for (i=1;i<=maxlevel;i++)
  {
  L=U->Lk+i;
  free(L->f);
  free(L->h);
  free(L->p);
  free(L->pold);
  free(L->pconv);
  }
free(U->Lk);
}


/**********FUNCTIONS**********/


double P_b(double x, double y)
{
/* take the boundary condition as zero */

return(0.0);
}


double P_i(double x, double y)
{
/* zero initial approximation */

return(0.0);
}


double H_i(double x, double y)
{
/* film thickness function */

return(1.0+epsilon*cos(x));
}


double F_i(double x, double y)
{
/* zero right hand side function */
```

```
return(0.0);
}


double Lu(double **p, double **h, double rhx2, double rhy2, int i, int j)
{
/* Computes operator in point i,j */
double hr, hl, hu, hd;

hr = 0.5*(h[i+1][j]+h[i][j]);
hl = 0.5*(h[i-1][j]+h[i][j]);
hu = 0.5*(h[i][j+1]+h[i][j]);
hd = 0.5*(h[i][j-1]+h[i][j]);

return(  rhx2*(hr*hr*hr*(p[i+1][j  ]-p[i  ][j  ])
             -hl*hl*hl*(p[i  ][j  ]-p[i-1][j  ]))
+RoL*RoL*rhy2*(hu*hu*hu*(p[i  ][j+1]-p[i  ][j  ])
             -hd*hd*hd*(p[i  ][j  ]-p[i  ][j-1]))
      -0.5*sqrt(rhx2)*(h[i+1][j]-h[i-1][j]));
}


/**********SINGLE GRID ROUTINES**********/


void init_uf(Stack *U, double (*p0)(double x, double y),
                       double (*pb)(double x, double y),
                       double (*h0)(double x, double y),
                       double (*f0)(double x, double y), int k)

/* initializes u, sets boundary condition for u, and  */
/* initializes right hand side values on grid level k */
{
int    i,j,ii,jj;
Level *L;
double x,y;

L =U->Lk+k;
ii=L->ii;
jj=L->jj;

for (i=0;i<=ii;i++)
  {
  x=U->xa+i*L->hx;
  for (j=0;j<=jj;j++)
    {
    y=U->ya+j*L->hy;
    if ((i==0)||(j==0)||(i==ii)||(j==jj))
      L->p[i][j]=pb(x,y);
    else
      L->p[i][j]=p0(x,y);
    L->h[i][j]=h0(x,y);
    L->f[i][j]=f0(x,y);
```

```
    }
  }
}


void prerelax1(Stack *U, int k)
{
/* perform Gauss-Seidel relaxation on gridlevel k */

Level *L;
double hx, hy, rhx2, rhy2, hu, hd, hl, hr, **p, **h, **f;
int    i,j,ii,jj;

L =U->Lk+k;
hx=L->hx;
hy=L->hy;
ii=L->ii;
jj=L->jj;
p =L->p;
h =L->h;
f =L->f;

rhx2=1.0/(hx*hx);
rhy2=1.0/(hy*hy);

for (i=ii/2;i<=3*ii/4;i++)
  for (j=1;j<=jj-1;j++)
    {
    hr = 0.5*(h[i+1][j]+h[i][j]);
    hl = 0.5*(h[i-1][j]+h[i][j]);
    hu = 0.5*(h[i][j+1]+h[i][j]);
    hd = 0.5*(h[i][j-1]+h[i][j]);
    p[i][j]+=(f[i][j]-Lu(p,h,rhx2,rhy2,i,j))/
             (-rhx2*(hr*hr*hr+hl*hl*hl)-RoL*RoL*rhy2*(hu*hu*hu+hd*hd*hd));
    if (p[i][j]<0.0) p[i][j]=0.0;
    }
}


void prerelax2(Stack *U, int k)
{
/* perform Gauss-Seidel relaxation on gridlevel k close to the cavitation boundary */

Level *L;
double hx, hy, rhx2, rhy2, hu, hd, hl, hr, **p, **h, **f;
int    i,i0,i1,i2,j,ii,jj;

L =U->Lk+k;
hx=L->hx;
hy=L->hy;
ii=L->ii;
jj=L->jj;
p =L->p;
h =L->h;
```

```
f =L->f;

rhx2=1.0/(hx*hx);
rhy2=1.0/(hy*hy);

for (j=1;j<=jj-1;j++)
 {
 i0=1;
 while ((p[i0][j] >0.0)&&(i0<ii)) i0++;
 i1=i0-3; if (i1<1 ) i1=1;
 i2=i0+2; if (i2>ii-1) i2=ii-1;
 for (i=i1;i<=i2;i++)
    {
    hr = 0.5*(h[i+1][j]+h[i][j]);
    hl = 0.5*(h[i-1][j]+h[i][j]);
    hu = 0.5*(h[i][j+1]+h[i][j]);
    hd = 0.5*(h[i][j-1]+h[i][j]);
    p[i][j]+=(f[i][j]-Lu(p,h,rhx2,rhy2,i,j))/
             (-rhx2*(hr*hr*hr+hl*hl*hl)-RoL*RoL*rhy2*(hu*hu*hu+hd*hd*hd));
    if (p[i][j]<0.0) p[i][j]=0.0;
    }
  }
}


void relax(Stack *U, int k)
{
/* perform Gauss-Seidel relaxation on gridlevel k */

Level *L;
double hx, hy, rhx2, rhy2, hu, hd, hl, hr, err, **p, **h, **f;
int    i,j,ii,jj;

prerelax2(U, k);
prerelax2(U, k);
prerelax2(U, k);

L =U->Lk+k;
hx=L->hx;
hy=L->hy;
ii=L->ii;
jj=L->jj;
p =L->p;
h =L->h;
f =L->f;

rhx2=1.0/(hx*hx);
rhy2=1.0/(hy*hy);

for (i=1;i<=ii-1;i++)
  for (j=1;j<=jj-1;j++)
    {
    hr = 0.5*(h[i+1][j]+h[i][j]);
    hl = 0.5*(h[i-1][j]+h[i][j]);
```

```
    hu = 0.5*(h[i][j+1]+h[i][j]);
    hd = 0.5*(h[i][j-1]+h[i][j]);
    p[i][j]+=(f[i][j]-Lu(p,h,rhx2,rhy2,i,j))/
             (-rhx2*(hr*hr*hr+hl*hl*hl)-RoL*RoL*rhy2*(hu*hu*hu+hd*hd*hd));
    if (p[i][j]<0.0) p[i][j]=0.0;
    }

err=0.0;
for (i=1;i<=ii-1;i++)
  for (j=1;j<=jj-1;j++)
    if (p[i][j]>0.0) err+=fabs(f[i][j]-Lu(p,h,rhx2,rhy2,i,j) );

U->wu += exp((U->maxlevel-k)*log(0.25));
printf("\nLevel %d Residual %8.5e Wu %8.5e",k,err/((ii-1)*(jj-1)),U->wu);
}


/**********INTER GRID ROUTINES**********/


void coarsen_p(Stack *U, int k)
{
/* compute initial approximation on level k-1 */
/* in coarsening step from level k            */

int     ic,jc,iic,jjc;
Level  *Lc, *L;
double **pc, **pco, **p ;

L =U->Lk+k;
p =L->p;

Lc =U->Lk+k-1;
iic=Lc->ii; jjc=Lc->jj;
pc =Lc->p;
pco=Lc->pold;

for (ic=1;ic<=iic-1;ic++)
  for (jc=1;jc<=jjc-1;jc++)
    pc[ic][jc]=       p[2*ic  ][2*jc  ];

/* store coarse grid solution in pco array */
for (ic=0;ic<=iic;ic++)
  for (jc=0;jc<=jjc;jc++)
    pco[ic][jc]=pc[ic][jc];
}


void coarsen_f(Stack *U, int k)
{
/* compute coarse grid right hand side on level k-1 */
/* in coarsening step from level k                  */

int     ic,jc,iic,jjc;
```

```
Level  *Lc,*L;
double **p,**pc,**h,**hc,**f,**fc ;
double hx,hy,rh2x,rh2y,hxc,hyc,rh2xc,rh2yc  ;

L  =U->Lk+k;
hx =L->hx; hy=L->hy;
p  =L->p;
h  =L->h;
f  =L->f;

Lc  =U->Lk+k-1;
iic =Lc->ii; jjc=Lc->jj;
hxc =Lc->hx; hyc=Lc->hy;
fc  =Lc->f;
hc  =Lc->h;
pc  =Lc->p;

rh2xc=1.0/(hxc*hxc);
rh2yc=1.0/(hyc*hyc);
rh2x =1.0/(hx*hx);
rh2y =1.0/(hy*hy);

for (ic=1;ic<=iic-1;ic++)
  for (jc=1;jc<=jjc-1;jc++)
    {

    /* FAS coarse grid right hand side   */
    /* with full weighting of residuals  */

    fc[ic][jc]=Lu(pc,hc,rh2xc,rh2yc,  ic,  jc)+f[2*ic  ][2*jc  ]
              -Lu(p ,h ,rh2x ,rh2y ,2*ic,2*jc);
    }
}


void refine(Stack *U, int k)
{
/* Interpolation and addition of coarse grid correction from grid k-1 */
/* to grid k                                                          */

int    i,j,ic,jc,iic,jjc,ii,jj;
Level  *Lc,*L;
double **pc,**pco,**p;

L  =U->Lk+k;
ii =L ->ii; jj =L->jj;
p  =L->p  ;

Lc =U->Lk+k-1;
iic=Lc->ii; jjc=Lc->jj;
pc =Lc->p ;
pco=Lc->pold ;

for (ic=1;ic<=iic;ic++)
```

```
  for (jc=1;jc<=jjc;jc++)
    {
    if (ic<iic)
      if (p[2*ic  ][2*jc  ]>0.0)
        p[2*ic  ][2*jc  ]+=(pc[ic][jc]-pco[ic][jc]);

    if (jc<jjc)
      if  (p[2*ic-1][2*jc  ]>0.0)
        p[2*ic-1][2*jc  ]+=(pc[ic  ][jc]-pco[ic  ][jc]+
                            pc[ic-1][jc]-pco[ic-1][jc])*0.5;

    if (ic<iic)
      if (p[2*ic  ][2*jc-1]>0.0)
        p[2*ic  ][2*jc-1]+=(pc[ic][jc  ]-pco[ic][jc  ]+
                            pc[ic][jc-1]-pco[ic][jc-1])*0.5;

    if (p[2*ic-1][2*jc-1]>0.0)
        p[2*ic-1][2*jc-1]+=(pc[ic  ][jc  ]-pco[ic  ][jc  ]+
                            pc[ic  ][jc-1]-pco[ic  ][jc-1]+
                            pc[ic-1][jc  ]-pco[ic-1][jc  ]+
                            pc[ic-1][jc-1]-pco[ic-1][jc-1])*0.25;
    }
for (i=1;i<=ii;i++)
  for (j=1;j<=jj;j++)
    if (p[i][j]<0.0) p[i][j]=0.0;
}


void fmg_interpolate(Stack *U, int k)
{
/* interpolation of coarse grid k-1 solution to fine grid k */
/* to serve as first approximation. bi-cubic interpolation  */

int    ic,jc,iic,jjc,i,j,ii,jj;
Level  *Lc,*L;
double **pc,**p, **pconv;

L  =U->Lk+k;
ii =L ->ii; jj =L->jj;
p  =L->p  ;

Lc    =U->Lk+k-1;
iic   =Lc->ii; jjc=Lc->jj;
pc    =Lc->p ;
pconv =Lc->pconv;

/* store grid k-1 solution for later use in convergence check */

for (ic=1;ic<=iic-1;ic++)
  for (jc=1;jc<=jjc-1;jc++)
    pconv[ic][jc]=pc[ic][jc];

/* first inject to points coinciding with coarse grid points */
```

```
for (ic=1;ic<=iic-1;ic++)
  for (jc=1;jc<=jjc-1;jc++)
    p[2*ic][2*jc]=pc[ic][jc];

/* interpolate intermediate y direction */

for (i=2;i<=ii-2;i+=2)
  {
  p[i][1]=(5.0*p[i][0]+15.0*p[i][2]-5.0*p[i][4]+p[i][6])*0.0625;

  for (j=3;j<=jj-3;j+=2)
    p[i][j]=(-p[i][j-3]+9.0*p[i][j-1]+9.0*p[i][j+1]-p[i][j+3])*0.0625;

  p[i][jj-1]=(5.0*p[i][jj]+15.0*p[i][jj-2]-5.0*p[i][jj-4]+p[i][jj-6])*0.0625;
  }

/* interpolate in x direction */

for (j=1;j<=jj-1;j++)
  {
  p[1][j]=(5.0*p[0][j]+15.0*p[2][j]-5.0*p[4][j]+p[6][j])*0.0625;

  for (i=3;i<=ii-3;i+=2)
    p[i][j]=(-p[i-3][j]+9.0*p[i-1][j]+9.0*p[i+1][j]-p[i+3][j])*0.0625;

  p[ii-1][j]=(5.0*p[ii][j]+15.0*p[ii-2][j]-5.0*p[ii-4][j]+p[ii-6][j])*0.0625;
  }
}


double conver(Stack *U, int k)
{
/* convergence check using converged solutions on level k and */
/* on next coarser grid k-1                                   */

int    ic,jc;
Level  *Lc,*L;
double **pc,**p;
double err;

L =U->Lk+k;

if (k==U->maxlevel) p=L->p;
  else p=L->pconv;

Lc =U->Lk+k-1;
pc =Lc->pconv;

err=0.0;
for (ic=1;ic<=Lc->ii-1;ic++)
  for (jc=1;jc<=Lc->jj-1;jc++)
    err+=fabs(pc[ic][jc]-p[2*ic][2*jc]);

return(err/((Lc->ii-1)*(Lc->jj-1)));
```

```
}


/**********MULTIGRID DRIVING ROUTINES**********/


void cycle(Stack *U, int k, int nu0, int nu1, int nu2, int gamma)

/* performs coarse grid correction cycle starting on level k */
/* nu1 pre-relaxations, nu2 postrelaxations, nu0 relaxations */
/* on the coarsest grid, cycleindex gamma=1 for Vcycle,      */
/* gamma=2 for Wcycle                                         */

{
int i,j;

if (k==1)
  for (i=1;i<=nu0;i++) relax(U,k);
else
  {
  for (i=1;i<=nu1;i++) relax(U,k);
  coarsen_p(U,k);
  coarsen_f(U,k);
  for (j=1;j<=gamma;j++) cycle(U,k-1,nu0,nu1,nu2,gamma);
  refine(U,k);
  for (i=1;i<=nu2;i++) relax(U,k);
  }
}


void fmg(Stack *U, int k, int nu0, int nu1, int nu2, int gamma, int ncy)
{
/* performs FMG with k levels and ncy cycles per level */

int i,j;

if (U->maxlevel==1)
  for (j=1;j< ncy;j++)
    for (i=1;i<=nu0;i++) relax(U,k);
if (k==1) for (i=1;i<=nu0;i++) relax(U,k);
else
  {
  fmg(U,k-1,nu0,nu1,nu2,gamma,ncy);
  fmg_interpolate(U,k);
  for (j=1;j<=ncy;j++)
    {
    cycle(U,k,nu0,nu1,nu2,gamma);
    printf("\n");
    }
  }
}


/**********OUTPUT ROUTINE *********/
```

```
void output(Stack *U, int k)
{
/* Writes output into a file PH2.dat */

Level *L;
double lx, ly, **p, **h;
int    i,j,ii,jj;
FILE *fp;

L =U->Lk+k;
p =L->p;
h =L->h;
ii=L->ii;
jj=L->jj;

fp = fopen("PH2.dat","w");
j = jj/2;
for (i=0;i<=ii;i++)
  fprintf(fp,"%e %e %e\n",i*L->hx,p[i][j],h[i][j]);
fclose(fp);

lx = ly =0.0;
for (i=1;i<=ii-1;i++)
  for (j=1;j<=jj-1;j++)
    if (p[i][j]>0.0)
      {
      ly += p[i][j]*cos(i*L->hx);
      lx += p[i][j]*sin(i*L->hx);
      }
lx *= L->hx*L->hy;
ly *= L->hx*L->hy;
printf("\n\nlx= %e ly = %e l = %e S0= %e angle = %e\n\n",
  lx,ly,sqrt(lx*lx+ly*ly),3.0*sqrt(lx*lx+ly*ly),atan(lx/ly));

}


/**********MAIN PROGRAM**********/


void main()
{
Stack  U;
int j,maxlev,ncy;

printf("\ngive maxlevel"); scanf("%d",&maxlev);
printf("\ngive ncy      "); scanf("%d",&ncy);
printf("\ngive epsilon "); scanf("%lf",&epsilon);

initialize(&U,24,4,maxlev,0.0,2.0*pi,0.0,1.0);

for (j=maxlev;j>=1;j--) init_uf(&U, P_i, P_b, H_i, F_i, j);
```

```
fmg(&U,maxlev,10,2,1,1,ncy);

if (maxlev>1)
  {
  printf("\n");
  for (j=2;j<=maxlev;j++) printf("\n aen(%2d,%2d)=%8.5e",j,j-1,conver(&U,j));
  }
output(&U,maxlev);
finalize(&U,maxlev);
}
```

Appendix G

Program Listing: DRY2d.c

```
#include <stdio.h>
#include <stdlib.h>
#include <math.h>


/* FMG solver for the 2d dry contact problem                          */
/* using intermediate grids for fast evaluation of deformation        */


#define pi  3.1415926535897931


typedef struct
{
double hx;             /* mesh size x                                       */
double hy;             /* mesh size y                                       */
int    ii;             /* number of nodes x                                 */
int    jj;             /* number of nodes y                                 */
double **p, **w, **f;  /* pressure, integral and right hand side values     */
                       /* For convenience:                                  */
double **K, **K1;      /* kernels                                           */
double **pconv;        /* converged pressure for convergence check in FMG   */
double **pold;         /* 'old' pressure for use in FAS coarsening          */
double **hfi, **hrhs;  /* film thickness and right hand side film thickness */
double rg;             /* right hand side of force balance equation         */
} Level;


typedef struct
{
int    nx0,ny0;         /* number of nodes coarsest grid     */
int    m1,m2,od;        /* number of correction points       */
int    maxlev, deep;    /* number of grid-levels, grids deep */
double xa,xb;           /* begin,end of computational domain */
double ya,yb;           /* begin,end of computational domain */
double h0,wu;           /* global constant and work unit     */
Level  *Lk ;            /* array of grid levels              */
} Stack;
```

```
/********** ROUTINES FOR DATASTRUCTURE **********/


double **matrix(Stack *U, int nx, int ny)
{
int i;

double **m=(double **)calloc(nx+1+4*U->od,sizeof(double*));
for (i=0;i<=nx+4*U->od;i++) m[i]=(double *)calloc(ny+1+4*U->od,sizeof(double));
return m;
}


double **smatri(Stack *U, int nx, int ny) /* shifted matrix */
{
int i;

double **m=(double **)calloc(nx+1+2*U->od,sizeof(double*));
for (i=0;i<=nx+2*U->od;i++) m[i]=(double *)calloc(ny+1+2*U->od,sizeof(double))+U->od;
return m+U->od;
}


void initialize(Stack *U, int nx0, int ny0, int maxl, int deepl, int ord,
                double xa, double xb, double ya, double yb)
{
/* initialize values in datastructure */

double  hx,hy;
Level  *L;
int     l,ii,jj;

U->xa=xa;
U->xb=xb;
U->ya=ya;
U->yb=yb;
U->maxlev=maxl;
U->wu=0.0;
U->deep=deepl;
U->od=ord-2;
U->h0= -0.0;

U->Lk=(Level *)calloc(maxl+1,sizeof(Level));

hx=(xb-ya)/nx0;
hy=(yb-ya)/ny0;
ii=nx0;
jj=ny0;
for (l=1;l<=maxl;l++)
  {
  L=U->Lk+l;
  L->hx=hx;
  L->hy=hy;
```

```
  L->ii=ii;
  L->jj=jj;
  L->p    =smatri(U,ii,jj);
  L->w    =smatri(U,ii,jj);
  L->f    =smatri(U,ii,jj);
  L->K    =matrix(U,ii,jj);
  L->K1   =matrix(U,ii,jj);
  L->pold =smatri(U,ii,jj);
  L->pconv=smatri(U,ii,jj);
  L->hfi  =smatri(U,ii,jj);
  L->hrhs =smatri(U,ii,jj);

  printf("\n level: %2d ii=%4d, jj=%4d hx=%f hy=%f",l,ii,jj,hx,hy);

  if ((2*(l/2)-l)!=0) {hx*=0.5; ii*=2;} else {hy*=0.5; jj*=2;}

  if ((L->p==NULL)||(L->w==NULL)||(L->f==NULL)||(L->K==NULL)||(L->pold==NULL)
                ||(L->pconv==NULL)||(L->hfi==NULL)||(L->hrhs==NULL))
    {
    printf("\nproblem allocating memory\n");
    exit(-1);
    }
  }
U->m1=1000; U->m2=1000;  /* for debugging */
U->m1=(int)(3+log(1.0*(U->Lk+maxl)->ii)); U->m2=2;
printf("\n m1=%2d, m2=%2d",U->m1,U->m2);
}


void finalize(Stack *U, int maxlevel)
{
/* free memory at end of program */

Level  *L;
int     i;
for (i=1;i<=U->maxlev;i++)
  {
  L=U->Lk+i;
  free(L->f-U->od);
  free(L->w-U->od);
  free(L->p-U->od);
  free(L->K);
  free(L->K1);
  free(L->pold-U->od);
  free(L->pconv-U->od);
  free(L->hfi-U->od);
  free(L->hrhs-U->od);
  }
free(U->Lk);
}


/********** SINGLE GRID ROUTINES **********/
```

```
void init_f(Stack *U, int l)
{
int    i,j;
Level *L;
double x,y;

L =U->Lk+l;

for (i=0; i<=L->ii; i++)
  {
  x=U->xa+i*L->hx;
  for (j=0; j<=L->jj; j++)
    {
    y=U->ya+j*L->hy;
    L->hrhs[i][j]= 0.5*x*x+0.5*y*y;
    }
  }
L->rg = -2.0*pi/3.0;
}


void init_p(Stack *U, int l)
{
int    i,j;
Level *L;
double x,y;
void calchi();

L =U->Lk+l;

for (i=0; i<=L->ii; i++)
  {
  x=U->xa+i*L->hx;
  for (j=0; j<=L->jj; j++)
    {
    y=U->ya+j*L->hy;
    if (x*x+y*y<1.0) L->p[i][j]=0.999*sqrt(1.0-x*x-y*y); else L->p[i][j]=0.0;
    L->p[i][j]=0.0;
    }
  }
calchi(U,l);
}


double resnorm(Stack *U, int l)
/* computes the absolute norm of the residual */
/* on level l in points where the pressure>0  */
{
int i,j;
Level *L;
double herr;

L =U->Lk+l;
```

```
herr = 0.0;

for (i=0; i<=L->ii; i++)
  for (j=0; j<=L->jj; j++)
    if (L->p[i][j]>0.0) herr += fabs(L->hfi[i][j]);
return(herr*L->hx*L->hy);
}


void relax(Stack *U, int l, int o)
/* relaxes the equation on level l, using distributive Jacobi */
/* and computes the work                                       */
{
int i,j;
Level *L;
double del,g,c0,ur=0.6;
void calchi();

L =U->Lk+l;
c0=2.0/pi/pi*(L->K[0][0]-0.5*(L->K[1][0]+L->K[0][1]));
calchi(U,l);

for (i=1; i<L->ii; i++)
  for (j=1; j<L->jj; j++)
    {
    del =  ur*L->hfi[i][j]/c0;
    if (del>L->p[i][j]) del=L->p[i][j];
          L->p[i  ][j  ] -=      del;
    if ((L->p[i+1][j  ]>0)&&(i<L->ii)) L->p[i+1][j  ] += 0.25*del;
    if  (L->p[i+1][j  ]<0)             L->p[i+1][j  ]  = 0.0;
    if ((L->p[i-1][j  ]>0)&&(i>0    )) L->p[i-1][j  ] += 0.25*del;
    if  (L->p[i-1][j  ]<0)             L->p[i-1][j  ]  = 0.0;
    if ((L->p[i  ][j+1]>0)&&(j<L->jj)) L->p[i  ][j+1] += 0.25*del;
    if  (L->p[i  ][j+1]<0)             L->p[i  ][j+1]  = 0.0;
    if ((L->p[i  ][j-1]>0)&&(j>0    )) L->p[i  ][j-1] += 0.25*del;
    if  (L->p[i  ][j-1]<0)             L->p[i  ][j-1]  = 0.0;
    }
g=0.0;
for (i=0; i<=L->ii; i++)
  for (j=0; j<=L->jj; j++)
    g += L->p[i][j];
g = g*L->hx*L->hy+L->rg;
U->wu+=pow(0.25,1.0*((U->maxlev+1)/2-(l+1)/2));
if (o==1) printf("\n k=%2d, resn=%10.3e, g=%10.3e, h0=%12.8f, wu=%7.3f",
 (l+1)/2,resnorm(U,l),g,U->h0,U->wu);
}


void relaxh0(Stack *U, int l)
/* relaxes the force-balance equation */
/*  hfact=0.2 give very stable convergence */
{
int i,j;
Level *L;
```

```
double g, hfact=0.3;

L =U->Lk+l;
g = 0.0;
for (i=0; i<=L->ii; i++)
  for (j=0; j<=L->jj; j++)
    g += L->p[i][j];
g = g*L->hx*L->hy + L->rg;
U->h0 += hfact*g;
}


/********** INTER GRID ROUTINES **********/


void coarsen_p(Stack *U, int l)
{
/* coarsen the solution from level l to level l-2 */
/* and store coarse grid solution in pold array   */

int     i,j;
Level  *Lc, *L;
void calchi();
void init_log();

calchi(U,l);
init_log(U,l-2);

L  =U->Lk+l;
Lc =U->Lk+l-2;

for (i=0; i<=Lc->ii; i++)
  for (j=0; j<=Lc->jj; j++)
    Lc->pold[i][j]=Lc->p[i][j]=L->p[2*i][2*j];
}


void coarsen_f(Stack *U, int l)
{
/* compute coarse grid right hand side on level l-2 */
/* in coarsening step from level l                  */

int     i,j;
Level  *Lc,*L;
double gf, gc;
void calchi();

L  =U->Lk+l;
Lc =U->Lk+l-2;

calchi(U,l-2);
for (i=0;i<=Lc->ii;i++)
  for (j=0;j<=Lc->jj;j++)
    Lc->hrhs[i][j]=L->hrhs[2*i][2*j]+2./pi/pi*(L->w[2*i][2*j]-Lc->w[i][j]);
```

```
gf=0.0;
for (i=0; i<=L->ii; i++)
  for (j=0; j<=L->jj; j++)
    gf += L->p[i][j];

gc=0.0;
for (i=0; i<=Lc->ii; i++)
  for (j=0; j<=Lc->jj; j++)
    gc += Lc->p[i][j];

Lc->rg=L->rg+gf*L->hx*L->hy-gc*Lc->hx*Lc->hy;
}


void refine(Stack *U, int k)
{
/* Interpolation and addition of coarse grid correction from grid k-2 */
/* to grid k                                                           */

int     i,j,ic,jc,iic,jjc,ii,jj;
Level  *Lc,*L;
double **pc,**pco,**p;

L  =U->Lk+k;
ii =L ->ii; jj =L->jj;
p  =L->p  ;

Lc =U->Lk+k-2;
iic=Lc->ii; jjc=Lc->jj;
pc =Lc->p ;
pco=Lc->pold ;

for (ic=1;ic<=iic;ic++)
  for (jc=1;jc<=jjc;jc++)
    {
    if (ic<iic)
      if (p[2*ic  ][2*jc  ]>0.0)
        p[2*ic  ][2*jc  ]+=(pc[ic][jc]-pco[ic][jc]);

    if (jc<jjc)
      if  (p[2*ic-1][2*jc  ]>0.0)
        p[2*ic-1][2*jc  ]+=(pc[ic  ][jc]-pco[ic  ][jc]+
                            pc[ic-1][jc]-pco[ic-1][jc])*0.5;

    if (ic<iic)
      if (p[2*ic  ][2*jc-1]>0.0)
        p[2*ic  ][2*jc-1]+=(pc[ic][jc  ]-pco[ic][jc  ]+
                            pc[ic][jc-1]-pco[ic][jc-1])*0.5;

    if (p[2*ic-1][2*jc-1]>0.0)
        p[2*ic-1][2*jc-1]+=(pc[ic  ][jc  ]-pco[ic  ][jc  ]+
                            pc[ic  ][jc-1]-pco[ic  ][jc-1]+
                            pc[ic-1][jc  ]-pco[ic-1][jc  ]+
                            pc[ic-1][jc-1]-pco[ic-1][jc-1])*0.25;
```

```
    }
for (i=0;i<=ii;i++)
  for (j=0;j<=jj;j++)
    if (p[i][j]<0.0) p[i][j]=0.0;
}


void fmg_interpolate(Stack *U, int k)
{
/* interpolation of coarse grid k-2 solution to fine grid k */
/* to serve as first approximation. bi-cubic interpolation  */

int    ic,jc,iic,jjc;
Level  *Lc,*L;
double **pc,**p, **pconv;

L  =U->Lk+k;
p  =L->p  ;

Lc    =U->Lk+k-2;
iic   =Lc->ii; jjc=Lc->jj;
pc    =Lc->p ;
pconv =Lc->pconv;

for (ic=1;ic<=iic;ic++)
  for (jc=1;jc<=jjc;jc++)
    { /* store grid k-2 solution for later use in convergence check */
    pconv[ic][jc]=pc[ic][jc];
    p[2*ic  ][2*jc  ]= pc[ic  ][jc  ];
    p[2*ic-1][2*jc  ]=(pc[ic  ][jc  ]+pc[ic-1][jc  ])*0.5;
    p[2*ic  ][2*jc-1]=(pc[ic  ][jc  ]+pc[ic  ][jc-1])*0.5;
    p[2*ic-1][2*jc-1]=(pc[ic  ][jc  ]+pc[ic  ][jc-1]+
                       pc[ic-1][jc  ]+pc[ic-1][jc-1])*0.25;
    }
}


double conver(Stack *U, int k)
{
/* convergence check using converged solutions on level k */
/* and on next coarser (solution) grid k-2                */

int    ic,jc;
Level  *Lc,*L;
double **pc,**p;
double err;

L =U->Lk+k;

if (k==U->maxlev) p=L->p;
  else p=L->pconv;

Lc =U->Lk+k-2;
pc =Lc->pconv;
```

```
err=0.0;
for (ic=1;ic<=Lc->ii-1;ic++)
  for (jc=1;jc<=Lc->jj-1;jc++)
    err+=fabs(pc[ic][jc]-p[2*ic][2*jc]);

return(err/((Lc->ii-1)*(Lc->jj-1)));
}


/********** MULTIGRID DRIVING ROUTINES **********/


void cycle(Stack *U, int k, int nu0, int nu1, int nu2, int gamma)
/* performs coarse grid correction cycle starting on level k */
/* nu1 pre-relaxations, nu2 postrelaxations, nu0 relaxations */
/* on the coarsest grid, cycleindex gamma=1 for Vcycle,      */
/* gamma=2 for Wcycle                                         */
{
int i,j;

if (k==1)
  {
  relax(U,k,1);
  for (i=1;i<=nu0/2;i++) relax(U,k,0);
  relaxh0(U,k);
  for (i=1;i<=nu0/2;i++) relax(U,k,0);
  relax(U,k,1);
  }
else
  {
  for (i=1;i<=nu1;i++) relax(U,k,1);
  coarsen_p(U,k);
  coarsen_f(U,k);
  for (j=1;j<=gamma;j++) cycle(U,k-2,nu0,nu1,nu2,gamma);
  refine(U,k);
  for (i=1;i<=nu2;i++) relax(U,k,1);
  }
}


void fmg(Stack *U, int k, int nu0, int nu1, int nu2, int gamma, int ncy)
{
/* performs FMG with k levels and ncy cycles per level */
int i,j;

if (U->maxlev==1)
  for (j=1;j<ncy;j++)
    {
    relax(U,1,1);
    for (i=1;i<=nu0/2;i++) relax(U,1,0);
    relaxh0(U,1);
    for (i=1;i<=nu0/2;i++) relax(U,1,0);
    relax(U,1,1);
```

```
    }
if (k==1)
  {
  relax(U,k,1);
  for (i=1;i<=nu0/2;i++) relax(U,k,0);
  relaxh0(U,k);
  for (i=1;i<=nu0/2;i++) relax(U,k,0);
  relax(U,k,1);
  }
else
  {
  fmg(U,k-2,nu0,nu1,nu2,gamma,ncy);
  fmg_interpolate(U,k);
  for (j=1;j<=ncy;j++)
    {
    cycle(U,k,nu0,nu1,nu2,gamma);
    printf("\n");
    }
  }
}


/* ++++++++++++++++++++++   Gap Height Computation    ++++++++++++++++++++*/
/* ++++++++++++++++++++++ Multi Level Multi Summation ++++++++++++++++++++*/


double ah(double a, double b)
/*calculates arcsinh(a/b)*/
{
if (a==0.0) return(0.0);
else
   if (b==0.0) return(1.0);
   else return(log(a/b+sqrt(a*a/b/b+1)));
}


void init_log(Stack *U, int l)
{
/* computes the kernel on level l */
int    i,j;
Level *L;
double xp,yp,xm,ym;

L =U->Lk+l;

for (i=0; i<=L->ii+4*U->od; i++)
  {
  xp=(i+0.5)*L->hx; xm=xp-L->hx;
  for (j=0; j<=L->jj+4*U->od; j++)
    {
    yp=(j+0.5)*L->hy; ym=yp-L->hy;
    L->K[i][j]=
      fabs(xp)*log(yp/xp+sqrt(1.0+yp*yp/xp/xp))
     -fabs(xp)*log(ym/xp+sqrt(1.0+ym*ym/xp/xp))
```

```
      +fabs(xm)*log(ym/xm+sqrt(1.0+ym*ym/xm/xm))
      -fabs(xm)*log(yp/xm+sqrt(1.0+yp*yp/xm/xm))
      +fabs(yp)*log(xp/yp+sqrt(1.0+xp*xp/yp/yp))
      -fabs(yp)*log(xm/yp+sqrt(1.0+xm*xm/yp/yp))
      +fabs(ym)*log(xm/ym+sqrt(1.0+xm*xm/ym/ym))
      -fabs(ym)*log(xp/ym+sqrt(1.0+xp*xp/ym/ym));
     }
  }
}


double kval(Stack *U, int lf, int lc, int i1, int j1)
/* calculates the value of k on the fine grid  lf in the point i1,j1*/
/* and injected to the coarse grid lc                               */
{
int l, i,j;
double x1,x2,y1,y2,help;
Level *Lf, *Lc;

Lf=U->Lk+lf;
Lc=U->Lk+lc;

i=i1; j=j1;
for (l=lf-1; l>=lc; l--)
    if ((2*(l/2)-l)!=0) i *= 2; else j *= 2;
x1=(i-0.5)*Lf->hx; x2=x1+Lf->hx; y1=(j-0.5)*Lf->hy; y2=y1+Lf->hy;

if ((x1==0.0)||(x2==0.0)||(y1==0.0)||(y2==0.0))
   printf("error in kval lf=%d lc=%d i=%d j=%d",lf,lc,i1,j1);
help=fabs(x2)*ah(y2,x2)+fabs(y2)*ah(x2,y2)-fabs(x2)*ah(y1,x2)-fabs(y2)*ah(x1,y2)
    -fabs(x1)*ah(y2,x1)-fabs(y1)*ah(x2,y1)+fabs(x1)*ah(y1,x1)+fabs(y1)*ah(x1,y1);
return(help*Lc->hx*Lc->hy/Lf->hx/Lf->hy);
}


void fillk(Stack *U, int l)
/* fills the matrix K with integral values          */
/* for matrix multiplication to obtain the integral*/
{
int l1, i,j;
Level *L, *Lc;

for (l1=l; l1>=2; l1--)
  {
  L =U->Lk+l1;
  Lc=U->Lk+l1-1;
  if ((2*(l1/2)-l1)==0)
    for (j=0; j<=L->jj+4*U->od; j++)
      {
      for (i=0; i<=Lc->ii+2*U->od; i++) Lc->K[i][j]=2.0*L->K[2*i][j];
      for (i=Lc->ii+2*U->od+1; i<=Lc->ii+4*U->od; i++) Lc->K[i][j]=kval(U,l,l1-1,i,j);
      }
  else
    for (i=0; i<=L->ii+4*U->od; i++)
```

```
      {
      for (j=0; j<=Lc->jj+2*U->od; j++) Lc->K[i][j]=2.0*L->K[i][2*j];
      for (j=Lc->jj+2*U->od+1; j<=Lc->jj+4*U->od; j++) Lc->K[i][j]=kval(U,l,l1-1,i,j);
      }
  }
}


void calcku(Stack *U, int l)
/*calculates the integral on grid l*/
{
int i1,j1,i,j;
double help;
Level *L;

L =U->Lk+l;
for (i1=-U->od; i1<=L->ii+U->od; i1++)
  for (j1=-U->od; j1<=L->jj+U->od; j1++)
    {
    help=0.0;
    for (i=-U->od; i<=L->ii+U->od; i++)
      for (j=-U->od; j<=L->jj+U->od; j++)
        help+=L->K[abs(i-i1)][abs(j-j1)]*L->p[i][j];
    L->w[i1][j1]=help;
    }
}


void sto6k1(Stack *U, int l)
/* stores values of k1 for refine6    */
/* k1 is the value of k minus the 6th  */
/* order coarse grid approximation to k*/
{
int i,j,maxx,maxy;
Level *L;

L =U->Lk+l;

if (4*U->m1<L->ii-1) maxx=4*U->m1+1; else maxx=L->ii;
if (4*U->m1<L->jj-1) maxy=4*U->m1+1; else maxy=L->jj;

if ((2*(l/2)-l)!=0)
  for (i=0; i<=maxx+4*U->od; i++)
    {
    for (j=0; j<=maxy+2*U->od; j++)
      {
      L->K1[i][j]=   L->K[i][j]
        -(  3*L->K[i][abs(5-j)]- 25*L->K[i][abs(3-j)]
         +150*L->K[i][abs(1-j)]+150*L->K[i][abs(1+j)]
          + 3*L->K[i][abs(5+j)]- 25*L->K[i][abs(3+j)])/256;
      }
    for (j=maxy+2*U->od+1; j<=maxy+4*U->od; j++)
      {
      L->K1[i][j]=   L->K[i][j]
```

```
          -(  3*L->K[i][abs(5-j)]- 25*L->K[i][abs(3-j)]
           +150*L->K[i][abs(1-j)] +150*kval(U,U->maxlev,l,i,j+1)
            + 3*kval(U,U->maxlev,l,i,j+5)- 25*kval(U,U->maxlev,l,i,j+3))/256;
        }
    }
else
  {
  for (j=0; j<=maxy+4*U->od; j++)
    {
    for (i=0; i<=maxx+2*U->od; i++)
      {
      L->K1[i][j]=  L->K[i][j]-
               (  3*L->K[abs(5-i)][j]-  25*L->K[abs(3-i)][j]
               +150*L->K[abs(1-i)][j]+ 150*L->K[abs(1+i)][j]
                + 3*L->K[abs(5+i)][j] - 25*L->K[abs(3+i)][j])/256;
      }
    for (i=maxx+2*U->od+1; i<=maxx+4*U->od; i++)
      {
      L->K1[i][j]=  L->K[i][j]-
          (  3*L->K[abs(5-i)][j]-  25*L->K[abs(3-i)][j]
          +150*L->K[abs(1-i)][j] +150*kval(U,U->maxlev,l,i+1,j)
           + 3*kval(U,U->maxlev,l,i+5,j)- 25*kval(U,U->maxlev,l,i+3,j))/256;
      }
    }
  }
}


void coarsenp6x(Stack *U, int l)
/* 6th order weighting in x direction                             */
/* actually the 6th order approximation is obtained in K          */
/* the stencil is 1/512*(3,0,-25,0,150,256,150,0,-25,0,3) in x   */
/* direction                                                      */

{
int i,j;
double q0,q1,q2,q3,q4,q5,q7,q9;
Level *L, *Lc;

L =U->Lk+l;
Lc=U->Lk+l-1;

if ((2*(l/2)-l)!=0) printf("\nwrong coarsening in coarsenp6x\n");


for (j=-4; j<=L->jj+4; j++)
  {
  q0=L->p[-4][j]; q1 =L->p[-3][j]; q2=L->p[-2][j]; q3 =L->p[-1][j];
  q4=L->p[ 0][j]; q5 =L->p[ 1][j]; q7=L->p[ 3][j]; q9 =L->p[ 5][j];

  Lc->p[   -4][j]=   (                                        3*q1)/512;
  Lc->p[   -3][j]=   (                                 -25*q1 +3*q3)/512;
  Lc->p[   -2][j]=   (                    256*q0 +150*q1 -25*q3 +3*q5)/512;
  Lc->p[   -1][j]=   (         150*q1 +256*q2 +150*q3 -25*q5 +3*q7)/512;
```

```
  Lc->p[    0][j]=    (-25*q1 +150*q3 +256*q4 +150*q5 -25*q7 +3*q9)/512;

  for (i=1; i<=Lc->ii-1; i++)
    Lc->p[i][j]=(3*L->p[2*i-5][j]-25*L->p[2*i-3][j]+150*L->p[2*i-1][j]
              +256*L->p[2*i  ][j]
                +3*L->p[2*i+5][j]-25*L->p[2*i+3][j]+150*L->p[2*i+1][j])/512;

  q0=L->p[L->ii+4][j]; q1=L->p[L->ii+3][j]; q2=L->p[L->ii+2][j]; q3=L->p[L->ii+1][j];
  q4=L->p[L->ii  ][j]; q5=L->p[L->ii-1][j]; q7=L->p[L->ii-3][j]; q9=L->p[L->ii-5][j];

  Lc->p[Lc->ii  ][j]=(-25*q1 +150*q3 +256*q4 +150*q5- 25*q7 +3*q9)/512;
  Lc->p[Lc->ii+1][j]=(        150*q1 +256*q2 +150*q3- 25*q5 +3*q7)/512;
  Lc->p[Lc->ii+2][j]=(                256*q0 +150*q1- 25*q3 +3*q5)/512;
  Lc->p[Lc->ii+3][j]=(                              - 25*q1 +3*q3)/512;
  Lc->p[Lc->ii+4][j]=(                                       3*q1)/512;
  }
}


void coarsenp6y(Stack *U, int l)
/*6th order weighting, in y-direction                          */
/*actually the 6th order approximation is obtained in K        */
/*the stencil is 1/512*93,0,-25,0,150,256,150,0,-25,0,3) in y */
/*direction                                                    */
{
int    i,j;
double q0,q1,q2,q3,q4,q5,q7,q9;
Level *L, *Lc;

L =U->Lk+l;
Lc=U->Lk+l-1;

if ((2*(l/2)-l)==0) printf("\nwarning:wrong coarsening in coarsenp6y\n");

for (i=-4; i<=L->ii+4; i++)
  {
  q0=L->p[i][-4]; q1=L->p[i][-3]; q2=L->p[i][-2]; q3=L->p[i][-1];
  q4=L->p[i][ 0]; q5=L->p[i][ 1]; q7=L->p[i][ 3]; q9=L->p[i][ 5];

  Lc->p[i][   -4]=(                                         3*q1)/512;
  Lc->p[i][   -3]=(                                  -25*q1 +3*q3)/512;
  Lc->p[i][   -2]=(                   256*q0 +150*q1 -25*q3 +3*q5)/512;
  Lc->p[i][   -1]=(           150*q1 +256*q2 +150*q3 -25*q5 +3*q7)/512;
  Lc->p[i][    0]=( -25*q1 +150*q3 +256*q4 +150*q5 -25*q7 +3*q9)/512;

  for (j=1; j<=Lc->jj-1; j++)
    {
    Lc->p[i][j]=(3*L->p[i][2*j-5]-25*L->p[i][2*j-3]+150*L->p[i][2*j-1]
              +256*L->p[i][2*j  ]
                +3*L->p[i][2*j+5]-25*L->p[i][2*j+3]+150*L->p[i][2*j+1])/512;
    }
  q0=L->p[i][L->jj+4]; q1=L->p[i][L->jj+3]; q2=L->p[i][L->jj+2]; q3=L->p[i][L->jj+1];
  q4=L->p[i][L->jj  ]; q5=L->p[i][L->jj-1]; q7=L->p[i][L->jj-3]; q9=L->p[i][L->jj-5];
```

```
  Lc->p[i][Lc->jj  ]=( -25*q1 +150*q3 +256*q4 +150*q5 -25*q7 +3*q9)/512;
  Lc->p[i][Lc->jj+1]=(         150*q1 +256*q2 +150*q3 -25*q5 +3*q7)/512;
  Lc->p[i][Lc->jj+2]=(                256*q0 +150*q1 -25*q3 +3*q5)/512;
  Lc->p[i][Lc->jj+3]=(                               -25*q1 +3*q3)/512;
  Lc->p[i][Lc->jj+4]=(                                      3*q1)/512;
  }
}


void refine6x(Stack *U, int l)
/* O(6) correction of values of hfi on the fine grid after interpolation */
/* in x direction                                                        */
{
int    i,j,i1,j1,ibe,ien,jbe,jen;
double help;
Level *L, *Lc;

L =U->Lk+l;
Lc=U->Lk+l-1;

if ((2*(l/2)-l)!=0) printf("\nwarning:error in refine6x: l=%d\n",l);

for (i=-4; i<=Lc->ii+4; i++) /* fill also the ghost points */
  {
  ibe=2*i-U->m1; if (ibe<     -4) ibe=     -4; if ((2*(ibe/2)-ibe)==0) ibe++;
  ien=2*i+U->m1; if (ien>L->ii+4) ien=L->ii+4; if ((2*(ien/2)-ien)==0) ien--;
  for (j=-4; j<=L->jj+4; j++)
    {
    jbe=j-U->m2; if (jbe<     -4) jbe=     -4;
    jen=j+U->m2; if (jen>L->jj+4) jen=L->jj+4;
    help=0.0;
    for (i1=ibe; i1<=ien; i1 += 2)
      for (j1=jbe; j1<=jen; j1++)
        help += L->p[i1][j1]*L->K1[abs(2*i-i1)][abs(j-j1)];
    Lc->w[i][j] += help;
    }
  }

for (i=-2; i<=Lc->ii+2; i++) /* fill also the ghost points */
  for (j=-4; j<=L->jj+4; j++)
    L->w[2*i][j]=Lc->w[i][j];

for (i=-2; i<=Lc->ii+1; i++)
  {
  ibe=2*i+1-U->m1; if (ibe<     -4) ibe=     -4;
  ien=2*i+1+U->m1; if (ien>L->ii+4) ien=L->ii+4;
  for (j=-4; j<=L->jj+4; j++)
    {
    jbe=j-U->m2; if (jbe<     -4) jbe=     -4;
    jen=j+U->m2; if (jen>L->jj+4) jen=L->jj+4;
    help=0.0;
    for (i1=ibe; i1<=ien; i1++)
      for (j1=jbe; j1<=jen; j1++)
        help += L->p[i1][j1]*L->K1[abs(2*i+1-i1)][abs(j-j1)];
```

```
    L->w[2*i+1][j]=(      3*Lc->w[i-2][j]-  25*Lc->w[i-1][j]
                       +150*Lc->w[i  ][j]+ 150*Lc->w[i+1][j]
                        + 3*Lc->w[i+3][j]-  25*Lc->w[i+2][j])/256+help;
    }
  }
}


void refine6y(Stack *U, int l)
/* O(6) correction of values of WI on the fine grid after interpolation*/
{
int i,j,i1,j1,ibe,ien,jbe,jen;
double help;
Level *L, *Lc;

L =U->Lk+l;
Lc=U->Lk+l-1;

if ((2*(l/2)-l)==0) printf("\nwarning:error in refine6y: %d\n",l);

for (j=-4; j<=Lc->jj+4; j++)
  {
  jbe=2*j-U->m1  ; if (jbe<     -4) jbe=     -4; if ((2*(jbe/2)-jbe)==0) jbe++;
  jen=2*j+U->m1  ; if (jen>L->jj+4) jen=L->jj+4; if ((2*(jen/2)-jen)==0) jen--;
  for (i=-4; i<=L->ii+4; i++)
    {
    ibe=i-U->m2; if (ibe<     -4) ibe=     -4;
    ien=i+U->m2; if (ien>L->ii+4) ien=L->ii+4;
    j1=jbe; help=0.0;
    for (j1=jbe; j1<=jen; j1 += 2)
      for (i1=ibe; i1<=ien; i1++)
        help+=L->p[i1][j1]*L->K1[abs(i-i1)][abs(2*j-j1)];
    Lc->w[i][j] += help;
    }
  }

for (i=-4; i<=L->ii+4; i++)
  for (j=-2; j<=Lc->jj+2; j++)
    L->w[i][2*j]=Lc->w[i][j];

for (j=-2; j<=Lc->jj+1; j++)
  {
  jbe=2*j+1-U->m1; if (jbe<     -4) jbe=     -4;
  jen=2*j+1+U->m1; if (jen>L->jj+4) jen=L->jj+4;
  for (i=-4; i<=L->ii+4; i++)
    {
    ibe=i-U->m2; if (ibe<     -4) ibe=     -4;
    ien=i+U->m2; if (ien>L->ii+4) ien=L->ii+4;
    help=0.0;
    for (j1=jbe; j1<=jen; j1++)
      for (i1=ibe; i1<=ien; i1++)
        help+=L->p[i1][j1]*L->K1[abs(i-i1)][abs(2*j+1-j1)];
    L->w[i][2*j+1]= (  3*Lc->w[i][j-2]- 25*Lc->w[i][j-1]
                     +150*Lc->w[i][j  ]+150*Lc->w[i][j+1]
```

```
                      + 3*Lc->w[i][j+3]- 25*Lc->w[i][j+2])/256+help;
    }
  }
}


void calchi(Stack *U, int lev)
/* calculates film thickness on level l */
{
int i,j,ldeep,l;
Level *L;
double x,y,amp=0.00,wav=0.2;

ldeep=U->deep;
if (ldeep<0) ldeep=0; if (ldeep>lev-1) ldeep=lev-1;
fillk(U,lev);
if (U->od==4) for (l=lev;l>=lev-ldeep+1;l--) sto6k1(U,l);
for (l=1; l<=ldeep; l+=2)
  {
  if (U->od==4) {coarsenp6y(U,lev-l+1); coarsenp6x(U,lev-l);}
  }
calcku(U,lev-ldeep);
for (l=ldeep; l>=1; l-=2)
  {
  if (U->od==4) {refine6x(U,lev-l+1); refine6y(U,lev-l+2);}
  }

L =U->Lk+lev;
for (i=0; i<=L->ii; i++)
  {
  x=U->xa+i*L->hx;
  for (j=0; j<=L->jj; j++)
    {
    y=U->ya+j*L->hy;
    L->hfi[i][j]=U->h0+amp*cos(2.*pi*x/wav)*cos(2.*pi*y/wav)
                 +2.0/pi/pi*L->w[i][j]+L->hrhs[i][j];
    }
  }
}


/**********OUTPUT ROUTINES**********/


void outputP(Stack *U)
/* writes an output file */
{
int i,j;
Level *L;
double x;
FILE *fp;

L =U->Lk+U->maxlev;
fp=fopen("px.dat","w");
```

```
j=L->jj/2;
for (i=0; i<=L->ii; i++)
  {
  x=U->xa+i*L->hx;
  fprintf(fp,"%f %f\n",x,L->p[i][j]);
  }
fclose(fp);
}


void outputH(Stack *U)
/* writes an output file */
{
int i,j;
Level *L;
double x;
FILE *fp;

L =U->Lk+U->maxlev;
fp=fopen("hx.dat","w");

j=L->jj/2;
for (i=0; i<=L->ii; i++)
  {
  x=U->xa+i*L->hx;
  fprintf(fp,"%f %f\n",x,L->hfi[i][j]);
  }
fclose(fp);
}


/********** MAIN PROGRAM **********/


void main()
{
Stack  U;
int l, ncy, maxl, deepl, order=6;

printf("\nhow many levels   ?");
scanf("%d",&maxl); maxl = 2*maxl-1; /* account for         */
printf("\nhow many l deep   ?");
scanf("%d",&deepl); deepl=2*deepl;  /* intermediate grids */
printf("\nhow many cycles   ?");
scanf("%d",&ncy);

initialize(&U,4,4,maxl,deepl,order,-2.0,2.0,-2.0,2.0);
for (l=1; l<=maxl; l+=2) {init_log(&U,l); init_f(&U,l); }

init_p(&U,1);
fmg(&U,maxl,10,2,1,1,ncy);

outputP(&U);
```

```
outputH(&U);

if (maxl>1)
  for (l=3;l<=maxl;l+=2) printf(" aen(%2d,%2d)=%8.5e \n",
       (l+1)/2,(l-1)/2,conver(&U,l));

finalize(&U,maxl);
}
```

Appendix H

Program Listing: EHL2d.c

```
#include <stdio.h>
#include <stdlib.h>
#include <math.h>
#include <time.h>

/* FMG Solver of the EHL circular contact  */

#define pi  3.1415926535897931


typedef struct
{
double hx;               /* mesh size x                                          */
double hy;               /* mesh size y                                          */
int    ii;               /* number of nodes x                                    */
int    jj;               /* number of nodes y                                    */
double **p,    **f;      /* pressure and right hand side values pressure         */
double **hfi,  **hrhs;   /* film thickness and right hand side film thickness    */
double **w;              /* elastic deformation integrals                        */
double **K,    **K1;     /* kernels                                              */
double **pconv;          /* converged pressure for convergence check in FMG      */
double **pold;           /* 'old' pressure for use in FAS coarsening             */
double **pjac;           /* 'old' pressure for use in jacobi relaxation          */
double **A, *X, *Y;      /* for line relaxation                                  */
double rg;               /* right hand side of force balance equation            */
double Hm, Hcp, Hc;      /* minimum and central film thickness for output        */
} Level;


typedef struct
{
int    nx0,ny0;           /* number of nodes coarsest grid     */
int    m1,m2,od;          /* number of correction points       */
int    maxlev, deep;      /* number of grid-levels, grids deep */
double xa,xb;             /* begin,end of computational domain */
double ya,yb;             /* begin,end of computational domain */
double h0,wu;             /* global constant and work unit     */
Level  *Lk ;              /* array of grid levels              */
} Stack;
```

```
/********** GLOBAL VARIABLES ********************/

double        MMoes, LMoes, H_0, rlambda, alphabar;
double        p0r, alpha, eta0, zr;
int           maxl,deepl,starl,order,typecy,ncy,outlev,currfilev;
unsigned long *cputimes,t0;
double        hfact,xi_l,urja,urgs;
int           typecy;


/********** ROUTINES FOR DATASTRUCTURE **********/


double **matrix(int nx, int ny, int shift)
{
int i;

double **m=(double **)calloc(nx+1,sizeof(double*));
for (i=0;i<=nx;i++) m[i]=(double *)calloc(ny+1,sizeof(double))+shift;
return m+shift;
}


void initialize(Stack *U, int nx0, int ny0, int maxl, int deepl, int ord,
                double xa, double xb, double ya, double yb, double h0)
{
/* initialize values in datastructure */

double  hx,hy;
Level  *L;
int     l,ii,jj;

U->xa=xa;
U->xb=xb;
U->ya=ya;
U->yb=yb;
U->maxlev=maxl;
U->deep=deepl;
U->wu=0.0;
U->od=ord-2;
U->h0=  h0;

cputimes=(unsigned long *)calloc(maxl+1,sizeof(unsigned long));

U->Lk=(Level *)calloc(maxl+1,sizeof(Level));

hx=(xb-xa)/nx0;
hy=(yb-ya)/ny0;
ii=nx0;
jj=ny0;
for (l=1;l<=maxl;l++)
```

```
  {
  L=U->Lk+l;
  L->hx=hx;
  L->hy=hy;
  L->ii=ii;
  L->jj=jj;
  L->p    =matrix(ii+2*U->od,jj+2*U->od,U->od);
  L->w    =matrix(ii+2*U->od,jj+2*U->od,U->od);
  L->f    =matrix(ii+2*U->od,jj+2*U->od,U->od);
  L->pold =matrix(ii+2*U->od,jj+2*U->od,U->od);
  L->pjac =matrix(ii+2*U->od,jj+2*U->od,U->od);
  L->pconv=matrix(ii+2*U->od,jj+2*U->od,U->od);
  L->hfi  =matrix(ii+2*U->od,jj+2*U->od,U->od);
  L->hrhs =matrix(ii+2*U->od,jj+2*U->od,U->od);

  L->K    =matrix(ii+4*U->od,jj+4*U->od,0);
  L->K1   =matrix(ii+4*U->od,jj+4*U->od,0);

  L->A=matrix(ii-1,11,0);
  L->X=(double *)calloc(ii-1,sizeof(double));
  L->Y=(double *)calloc(ii-1,sizeof(double));

  printf("\n level: %2d ii=%4d, jj=%4d hx=%f hy=%f",l,ii,jj,hx,hy);

  if ((2*(l/2)-l)!=0) {hx*=0.5; ii*=2;} else {hy*=0.5; jj*=2;}

  if ((L->p==NULL)||(L->w==NULL)||(L->f==NULL)||(L->K==NULL)||(L->pold==NULL)
                 ||(L->pconv==NULL)||(L->hfi==NULL)||(L->hrhs==NULL)||(L->K==NULL)
                 ||(L->K1==NULL)||(L->A==NULL)||(L->X==NULL)||(L->Y==NULL))
    {
    printf("\nproblem allocating memory\n");
    exit(-1);
    }
  }
U->m1=(int)(3+log(1.0*(U->Lk+maxl)->ii)); U->m2=2;
printf("\ncorrection patch in multi-integration: m1=%2d, m2=%2d",U->m1,U->m2);
}


void finalize(Stack *U, int maxlevel)
{
/* free memory at end of program */

Level  *L;
int     i;
for (i=1;i<=U->maxlev;i++)
  {
  L=U->Lk+i;
  free(L->f-U->od);
  free(L->w-U->od);
  free(L->p-U->od);
  free(L->K);
  free(L->K1);
  free(L->pold-U->od);
```

```
  free(L->pconv-U->od);
  free(L->hfi-U->od);
  free(L->hrhs-U->od);
  free(L->A);
  free(L->X);
  free(L->Y);
  }
free(cputimes);
free(U->Lk);
}


/********** SPECIAL FUNCTIONS    **********/


double reta(double p)
{
/* Barus */

return(exp(-alphabar*p));

/* Roelands */

/* return(exp(-alpha*p0r/zr*(-1.0+pow(1.0+(p/p0r)*(alphabar/alpha),zr)))); */
}


double rho(double p)
{
/* Incompressible */

return(1.0);

/* Compressible */

/* return((5.9e8+1.34*(alphabar/alpha)*p)/(5.9e8+(alphabar/alpha)*p)); */
}


double Lu(double **H, double **P, double rhx, double rhx2, double rhy2, int i, int j)
/* computes operator in point i,j */
{
double H3, Hx, Qx, Qy, xi_n,xi_s,xi_e,xi_w,r0,r1,r2;

r0=rho(P[i][j]);
H3=r0*H[i][j]*H[i][j]*H[i][j]*reta(P[i][j]);
xi_n=0.5*(rho(P[i  ][j+1])*H[i  ][j+1]*H[i  ][j+1]*H[i  ][j+1]*reta(P[i][j+1]) + H3)
                                                                  *rlambda*rhy2;
xi_s=0.5*(rho(P[i  ][j-1])*H[i  ][j-1]*H[i  ][j-1]*H[i  ][j-1]*reta(P[i][j-1]) + H3)
                                                                  *rlambda*rhy2;
xi_e=0.5*(rho(P[i+1][j  ])*H[i+1][j  ]*H[i+1][j  ]*H[i+1][j  ]*reta(P[i+1][j]) + H3)
                                                                  *rlambda*rhx2;
xi_w=0.5*(rho(P[i-1][j  ])*H[i-1][j  ]*H[i-1][j  ]*H[i-1][j  ]*reta(P[i-1][j]) + H3)
                                                                  *rlambda*rhx2;
```

```
Qx=(xi_e*P[i+1][j  ] -(xi_e+xi_w)*P[i][j]  + xi_w*P[i-1][j  ]);
Qy=(xi_n*P[i  ][j+1] -(xi_n+xi_s)*P[i][j]  + xi_s*P[i  ][j-1]);

r1=rho(P[i-1][j]);
if (i==1) Hx=rhx*(r0*H[i][j]-r1*H[i-1][j]);
else
  {
  r2=rho(P[i-2][j]);
  Hx=rhx*(1.5*r0*H[i][j]-2*r1*H[i-1][j]+0.5*r2*H[i-2][j]);
  }
return(Qx+Qy-Hx);
}


/********** SINGLE GRID ROUTINES **********/


void init_f(Stack *U, int l)
{
int    i,j;
Level *L;
double x,y;

L =U->Lk+l;

for (i=-1; i<=L->ii; i++)
  {
  x=U->xa+i*L->hx;
  for (j=0; j<=L->jj; j++)
    {
    y=U->ya+j*L->hy;
    L->hrhs[i][j]= 0.5*x*x+0.5*y*y;
    }
  }
L->rg = -2.0*pi/3.0;
}


void init_p(Stack *U, int l)
{
int    i,j;
Level *L;
double x,y;
void calchi();

L =U->Lk+l;

for (i=0; i<=L->ii; i++)
  {
  x=U->xa+i*L->hx;
  for (j=0; j<=L->jj; j++)
    {
    y=U->ya+j*L->hy;
```

```
    L->p[i][j]=0.0;
    if (x*x+y*y<1.0) L->p[i][j]=sqrt(1.0-x*x-y*y); else L->p[i][j]=0.0;
    }
  }
calchi(U,l);
}


double resnorm(Stack *U, int l)
/* computes the absolute norm of the residual */
/* on level l in points where the pressure>0  */
{
int i,j;
Level *L;
double herr,rhx2,rhy2,rhx;

L =U->Lk+l;
rhx=1.0/L->hx;
rhx2=rhx*rhx;
rhy2=1/(L->hy*L->hy);
herr = 0.0;

for (i=1; i<=L->ii-1; i++)
  for (j=1; j<=L->jj-1; j++)
    {
    if (L->p[i][j]>0.0)
    herr+=fabs(L->f[i][j]-Lu(L->hfi,L->p,rhx,rhx2,rhy2,i,j));
    }
 return(herr/(L->ii-1)/(L->jj-1));
}


void relax(Stack *U, int l)
/* relaxes the equation on level l */
{
int    i,j;
Level  *L;
double **P, **Pj;
double g;

void calchi();
void solve_line();

L =U->Lk+l;
P =L->p;
Pj=L->pjac;

calchi(U,l);

for (i=0; i<=L->ii; i++)
  for (j=0; j<=L->jj; j++)
      Pj[i][j]=P[i][j];

for (j=1; j<L->jj; j++)
```

```
  solve_line(U,l,j);

for (i=0; i<=L->ii; i++)
  for (j=0; j<=L->jj; j++)
      P[i][j]=Pj[i][j];

g=0.0;
for (i=0; i<=L->ii; i++)
  for (j=0; j<=L->jj; j++)
    g += L->p[i][j];

g = g*L->hx*L->hy+L->rg;

U->wu+=pow(0.25,1.0*((U->maxlev+1)/2-(l+1)/2));
if (l==outlev)
printf("\n k=%2d, resn=%10.3e, g=%10.3e, h0= %12.8f, wu= %7.3f",(l+1)/2,resnorm(U,l),
                                                                g,U->h0,U->wu);
}


void relaxh0(Stack *U, int l, int lf)
/* relaxes the force-balance equation              */
/*  hfact <0.05 gives stable convergence for W cycles */
{
int i,j;
Level *L;
double g;

L =U->Lk+l;
g = 0.0;
for (i=0; i<=L->ii; i++)
  for (j=0; j<=L->jj; j++)
      g += L->p[i][j];
g = g*L->hx*L->hy + L->rg;
U->h0 += hfact*g;
}


/********** INTER GRID ROUTINES **********/


void coarsen_p(Stack *U, int l)
{
/* coarsen the solution from level l to level l-2 */
/* and store coarse grid solution in pold array   */

int     i,j,iff,jff;
Level  *Lc, *L;
double **p ,**hfi;
void calchi();
void init_log();

L  =U->Lk+l;
Lc =U->Lk+l-2;
```

```
p  =L->p;
hfi=L->hfi;

calchi(U,l  );
init_log(U,l-2);

for (i=0; i<=Lc->ii; i++)
  {
  iff=2*i;
  for (j=0; j<=Lc->jj; j++)
    {
    jff=2*j;
    if ((i==0)||(j==0)||(i==Lc->ii)||(j==Lc->jj))
    Lc->p[i][j]=0.0;
    else
    {
    if ((p[iff  ][jff  ]==0)||
        (p[iff+1][jff+1]==0)||(p[iff+1][jff-1]==0)||
        (p[iff-1][jff+1]==0)||(p[iff-1][jff-1]==0)||
        (p[iff+1][jff  ]==0)||(p[iff-1][jff  ]==0)||
        (p[iff  ][jff+1]==0)||(p[iff  ][jff-1]==0))
      Lc->p[i][j]=p[iff  ][jff  ];
    else
      Lc->p[i][j]=
    (4.0*p[iff  ][jff  ]+
     2.0*(p[iff+1][jff  ]+p[iff-1][jff  ]+p[iff  ][jff+1]+p[iff  ][jff-1])+
          p[iff+1][jff+1]+p[iff+1][jff-1]+p[iff-1][jff+1]+p[iff-1][jff-1]  )/16.0;
    }
  }
}

for (i=0; i<=Lc->ii; i++)
  {
  iff=2*i;
  for (j=0; j<=Lc->jj; j++)
    {
    jff=2*j;
    Lc->hfi[i][j]=hfi[iff][jff];
    }
  }

for (i=0; i<=Lc->ii; i++)
  for (j=0; j<=Lc->jj; j++)
    Lc->pold[i][j]=Lc->p[i][j];
}


void coarsen_f(Stack *U, int l)
{
/* compute coarse grid right hand side on level l-2 */
/* in coarsening step from level l                  */

int     i,j,iff,jff;
```

```
Level  *Lc,*L;
double **f, **p, gf, gc, rhx, rhy, rhx2, rhy2, rhxc, rhyc, rhxc2, rhyc2;
double r0, rn, re, rs, rw, rne, rnw, rse, rsw;
void calchi();

L  =U->Lk+l;
Lc =U->Lk+l-2;

p=L->p;
f=L->f;

rhx=1.0/L->hx;
rhy=1.0/L->hy;
rhx2=rhx*rhx;
rhy2=rhy*rhy;

rhxc=1.0/Lc->hx;
rhyc=1.0/Lc->hy;
rhxc2=rhxc*rhxc;
rhyc2=rhyc*rhyc;

calchi(U,l-2);

/* right hand side film thickness equation */

for (i=0;i<=Lc->ii;i++)
  for (j=0;j<=Lc->jj;j++)
    Lc->hrhs[i][j]=-U->h0-2./pi/pi*Lc->w[i][j]+Lc->hfi[i][j];

/* right hand side Reynolds equation */

for (i=1; i<=Lc->ii-1; i++)
    {
    iff=2*i;
    for (j=1; j<=Lc->jj-1; j++)
    {
    jff=2*j;
    r0=f[iff][jff]-Lu(L->hfi,L->p,rhx,rhx2,rhy2,iff,jff);
    if (p[iff][jff+1]>0)
     rn= f[iff  ][jff+1]-Lu(L->hfi,L->p,rhx,rhx2,rhy2,iff  ,jff+1);
    else rn=0.0;
    if (p[iff+1][jff]>0)
     re= f[iff+1][jff  ]-Lu(L->hfi,L->p,rhx,rhx2,rhy2,iff+1,jff  );
    else re=0.0;
    if (p[iff][jff-1]>0)
     rs= f[iff  ][jff-1]-Lu(L->hfi,L->p,rhx,rhx2,rhy2,iff  ,jff-1);
    else rs=0.0;
    if (p[iff-1][jff]>0)
     rw= f[iff-1][jff  ]-Lu(L->hfi,L->p,rhx,rhx2,rhy2,iff-1,jff  );
    else rw=0.0;
    if (p[iff+1][jff+1]>0)
     rne=f[iff+1][jff+1]-Lu(L->hfi,L->p,rhx,rhx2,rhy2,iff+1,jff+1);
    else rne=0.0;
    if (p[iff-1][jff+1]>0)
```

```
     rnw=f[iff-1][jff+1]-Lu(L->hfi,L->p,rhx,rhx2,rhy2,iff-1,jff+1);
    else rnw=0.0;
    if (p[iff+1][jff-1]>0)
     rse=f[iff+1][jff-1]-Lu(L->hfi,L->p,rhx,rhx2,rhy2,iff+1,jff-1);
    else rse=0.0;
    if (p[iff-1][jff-1]>0)
     rsw=f[iff-1][jff-1]-Lu(L->hfi,L->p,rhx,rhx2,rhy2,iff-1,jff-1);
    else rsw=0.0;

    Lc->f[i][j]=(4.0*r0+2.0*(rn+rs+re+rw)+(rne+rse+rsw+rnw))/16.0;
    Lc->f[i][j]+=Lu(Lc->hfi,Lc->p,rhxc,rhxc2,rhyc2,i,j);
    }
  }

/* right hand side force balance equation */

gf=0.0;
for (i=0; i<=L->ii; i++)
  for (j=0; j<=L->jj; j++)
     gf += L->p[i][j];

gc=0.0;
for (i=0; i<=Lc->ii; i++)
  for (j=0; j<=Lc->jj; j++)
    gc += Lc->p[i][j];

Lc->rg=(L->rg+gf*L->hx*L->hy)-gc*Lc->hx*Lc->hy;
}


void refine(Stack *U, int k)
{
/* Interpolation and addition of coarse grid correction from grid k-2 */
/* to grid k                                                          */

int     i,j,ic,jc,iic,jjc,ii,jj;
Level  *Lc,*L;
double **pc,**pco,**p;

void calchi();

L  =U->Lk+k;
ii =L ->ii; jj =L->jj;
p  =L->p  ;

Lc =U->Lk+k-2;
iic=Lc->ii; jjc=Lc->jj;
pc =Lc->p ;
pco=Lc->pold ;

for (ic=1;ic<=iic;ic++)
  for (jc=1;jc<=jjc;jc++)
    {
    if (p[2*ic  ][2*jc  ]>0)
```

```
     p[2*ic  ][2*jc  ]+=(pc[ic][jc]-pco[ic][jc]);

    if ((jc<jjc)&&(p[2*ic-1][2*jc  ]>0))
     p[2*ic-1][2*jc  ]+=(pc[ic  ][jc]-pco[ic  ][jc]+
                         pc[ic-1][jc]-pco[ic-1][jc])*0.5;

    if ((ic<iic)&&(p[2*ic  ][2*jc-1]>0))
     p[2*ic  ][2*jc-1]+=(pc[ic][jc  ]-pco[ic][jc  ]+
                         pc[ic][jc-1]-pco[ic][jc-1])*0.5;

    if (p[2*ic-1][2*jc-1]>0)
     p[2*ic-1][2*jc-1]+=(pc[ic  ][jc  ]-pco[ic  ][jc  ]+
                         pc[ic  ][jc-1]-pco[ic  ][jc-1]+
                         pc[ic-1][jc  ]-pco[ic-1][jc  ]+
                         pc[ic-1][jc-1]-pco[ic-1][jc-1])*0.25;
    }

for (i=0;i<=ii;i++)
  for (j=0;j<=jj;j++)
    if (p[i][j]<0.0) p[i][j]=0.0;

calchi(U,k);
}


void fmg_interpolate(Stack *U, int k)
{
/* interpolation of coarse grid k-2 solution to fine grid k */
/* to serve as first approximation. bi-cubic interpolation  */

int    ic,jc,iic,jjc,i,j,ii,jj;
Level  *Lc,*L;
double **pc,**p, **pconv, x,y;

/* set time */

cputimes[k-2]=clock();

L  =U->Lk+k;
ii =L ->ii; jj =L->jj;
p  =L->p ;

Lc    =U->Lk+k-2;
iic   =Lc->ii; jjc=Lc->jj;
pc    =Lc->p ;
pconv =Lc->pconv;

/* store coarse grid solution for later use in convergence check */
/* and store minimum and central film thickness for output */

jc=jjc/2;
ic=1;
while ((Lc->p[ic][jc]>Lc->p[ic-1][jc])&&(ic<iic)) ic++;
Lc->Hcp=Lc->hfi[ic][jc];
```

```
Lc->Hm=1e5; /* arbitrary large value */

for (ic=1;ic<=iic-1;ic++)
    {
    x=U->xa+ic*Lc->hx;
    for (jc=1;jc<=jjc-1;jc++)
       {
       y=U->ya+jc*Lc->hy;
       if (Lc->hfi[ic][jc]<Lc->Hm) Lc->Hm=Lc->hfi[ic][jc];
       if ((x==0)&&(y==0)) Lc->Hc=Lc->hfi[ic][jc];
       pconv[ic][jc]=pc[ic][jc];
       }
    }

/* interpolation                                        */
/* first inject to points coinciding with coarse grid points */

for (ic=1;ic<=iic-1;ic++)
  for (jc=1;jc<=jjc-1;jc++)
    p[2*ic][2*jc]=pc[ic][jc];

/* interpolate intermediate y direction */

for (i=2;i<=ii-2;i+=2)
  {
  p[i][1]=(5.0*p[i][0]+15.0*p[i][2]-5.0*p[i][4]+p[i][6])*0.0625;

  for (j=3;j<=jj-3;j+=2)
    p[i][j]=(-p[i][j-3]+9.0*p[i][j-1]+9.0*p[i][j+1]-p[i][j+3])*0.0625;

  p[i][jj-1]=(5.0*p[i][jj]+15.0*p[i][jj-2]-5.0*p[i][jj-4]+p[i][jj-6])*0.0625;
  }

/* interpolate in x direction */

for (j=1;j<=jj-1;j++)
  {
  p[1][j]=(5.0*p[0][j]+15.0*p[2][j]-5.0*p[4][j]+p[6][j])*0.0625;

  for (i=3;i<=ii-3;i+=2)
    p[i][j]=(-p[i-3][j]+9.0*p[i-1][j]+9.0*p[i+1][j]-p[i+3][j])*0.0625;

  p[ii-1][j]=(5.0*p[ii][j]+15.0*p[ii-2][j]-5.0*p[ii-4][j]+p[ii-6][j])*0.0625;
  }
for (i=1;i<=ii-1;i++)
   for (j=1;j<=jj-1;j++)
       if (p[i][j]<0) p[i][j]=0.0;
}


double conver(Stack *U, int k)
{
/* convergence check using converged solutions on level k and  */
```

```
/* on next coarser (solution) grid k-2                        */

int    ic,jc;
Level  *Lc,*L;
double **pc,**p;
double err;

L =U->Lk+k;

if (k==U->maxlev) p=L->p;
  else p=L->pconv;

Lc =U->Lk+k-2;
pc =Lc->pconv;

err=0.0;
for (ic=1;ic<=Lc->ii-1;ic++)
  for (jc=1;jc<=Lc->jj-1;jc++)
    err+=fabs(pc[ic][jc]-p[2*ic][2*jc]);

return(err/((Lc->ii-1)*(Lc->jj-1)));
}


/********** MULTIGRID DRIVING ROUTINES **********/


void cycle(Stack *U, int k, int nu0, int nu1, int nu2, int gamma)
/* performs coarse grid correction cycle starting on level k */
/* nu1 pre-relaxations, nu2 postrelaxations, nu0 relaxations */
/* on the coarsest grid, cycleindex gamma=1 for Vcycle,      */
/* gamma=2 for Wcycle                                        */
{
int i,j;

if (k==1)
  {
  for (i=1;i<=nu0/2;i++) relax(U,k);
  relaxh0(U,k,currfilev);
  for (i=1;i<=nu0/2;i++) relax(U,k);
  }
else
  {
  for (i=1;i<=nu1;i++) relax(U,k);
  coarsen_p(U,k);
  coarsen_f(U,k);
  for (j=1;j<=gamma;j++) cycle(U,k-2,nu0,nu1,nu2,gamma);
  refine(U,k);
  for (i=1;i<=nu2;i++) relax(U,k);
  }
}


void fmg(Stack *U, int k, int ks, int nu0, int nu1, int nu2, int gamma, int ncy)
```

```
{
/* performs FMG with k levels and ncy cycles per level */
int i,j;
void init_log();

if (k==ks)
   {
   if (ks==1)
    {
    for (i=1;i<=nu0/2;i++) relax(U,k);
    relaxh0(U,k,currfilev);
    for (i=1;i<=nu0/2;i++) relax(U,k);
    }
  else
  for (j=1;j<=ncy;j++)
    {
    cycle(U,ks,nu0,nu1,nu2,gamma);
    printf("\n");
    }
  }
else
  if (k>ks)
  {
  fmg(U,k-2,ks,nu0,nu1,nu2,gamma,ncy);
  fmg_interpolate(U,k);
  outlev=currfilev=k;
  for (j=1;j<=ncy;j++)
    {
    cycle(U,k,nu0,nu1,nu2,gamma);
    printf("\n");
    }
  }
}


/********* INPUT ROUTINES *****************/


void input_loadpar()
/* input of the load conditions for the contact */
{

printf("\ngive M            ?")    ; scanf("%lf",&MMoes);
printf("\ngive L            ?")    ; scanf("%lf",&LMoes);

/* conversion to parameters appearing in equations */

rlambda=1.0/(pi*pow(128.0/3.0/pow(MMoes,4.0),1.0/3.0));
alphabar=LMoes/pi*pow(1.5*MMoes,1.0/3.0);

printf("\nrlambda=%8.5e   alpha*p_h=%8.5e\n",rlambda,alphabar);

/* computation initial guess H0 */
```

```
H_0=1.67*pow(MMoes,-1.0/9.0)-1.897+0.2*LMoes/50;
if (H_0<-0.99) H_0=-0.99;

printf("First approximation H0=%8.5e",H_0);

/* Optional use of hand-input H0 */
/*
printf("\ngive H0              ?");
scanf("%lf",&H_0);
*/

/* parameters Roelands equation */

p0r=1.96e8;
alpha=2.2e-8;
eta0=40.0e-3;
zr=(alpha*p0r)/(log(eta0)+9.67);
}


void input_solvepar()
/* input of parameters numerical process */
{
/* account for intermediate grids */
printf("\nhow many levels   ?"); scanf("%d",&maxl);  maxl = 2*maxl-1;
printf("\nstartlevel        ?"); scanf("%d",&starl); starl=2*starl-1;

printf("\nhow many cycles   ?"); scanf("%d",&ncy);
printf("\ntype of cycle     ?"); scanf("%d",&typecy);

deepl=2*maxl-2;/* number of coarse grids in multi-integration */
order=6;       /* order of transfer multi-integration         */

xi_l=0.3;      /* relaxation switch parameter                 */
urja=0.2;      /* underrelaxation jacobi part                 */
urgs=0.4;      /* underrelaxation Gauss-Seidel part           */

/* factor for relaxation of force balance equation */

if (typecy==2)
 hfact=0.05;
else
 hfact=0.1;
if ((LMoes>10)||(MMoes>1000)) hfact*=0.25;
}


/********** OUTPUT ROUTINES **********/


void output(Stack *U)
/* writes an output file of p and h */
/* output of Hm and Hc to screen    */
{
```

```
int i,j;
Level *L;
double x,y,dum;
FILE *fp,*fh;

L =U->Lk+U->maxlev;
fp=fopen("P.dat","w");
fh=fopen("H.dat","w");

L->Hm=1e5; /* arbitrary large value */

/* determine central film thickness */
j=L->jj/2;
i=1;
while ((L->p[i][j]>L->p[i-1][j])&&(i<L->ii)) i++;
L->Hcp=L->hfi[i][j];

for (i=0; i<=L->ii; i++)
  {
  x=U->xa+i*L->hx;
  for (j=0; j<=L->jj; j++)
    {
    y=U->ya+j*L->hy;
    if (L->hfi[i][j]<L->Hm) L->Hm=L->hfi[i][j];
    if ((x==0)&&(y==0)) L->Hc=L->hfi[i][j];
    fprintf(fp,"%f %f %f\n",x,y,L->p[i][j]);
    fprintf(fh,"%f %f %f\n",x,y,2.0-L->hfi[i][j]);
    }
  fprintf(fp,"\n");
  fprintf(fh,"\n");
  }
fclose(fp);
fclose(fh);

/* output of film thickness values */;

dum=sqrt(6*pi*rlambda);
printf("\n\nLevel      Hm            Hm(Moes)       ");
printf("Hc             Hc(Moes)      Hc(Moes)(p_x=0)");

printf("\n\n");
for (i=starl;i<=U->maxlev;i+=2)
    {
    L=U->Lk+i;
    printf("%d      %8.5e     %8.5e   %8.5e     %8.5e     %8.5e\n",
           (i+1)/2, L->Hm,L->Hm*dum,L->Hc,L->Hc*dum,L->Hcp*dum);
    }
printf("\n\n");
}


/* ++++++++++++++++++++++ Film Thickness Computation  ++++++++++++++++++++*/
/* ++++++++++++++++++++++ Multi Level Multi Summation ++++++++++++++++++++*/
```

```
double ah(double a, double b)
/*calculates arcsinh(a/b)*/
{
if (a==0.0) return(0.0);
else
   if (b==0.0) return(1.0);
   else return(log(a/b+sqrt(a*a/b/b+1)));
}


void init_log(Stack *U, int l)
{
/* computes the kernel on level l */
int    i,j;
Level *L;
double xp,yp,xm,ym;

L =U->Lk+l;

for (i=0; i<=L->ii+4*U->od; i++)
  {
  xp=(i+0.5)*L->hx; xm=xp-L->hx;
  for (j=0; j<=L->jj+4*U->od; j++)
    {
    yp=(j+0.5)*L->hy; ym=yp-L->hy;
    L->K[i][j]=
      fabs(xp)*log(yp/xp+sqrt(1.0+yp*yp/xp/xp))
     -fabs(xp)*log(ym/xp+sqrt(1.0+ym*ym/xp/xp))
     +fabs(xm)*log(ym/xm+sqrt(1.0+ym*ym/xm/xm))
     -fabs(xm)*log(yp/xm+sqrt(1.0+yp*yp/xm/xm))
     +fabs(yp)*log(xp/yp+sqrt(1.0+xp*xp/yp/yp))
     -fabs(yp)*log(xm/yp+sqrt(1.0+xm*xm/yp/yp))
     +fabs(ym)*log(xm/ym+sqrt(1.0+xm*xm/ym/ym))
     -fabs(ym)*log(xp/ym+sqrt(1.0+xp*xp/ym/ym));
    }
  }
}


double kval(Stack *U, int lf, int lc, int i1, int j1)
/* calculates the value of k on the fine grid  lf in the point i1,j1*/
/* and injected to the coarse grid lc                                */
{
int l, i,j;
double x1,x2,y1,y2,help;
Level *Lf, *Lc;

Lf=U->Lk+lf;
Lc=U->Lk+lc;

i=i1; j=j1;
for (l=lf-1; l>=lc; l--)
    if ((2*(l/2)-l)!=0) i *= 2; else j *= 2;
```

```
x1=(i-0.5)*Lf->hx; x2=x1+Lf->hx; y1=(j-0.5)*Lf->hy; y2=y1+Lf->hy;

if ((x1==0.0)||(x2==0.0)||(y1==0.0)||(y2==0.0))
   printf("error in kval lf=%d lc=%d i=%d j=%d",lf,lc,i1,j1);
help=fabs(x2)*ah(y2,x2)+fabs(y2)*ah(x2,y2)-fabs(x2)*ah(y1,x2)-fabs(y2)*ah(x1,y2)
    -fabs(x1)*ah(y2,x1)-fabs(y1)*ah(x2,y1)+fabs(x1)*ah(y1,x1)+fabs(y1)*ah(x1,y1);
return(help*Lc->hx*Lc->hy/Lf->hx/Lf->hy);
}


void fillk(Stack *U, int l)
/* fills the matrix K with integral values          */
/* for matrix multiplication to obtain the integral*/
{
int l1, i,j;
Level *L, *Lc;

for (l1=l; l1>=2; l1--)
  {
  L =U->Lk+l1;
  Lc=U->Lk+l1-1;
  if ((2*(l1/2)-l1)==0)
    for (j=0; j<=L->jj+4*U->od; j++)
      {
      for (i=0; i<=Lc->ii+2*U->od; i++) Lc->K[i][j]=2.0*L->K[2*i][j];
      for (i=Lc->ii+2*U->od+1; i<=Lc->ii+4*U->od; i++) Lc->K[i][j]=kval(U,l,l1-1,i,j);
      }
  else
    for (i=0; i<=L->ii+4*U->od; i++)
      {
      for (j=0; j<=Lc->jj+2*U->od; j++) Lc->K[i][j]=2.0*L->K[i][2*j];
      for (j=Lc->jj+2*U->od+1; j<=Lc->jj+4*U->od; j++) Lc->K[i][j]=kval(U,l,l1-1,i,j);
      }
  }
}


void calcku(Stack *U, int l)
/*calculates the integral on grid l*/
{
int i1,j1,i,j;
double help;
Level *L;

L =U->Lk+l;
for (i1=-U->od; i1<=L->ii+U->od; i1++)
  for (j1=-U->od; j1<=L->jj+U->od; j1++)
    {
    help=0.0;
    for (i=-U->od; i<=L->ii+U->od; i++)
      for (j=-U->od; j<=L->jj+U->od; j++)
        help+=L->K[abs(i-i1)][abs(j-j1)]*L->p[i][j];
    L->w[i1][j1]=help;
    }
```

```
}


void sto6k1(Stack *U, int l)
/* stores values of k1 for correct6    */
/* k1 is the value of k minus the 6th  */
/* order coarse grid approximation to k*/
{
int i,j,maxx,maxy;
Level *L;

L =U->Lk+l;

if (4*U->m1<L->ii-1) maxx=4*U->m1+1; else maxx=L->ii;
if (4*U->m1<L->jj-1) maxy=4*U->m1+1; else maxy=L->jj;

if ((2*(l/2)-l)!=0)
  for (i=0; i<=maxx+4*U->od; i++)
    {
    for (j=0; j<=maxy+2*U->od; j++)
      {
      L->K1[i][j]=   L->K[i][j]
        -(  3*L->K[i][abs(5-j)]- 25*L->K[i][abs(3-j)]
         +150*L->K[i][abs(1-j)]+150*L->K[i][abs(1+j)]
         - 25*L->K[i][abs(3+j)]+  3*L->K[i][abs(5+j)])/256;
      }
    for (j=maxy+2*U->od+1; j<=maxy+4*U->od; j++)
      {
      L->K1[i][j]=   L->K[i][j]
        -(  3*L->K[i][abs(5-j)]- 25*L->K[i][abs(3-j)]
         +150*L->K[i][abs(1-j)] +150*kval(U,U->maxlev,l,i,j+1)
         - 25*kval(U,U->maxlev,l,i,j+3)+  3*kval(U,U->maxlev,l,i,j+5))/256;
      }
    }
else
  {
  for (j=0; j<=maxy+4*U->od; j++)
    {
    for (i=0; i<=maxx+2*U->od; i++)
      {
      L->K1[i][j]=  L->K[i][j]-
         (  3*L->K[abs(5-i)][j]-  25*L->K[abs(3-i)][j]
         +150*L->K[abs(1-i)][j]+ 150*L->K[abs(1+i)][j]
         - 25*L->K[abs(3+i)][j]+   3*L->K[abs(5+i)][j])/256;
      }
    for (i=maxx+2*U->od+1; i<=maxx+4*U->od; i++)
      {
      L->K1[i][j]=  L->K[i][j]-
         (  3*L->K[abs(5-i)][j]-  25*L->K[abs(3-i)][j]
         +150*L->K[abs(1-i)][j] +150*kval(U,U->maxlev,l,i+1,j)
         - 25*kval(U,U->maxlev,l,i+3,j)  +  3*kval(U,U->maxlev,l,i+5,j))/256;
      }
    }
  }
```

```
}


void coarsenp6x(Stack *U, int l)
/* 6th order weighting in x direction                            */
/* actually the 6th order approximation is obtained in K         */
/* the stencil is 1/512*(3,0,-25,0,150,256,150,0,-25,0,3) in x  */
/* direction                                                     */

{
int i,j;
double q0,q1,q2,q3,q4,q5,q7,q9;
Level *L, *Lc;

L =U->Lk+l;
Lc=U->Lk+l-1;

if ((2*(l/2)-l)!=0) printf("\nwrong coarsening in coarsenp6x\n");

for (j=-4; j<=L->jj+4; j++)
  {
  q0=L->p[-4][j]; q1 =L->p[-3][j]; q2=L->p[-2][j]; q3 =L->p[-1][j];
  q4=L->p[ 0][j]; q5 =L->p[ 1][j]; q7=L->p[ 3][j]; q9 =L->p[ 5][j];

  Lc->p[   -4][j]=  (                                          3*q1)/512;
  Lc->p[   -3][j]=  (                                   -25*q1 +3*q3)/512;
  Lc->p[   -2][j]=  (                  256*q0 +150*q1 -25*q3 +3*q5)/512;
  Lc->p[   -1][j]=  (          150*q1 +256*q2 +150*q3 -25*q5 +3*q7)/512;
  Lc->p[    0][j]=  (-25*q1 +150*q3 +256*q4 +150*q5 -25*q7 +3*q9)/512;

  for (i=1; i<=Lc->ii-1; i++)
    Lc->p[i][j]=(3*L->p[2*i-5][j]-25*L->p[2*i-3][j]+150*L->p[2*i-1][j]
              +256*L->p[2*i  ][j]
                +3*L->p[2*i+5][j]-25*L->p[2*i+3][j]+150*L->p[2*i+1][j])/512;

  q0=L->p[L->ii+4][j]; q1=L->p[L->ii+3][j]; q2=L->p[L->ii+2][j]; q3=L->p[L->ii+1][j];
  q4=L->p[L->ii  ][j]; q5=L->p[L->ii-1][j]; q7=L->p[L->ii-3][j]; q9=L->p[L->ii-5][j];

  Lc->p[Lc->ii  ][j]=(-25*q1 +150*q3 +256*q4 +150*q5- 25*q7 +3*q9)/512;
  Lc->p[Lc->ii+1][j]=(        150*q1 +256*q2 +150*q3- 25*q5 +3*q7)/512;
  Lc->p[Lc->ii+2][j]=(                256*q0 +150*q1- 25*q3 +3*q5)/512;
  Lc->p[Lc->ii+3][j]=(                               - 25*q1 +3*q3)/512;
  Lc->p[Lc->ii+4][j]=(                                        3*q1)/512;
  }
}


void coarsenp6y(Stack *U, int l)
/*6th order weighting, in y-direction                            */
/*actually the 6th order approximation is obtained in K          */
/*the stencil is 1/512*93,0,-25,0,150,256,150,0,-25,0,3) in y */
/*direction                                                      */

{
```

```
int    i,j;
double q0,q1,q2,q3,q4,q5,q7,q9;
Level *L, *Lc;

L =U->Lk+l;
Lc=U->Lk+l-1;

if ((2*(l/2)-l)==0) printf("\nwarning:wrong coarsening in coarsenp6y\n");

for (i=-4; i<=L->ii+4; i++)
  {
  q0=L->p[i][-4]; q1=L->p[i][-3]; q2=L->p[i][-2]; q3=L->p[i][-1];
  q4=L->p[i][ 0]; q5=L->p[i][ 1]; q7=L->p[i][ 3]; q9=L->p[i][ 5];

  Lc->p[i][   -4]=(                                          3*q1)/512;
  Lc->p[i][   -3]=(                                    -25*q1 +3*q3)/512;
  Lc->p[i][   -2]=(                      256*q0 +150*q1 -25*q3 +3*q5)/512;
  Lc->p[i][   -1]=(               150*q1 +256*q2 +150*q3 -25*q5 +3*q7)/512;
  Lc->p[i][    0]=( -25*q1 +150*q3 +256*q4 +150*q5 -25*q7 +3*q9)/512;

  for (j=1; j<=Lc->jj-1; j++)
    {
    Lc->p[i][j]=(3*L->p[i][2*j-5]-25*L->p[i][2*j-3]+150*L->p[i][2*j-1]
              +256*L->p[i][2*j]
                +3*L->p[i][2*j+5]-25*L->p[i][2*j+3]+150*L->p[i][2*j+1])/512;
    }
  q0=L->p[i][L->jj+4]; q1=L->p[i][L->jj+3]; q2=L->p[i][L->jj+2]; q3=L->p[i][L->jj+1];
  q4=L->p[i][L->jj  ]; q5=L->p[i][L->jj-1]; q7=L->p[i][L->jj-3]; q9=L->p[i][L->jj-5];

  Lc->p[i][Lc->jj  ]=( -25*q1 +150*q3 +256*q4 +150*q5 -25*q7 +3*q9)/512;
  Lc->p[i][Lc->jj+1]=(        150*q1 +256*q2 +150*q3 -25*q5 +3*q7)/512;
  Lc->p[i][Lc->jj+2]=(               256*q0 +150*q1 -25*q3 +3*q5)/512;
  Lc->p[i][Lc->jj+3]=(                             -25*q1 +3*q3)/512;
  Lc->p[i][Lc->jj+4]=(                                   3*q1)/512;
  }
}


void refine6x(Stack *U, int l)
/* O(6) correction of values of hfi on the fine grid after interpolation */
/* in x direction                                                         */
{
int    i,j,i1,j1,ibe,ien,jbe,jen;
double help;
Level *L, *Lc;

L =U->Lk+l;
Lc=U->Lk+l-1;

if ((2*(l/2)-l)!=0) printf("\nwarning:error in refine6x: l=%d\n",l);

for (i=-4; i<=Lc->ii+4; i++) /* fill also the ghost points */
  {
  ibe=2*i-U->m1; if (ibe<-4)      ibe=-4;      if ((2*(ibe/2)-ibe)==0) ibe++;
```

```
  ien=2*i+U->m1; if (ien>L->ii+4) ien=L->ii+4; if ((2*(ien/2)-ien)==0) ien--;
  for (j=-4; j<=L->jj+4; j++)
    {
    jbe=j-U->m2; if (jbe<-4)      jbe=-4;
    jen=j+U->m2; if (jen>L->jj+4) jen=L->jj+4;
    help=0.0;
    for (i1=ibe; i1<=ien; i1 += 2)
      for (j1=jbe; j1<=jen; j1++)
        help += L->p[i1][j1]*L->K1[abs(2*i-i1)][abs(j-j1)];
    Lc->w[i][j] += help;
    }
  }

for (i=-2; i<=Lc->ii+2; i++) /* fill also the ghost points */
  for (j=-4; j<=L->jj+4; j++)
    L->w[2*i][j]=Lc->w[i][j];

for (i=-2; i<=Lc->ii+1; i++)
  {
  ibe=2*i+1-U->m1; if (ibe<-4)     ibe=-4;
  ien=2*i+1+U->m1; if (ien>L->ii+4) ien=L->ii+4;
  for (j=-4; j<=L->jj+4; j++)
    {
    jbe=j-U->m2; if (jbe<-4)      jbe=-4;
    jen=j+U->m2; if (jen>L->jj+4) jen=L->jj+4;
    help=0.0;
    for (i1=ibe; i1<=ien; i1++)
      for (j1=jbe; j1<=jen; j1++)
        help += L->p[i1][j1]*L->K1[abs(2*i+1-i1)][abs(j-j1)];
    L->w[2*i+1][j]=(     3*Lc->w[i-2][j]-  25*Lc->w[i-1][j]
                    +150*Lc->w[i  ][j]+ 150*Lc->w[i+1][j]
                    - 25*Lc->w[i+2][j]+   3*Lc->w[i+3][j])/256+help;
    }
  }
}


void refine6y(Stack *U, int l)
/* O(6) correction of values of WI on the fine grid after interpolation*/
{
int i,j,i1,j1,ibe,ien,jbe,jen;
double help;
Level *L, *Lc;

L =U->Lk+l;
Lc=U->Lk+l-1;

if ((2*(l/2)-l)==0) printf("\nwarning:error in refine6y: %d\n",l);

for (j=-4; j<=Lc->jj+4; j++)
  {
  jbe=2*j-U->m1  ; if (jbe<-4)      jbe=-4;      if ((2*(jbe/2)-jbe)==0) jbe++;
  jen=2*j+U->m1  ; if (jen>L->jj+4) jen=L->jj+4; if ((2*(jen/2)-jen)==0) jen--;
  for (i=-4; i<=L->ii+4; i++)
```

```
    {
    ibe=i-U->m2; if (ibe<-4)      ibe=-4;
    ien=i+U->m2; if (ien>L->ii+4) ien=L->ii+4;
    j1=jbe; help=0.0;
    for (j1=jbe; j1<=jen; j1 += 2)
      for (i1=ibe; i1<=ien; i1++)
        help+=L->p[i1][j1]*L->K1[abs(i-i1)][abs(2*j-j1)];
    Lc->w[i][j] += help;
    }
  }

for (i=-4; i<=L->ii+4; i++)
  for (j=-2; j<=Lc->jj+2; j++)
    L->w[i][2*j]=Lc->w[i][j];

for (j=-2; j<=Lc->jj+1; j++)
  {
  jbe=2*j+1-U->m1; if (jbe<-4)      jbe=-4;
  jen=2*j+1+U->m1; if (jen>L->jj+4) jen=L->jj+4;
  for (i=-4; i<=L->ii+4; i++)
    {
    ibe=i-U->m2; if (ibe<-4)      ibe=-4;
    ien=i+U->m2; if (ien>L->ii+4) ien=L->ii+4;
    help=0.0;
    for (j1=jbe; j1<=jen; j1++)
      for (i1=ibe; i1<=ien; i1++)
        help+=L->p[i1][j1]*L->K1[abs(i-i1)][abs(2*j+1-j1)];
    L->w[i][2*j+1]=
        (  3*Lc->w[i][j-2]- 25*Lc->w[i][j-1]
        +150*Lc->w[i][j  ]+150*Lc->w[i][j+1]
        - 25*Lc->w[i][j+2]+  3*Lc->w[i][j+3])/256+help;
    }
  }
}


void calchi(Stack *U, int lev)
/* calculates film thickness on level l */
{
int i,j,ldeep,l;
Level *L;

ldeep=U->deep;
if (ldeep<0) ldeep=0; if (ldeep>lev-1) ldeep=lev-1;
fillk(U,lev);
if (U->od==4) for (l=lev;l>=lev-ldeep+1;l--) sto6k1(U,l);
for (l=1; l<=ldeep; l+=2)
  {
  if (U->od==4) {coarsenp6y(U,lev-l+1); coarsenp6x(U,lev-l);}
  }
calcku(U,lev-ldeep);
for (l=ldeep; l>=1; l-=2)
  {
  if (U->od==4) {refine6x(U,lev-l+1); refine6y(U,lev-l+2);}
```

```
  }

L =U->Lk+lev;

/* compute extra point for 2nd order discretization at first line */

for (i=-1; i<=L->ii; i++)
  for (j=0; j<=L->jj; j++)
    L->hfi[i][j]=U->h0+2.0/pi/pi*L->w[i][j]+L->hrhs[i][j];
}


/********** ROUTINES FOR LINE RELAXATION **********/


void init_line(Stack *U, int k, int j, int ml, int mr)
/* prepares system to be solved */
{

double H3, Hx, Qx, Qy, xi_n,xi_s,xi_e,xi_w, r0, r1, r2;
double dHx,dHxm,dHxp,dHxmm,dHxpp,dHxm3;
double dK0, dK1, dK2,dK3,dK4,rhx,rhy,rhx2,rhy2,rpi2;
double **P,**Pj, **H,**f, **A, *Y, *X, **K;
Level *L;
int ii,i,m,gs;

L=U->Lk+k;
ii=L->ii;

rhx=1.0/L->hx;
rhy=1.0/L->hy;
rhx2=rhx*rhx;
rhy2=rhy*rhy;
rpi2=1.0/(pi*pi);

P=L->p;
Pj=L->pjac;
K=L->K;
H=L->hfi;
f=L->f;
A=L->A;
Y=L->Y;
X=L->X;

for (i=0;i<=L->ii-2;i++)
   {
   for (m=0;m<=3*ml+mr;m++)
       A[i][m]=0.0;
   X[i]=0.0;
   Y[i]=0.0;
   }

for (i=1;i<=ii-1;i++)
  {
```

```
  r0=rho(P[i][j]);
  H3=r0*H[i][j]*H[i][j]*H[i][j]*reta(P[i][j]);
  xi_n=0.5*(rho(P[i  ][j+1])*H[i  ][j+1]*H[i  ][j+1]*H[i  ][j+1]
         *reta(P[i][j+1]) + H3)*rlambda*rhy2;
  xi_s=0.5*(rho(P[i  ][j-1])*H[i  ][j-1]*H[i  ][j-1]*H[i  ][j-1]
         *reta(P[i][j-1]) + H3)*rlambda*rhy2;
  xi_e=0.5*(rho(P[i+1][j  ])*H[i+1][j  ]*H[i+1][j  ]*H[i+1][j  ]
         *reta(P[i+1][j]) + H3)*rlambda*rhx2;
  xi_w=0.5*(rho(P[i-1][j  ])*H[i-1][j  ]*H[i-1][j  ]*H[i-1][j  ]
         *reta(P[i-1][j]) + H3)*rlambda*rhx2;

  gs=((fabs(xi_n)>xi_l)||(fabs(xi_w)>xi_l)||(fabs(xi_e)>xi_l)||(fabs(xi_s)>xi_l));

if (gs==1)
  {
  dK0=K[0][0];
  dK1=K[1][0];
  dK2=K[2][0];
  dK3=K[3][0];
  dK4=K[4][0];
  }
  else
  {
  dK0=K[0][0]-0.25*(2*K[1][0]+2*K[0][1]);
  dK1=K[1][0]-0.25*(K[2][0]+K[0][0]+2*K[1][1]);
  dK2=K[2][0]-0.25*(K[3][0]+K[1][0]+2*K[2][1]);
  dK3=K[3][0]-0.25*(K[4][0]+K[2][0]+2*K[3][1]);
  dK4=K[4][0]-0.25*(K[5][0]+K[3][0]+2*K[4][1]);
  }
  r1=rho(P[i-1][j]);
  if (i==1) Hx=rhx*(r0*H[i][j]-r1*H[i-1][j]);
  else
   {
   r2=rho(P[i-2][j]);
   Hx=rhx*(1.5*r0*H[i][j]-2*r1*H[i-1][j]+0.5*r2*H[i-2][j]);
   }

  dHx=dHxm=dHxmm=dHxm3=0.0;
  dHxp=dHxpp=0.0;

  if (i==1)
  {
  dHxm3=0.0*rhx*(r0*dK3-r1*dK2);
  dHxmm=0.0*rhx*(r0*dK2-r1*dK1);
  dHxm= 0.0*rhx*(r0*dK1-r1*dK0);
  dHx=  rhx*(r0*dK0-r1*dK1);
  dHxp= rhx*(r0*dK1-r1*dK2);
  dHxpp=rhx*(r0*dK2-r1*dK3);
  }
 else
  {
  if (i>3)   dHxm3=rhx*(1.5*r0*dK3-2.0*r1*dK2+0.5*r2*dK1);
  if (i>2)   dHxmm=rhx*(1.5*r0*dK2-2.0*r1*dK1+0.5*r2*dK0);
  if (i>1)   dHxm= rhx*(1.5*r0*dK1-2.0*r1*dK0+0.5*r2*dK1);
```

```
  dHx=  rhx*(1.5*r0*dK0-2.0*r1*dK1+0.5*r2*dK2);
  if (i<ii-1) dHxp= rhx*(1.5*r0*dK1-2.0*r1*dK2+0.5*r2*dK3);
  if (i<ii-2) dHxpp= rhx*(1.5*r0*dK2-2.0*r1*dK3+0.5*r2*dK4);
  }

if (gs==1)
   {
   Qx=(xi_e*Pj[i+1][j  ] -(xi_e+xi_w)*Pj[i][j]  + xi_w*Pj[i-1][j  ]);
   Qy=(xi_n*Pj[i  ][j+1] -(xi_n+xi_s)*Pj[i][j]  + xi_s*Pj[i  ][j-1]);

   Y[i-1]=f[i][j] -Qx -Qy + Hx;
   A[i-1][ml]=-(xi_w+xi_e)-(xi_s+xi_n)-2*rpi2*dHx;

   if ((Pj[i-2][j]>0)&&(Pj[i-1][j]>0)&&(Pj[i+1][j]>0)&&(Pj[i+2][j]>0))
    {
    if (i>1) A[i-1][ml-1]=xi_w-2*rpi2*dHxm;
    if (i>2) A[i-1][ml-2]=-2*rpi2*dHxmm;
    if (i>3) A[i-1][ml-3]=-2*rpi2*dHxm3;
    if (i<ii-1) A[i-1][ml+1]=xi_e-2*rpi2*dHxp;
    if (i<ii-2) A[i-1][ml+2]=-2*rpi2*dHxpp;
    }
   }
else
  {
  Qx=(xi_e*P[i+1][j  ] -(xi_e+xi_w)*P[i][j]  + xi_w*P[i-1][j  ]);
  Qy=(xi_n*P[i  ][j+1] -(xi_n+xi_s)*P[i][j]  + xi_s*P[i  ][j-1]);

  Y[i-1]=f[i][j] -Qx -Qy + Hx;
  A[i-1][ml]=-1.25*(xi_w+xi_e+xi_s+xi_w)-2*rpi2*dHx;
  if ((Pj[i-2][j]>0)&&(Pj[i-1][j]>0)&&(Pj[i+1][j]>0)&&(Pj[i+2][j]>0))
   {
   if (i>1) A[i-1][ml-1]=      xi_w+(xi_n+xi_e+xi_e+xi_w)*0.25-2*rpi2*dHxm;
   if (i>2) A[i-1][ml-2]=-0.25*xi_w-2*rpi2*dHxmm;
   if (i>3) A[i-1][ml-3]=-2*rpi2*dHxm3;
   if (i<ii-1) A[i-1][ml+1]=      xi_e+(xi_n+xi_e+xi_e+xi_w)*0.25-2*rpi2*dHxp;
   if (i<ii-2) A[i-1][ml+2]=-0.25*xi_e-2*rpi2*dHxpp;
   }
  }
 }
}


int pivot(double **A, int ri, int ii, int ml)
{
int i,j,I,end;
double piv;

piv=fabs(A[ri][ml]);
I=ri;

end=ml; if ((ri+ml)>ii) end=ii-ri;
for (i=1;i<=end;i++)
  {
  j=ml-i;
```

```
  if (fabs(A[ri+i][j]) > piv)
    I=ri+i;
  }
return(I);
}


void swap(double **A, double *y, int ri, int rI, int ml, int m)
{
int j,cJ;
double temp;

for (j=ml;j<=m;j++)
  {
  cJ=j-(rI-ri);
  if (cJ<0) continue;
  temp=A[ri][j]; A[ri][j]=A[rI][cJ]; A[rI][cJ]=temp;
  }

temp=y[ri]; y[ri]=y[rI]; y[rI]=temp;
}


void bcksb(double **A,double *Y,double *X, int mm, int ml, int N)
{
int i,j;
double sum;

X[N]=Y[N]/(A[N][ml]);
for (i=N-1;i>=0;i--)
  {
  sum=0.0;
  for (j=1;j<=mm;j++)
    if (i+j<=N) sum+=A[i][ml+j]*X[i+j];
  X[i]=(Y[i]-sum)/(A[i][ml]);
  }
}


void solve_system(Stack *U, int l, int m, int ml, int mr)
{
int N,end,i,j,k,rI,J;
double dum;
Level *L;
double **A, *X, *Y;

L=U->Lk+l;
A=L->A;
X=L->X;
Y=L->Y;
N=L->ii-2;

for (i=0;i<=N;i++)
  {
```

```
  rI=pivot(A,i,N,ml);
  if (i!=rI) swap(A,Y,i,rI,ml,m);

  end=i+ml; if (end>N) end=N;
  for (k=i+1;k<=end;k++)
    {
    J=ml-(k-i);
    if (J<0) continue;
    dum=-(A[k][J]/A[i][ml]);
    for (j=J+1;j<=J+(mr+ml);j++)
      A[k][j]+=dum*A[i][j+(ml-J)];
    A[k][J]=0.0;
    Y[k]+=dum*Y[i];
    }
  }
bcksb(A,Y,X,m,ml,N);
}


void return_line(Stack *U, int k, int j)
{
double **H,H3,xi_n,xi_s,xi_e,xi_w,r0;
int ii,jj,i,gs;
double rhx,rhy,rhx2,rhy2,*X,del0,p0;
Level *L;

L=U->Lk+k;
ii=L->ii;
jj=L->jj;

rhx=1.0/L->hx;
rhy=1.0/L->hy;
rhx2=rhx*rhx;
rhy2=rhy*rhy;
X=L->X;
H=L->hfi;

for (i=1;i<=ii-1;i++)
    {
    r0=rho(L->p[i][j]);
    H3=r0*H[i][j]*H[i][j]*H[i][j]*reta(L->p[i][j]);
    xi_n=0.5*(rho(L->p[i  ][j+1])*H[i  ][j+1]*H[i  ][j+1]*H[i  ][j+1]
            *reta(L->p[i][j+1]) + H3)*rlambda*rhy2;
    xi_s=0.5*(rho(L->p[i  ][j-1])*H[i  ][j-1]*H[i  ][j-1]*H[i  ][j-1]
            *reta(L->p[i][j-1]) + H3)*rlambda*rhy2;
    xi_e=0.5*(rho(L->p[i+1][j  ])*H[i+1][j  ]*H[i+1][j  ]*H[i+1][j  ]
            *reta(L->p[i+1][j]) + H3)*rlambda*rhx2;
    xi_w=0.5*(rho(L->p[i-1][j  ])*H[i-1][j  ]*H[i-1][j  ]*H[i-1][j  ]
            *reta(L->p[i-1][j]) + H3)*rlambda*rhx2;

    gs=((fabs(xi_n)>xi_l)||(fabs(xi_w)>xi_l)||(fabs(xi_e)>xi_l)||(fabs(xi_s)>xi_l));

    if (gs==1)
    {
```

```
    p0=L->pjac[i][j];
    L->pjac[i][j]+=urgs*X[i-1];
    if (L->pjac[i][j]<0.0) L->pjac[i][j]=0.0;
    del0=L->pjac[i][j]-p0;
    }
    else
    {
    p0=L->pjac[i][j];
    L->pjac[i][j]+=urja*X[i-1];
    if (L->pjac[i][j]<0) L->pjac[i][j]=0.0;
    del0=L->pjac[i][j]-p0;
    if ((L->pjac[i][j]>0)&&(L->pjac[i-1][j]>0)&&(L->pjac[i+1][j]>0)
                       &&(L->pjac[i][j-1]>0)&&(L->pjac[i][j+1]>0))
    {
    if (i>1)
       {
       L->pjac[i-1][j]-=0.25*del0;
       if (L->pjac[i-1][j]<0) L->pjac[i-1][j]=0.0;
       }
    if (i<ii-1)
       {
       L->pjac[i+1][j]-=0.25*del0;
       if (L->pjac[i+1][j]<0) L->pjac[i+1][j]=0.0;
       }

    if (j>1)
       {
       L->pjac[i][j-1]-=0.25*del0;
       if (L->pjac[i][j-1]<0) L->pjac[i][j-1]=0.0;
       }

    if (j<jj-1)
       {
       L->pjac[i][j+1]-=0.25*del0;
       if (L->pjac[i][j+1]<0) L->pjac[i][j+1]=0.0;
       }
     }
    }
  }
}


void solve_line(Stack *U, int k, int j)
{
int ml,m,mr;
ml=3;
mr=2;
m=2*ml+mr;
init_line(U, k, j, ml,mr);
solve_system(U,k,m,ml,mr);
return_line(U,k,j);
}
```

```
/********** MAIN PROGRAM **********/


void main()
{
int    l;
Stack  U;

input_loadpar();
input_solvepar();

initialize(&U,16,16,maxl,deepl,order,-4.5,1.5,-3.0,3.0,H_0);

for (l=1; l<=maxl; l+=2) { init_log(&U,l); init_f(&U,l); }
outlev=currfilev=starl;
t0=clock();
init_p(&U,starl);

fmg(&U,maxl,starl,40,2,1,typecy,ncy);
cputimes[maxl]=clock();

output(&U);

if (maxl>1)
  for (l=starl+2;l<=maxl;l+=2) printf(" aen(%2d,%2d)=%8.5e \n",(l+1)/2,(l-1)/2,
                                   conver(&U,l));
printf("\n\n");
for (l=starl;l<=maxl;l+=2) printf("cpu times level %d = %8.5e \n",(l+1)/2,
                                   (cputimes[l]-t0)/1000000.0);
finalize(&U,maxl);
}
```

Appendix I

Program Listing: Second Order

```
void sto2k1(Stack *U, int l)
/* stores values of k1 for refine2    */
/* k1 is the value of k minus the 2nd  */
/* order coarse grid approximation to k*/
{
int i,j,maxx,maxy;
Level *L;

L =U->Lk+l;

if (4*U->m1<L->ii-1) maxx=4*U->m1+1; else maxx=L->ii;
if (4*U->m1<L->jj-1) maxy=4*U->m1+1; else maxy=L->jj;

if ((2*(l/2)-l)!=0)
  for (i=0; i<=maxx; i++)
    for (j=0; j<=maxy; j++)
      L->K1[i][j]= L->K[i][j]-(L->K[i][abs(1-j)]+L->K[i][1+j])/2.;
else
  for (j=0; j<=maxy; j++)
    for (i=0; i<=maxx; i++)
      L->K1[i][j]=L->K[i][j]-(L->K[abs(1-i)][j]+L->K[1+i][j])/2.;
}


void coarsenp2x(Stack *U, int l)
/* 2nd order weighting in x direction                           */
/* actually the 2nd order approximation is obtained in K        */
/* the stencil is 1/4*(1,2,1) in x  direction                   */
{
int i,j;
Level *L, *Lc;

L =U->Lk+l;
Lc=U->Lk+l-1;

if ((2*(l/2)-l)!=0) printf("\nwrong coarsening in coarsenp2x\n");

for (j=0; j<=L->jj; j++)
  {
```

```
  Lc->p[0     ][j]=(                  2*L->p[      0][j]+L->p[       1][j])/4;
  for (i=1; i<=Lc->ii-1; i++)
    Lc->p[i][j]=(L->p[2*i-1][j]+2*L->p[2*i][j]+L->p[2*i+1][j])/4;
  Lc->p[Lc->ii][j]=(L->p[L->ii-1][j]+2*L->p[L->ii+0][j]                 )/4;
  }
}


void coarsenp2y(Stack *U, int l)
/* 2nd order weighting in y direction                              */
/* actually the 2th order approximation is obtained in K           */
/* the stencil is 1/4*(1,2,1) in y  direction                      */
{
int i,j;
Level *L, *Lc;

L =U->Lk+l;
Lc=U->Lk+l-1;

if ((2*(l/2)-l)==0) printf("\nwrong coarsening in coarsenp2y\n");

for (i=0; i<=L->ii; i++)
  {
  Lc->p[i][     0]=(                  2*L->p[i][      0]+L->p[i][       1])/4;
  for (j=1; j<=Lc->jj-1; j++)
    Lc->p[i][j]=(L->p[i][2*j-1]+2*L->p[i][2*j]+L->p[i][2*j+1])/4;
  Lc->p[i][Lc->jj]=(L->p[i][L->jj-1]+2*L->p[i][L->jj+0]                 )/4;
  }
}


void refine2x(Stack *U, int l)
/* O(2) correction of values of hfi on the fine grid after interpolation */
/* in x direction                                                        */
{
int    i,j,i1,j1,ibe,ien,jbe,jen;
double help;
Level *L, *Lc;

L =U->Lk+l;
Lc=U->Lk+l-1;

if ((2*(l/2)-l)!=0) printf("\nwarning:error in refine2x: l=%d\n",l);

for (i=0; i<=Lc->ii; i++)
  {
  ibe=2*i-U->m1; if (ibe<0)     ibe=0;     if ((2*(ibe/2)-ibe)==0) ibe++;
  ien=2*i+U->m1; if (ien>L->ii) ien=L->ii; if ((2*(ien/2)-ien)==0) ien--;
  for (j=0; j<=L->jj; j++)
    {
    jbe=j-U->m2; if (jbe<0)     jbe=0;
    jen=j+U->m2; if (jen>L->jj) jen=L->jj;
    help=0.0;
    for (i1=ibe; i1<=ien; i1 += 2)
```

```
      for (j1=jbe; j1<=jen; j1++)
        help += L->p[i1][j1]*L->K1[abs(2*i-i1)][abs(j-j1)];
    Lc->w[i][j] += help;
    }
  }

for (i=0; i<=Lc->ii; i++)
  for (j=0; j<=L->jj; j++)
    L->w[2*i][j]=Lc->w[i][j];

for (i=0; i<=Lc->ii-1; i++)
  {
  ibe=2*i+1-U->m1; if (ibe<0)     ibe=0;
  ien=2*i+1+U->m1; if (ien>L->ii) ien=L->ii;
  for (j=0; j<=L->jj; j++)
    {
    jbe=j-U->m2; if (jbe<0)     jbe=0;
    jen=j+U->m2; if (jen>L->jj) jen=L->jj;
    help=0.0;
    for (i1=ibe; i1<=ien; i1++)
      for (j1=jbe; j1<=jen; j1++)
        help += L->p[i1][j1]*L->K1[abs(2*i+1-i1)][abs(j-j1)];
    L->w[2*i+1][j]=(Lc->w[i+0][j]+Lc->w[i+1][j])/2+help;
    }
  }
}


void refine2y(Stack *U, int l)
/* O(2) correction of values of hfi on the fine grid after interpolation */
/* in y direction                                                        */
{
int    i,j,i1,j1,ibe,ien,jbe,jen;
double help;
Level *L, *Lc;

L =U->Lk+l;
Lc=U->Lk+l-1;

if ((2*(l/2)-l)==0) printf("\nwarning:error in refine2y: l=%d\n",l);

for (j=0; j<=Lc->jj; j++)
  {
  jbe=2*j-U->m1; if (jbe<0)     jbe=0;     if ((2*(jbe/2)-jbe)==0) jbe++;
  jen=2*j+U->m1; if (jen>L->jj) jen=L->jj; if ((2*(jen/2)-jen)==0) jen--;
  for (i=0; i<=L->ii; i++)
    {
    ibe=i-U->m2; if (ibe<0)     ibe=0;
    ien=i+U->m2; if (ien>L->ii) ien=L->ii;
    help=0.0;
    for (j1=jbe; j1<=jen; j1 += 2)
      for (i1=ibe; i1<=ien; i1++)
        help += L->p[i1][j1]*L->K1[abs(i-i1)][abs(2*j-j1)];
    Lc->w[i][j] += help;
```

```
    }
  }

for (j=0; j<=Lc->jj; j++)
  for (i=0; i<=L->ii; i++)
    L->w[i][2*j]=Lc->w[i][j];

for (j=0; j<=Lc->jj-1; j++)
  {
  jbe=2*j+1-U->m1; if (jbe<0)     jbe=0;
  jen=2*j+1+U->m1; if (jen>L->ii) jen=L->jj;
  for (i=0; i<=L->ii; i++)
    {
    ibe=i-U->m2; if (ibe<0)     ibe=0;
    ien=i+U->m2; if (ien>L->ii) ien=L->ii;
    help=0.0;
    for (i1=ibe; i1<=ien; i1++)
      for (j1=jbe; j1<=jen; j1++)
        help += L->p[i1][j1]*L->K1[abs(i-i1)][abs(2*j+1-j1)];
    L->w[i][2*j+1]=(Lc->w[i][j+0]+Lc->w[i][j+1])/2+help;
    }
  }
}
```

Appendix J

Program Listing: Fourth Order

```
void sto4k1(Stack *U, int l)
/* stores values of k1 for refine4      */
/* k1 is the value of k minus the 4th   */
/* order coarse grid approximation to k*/
{
int i,j,maxx,maxy;
Level *L;

L =U->Lk+l;

if (4*U->m1<L->ii-1) maxx=4*U->m1+1; else maxx=L->ii;
if (4*U->m1<L->jj-1) maxy=4*U->m1+1; else maxy=L->jj;

if ((2*(l/2)-l)!=0)
  for (i=0; i<=maxx+4*U->od; i++)
    {
    for (j=0; j<=maxy+2*U->od; j++)
      L->K1[i][j]=L->K[i][j]-(-L->K[i][abs(3-j)]+ 9*L->K[i][abs(1-j)]
                             -L->K[i][    3+j ]+ 9*L->K[i][    1+j ])/16;
    for (j=maxy+2*U->od+1; j<=maxy+4*U->od; j++)
      L->K1[i][j]=L->K[i][j]-(-L->K[i][abs(3-j)]+ 9*L->K[i][abs(1-j)]
                             +9*kval(U,U->maxlev,l,i,j+1)
                             -  kval(U,U->maxlev,l,i,j+3))/16;
    }
else
  for (j=0; j<=maxy+4*U->od; j++)
    {
    for (i=0; i<=maxx+2*U->od; i++)
      L->K1[i][j]=L->K[i][j]-(-L->K[abs(3-i)][j]+ 9*L->K[abs(1-i)][j]
                             -L->K[    3+i ][j]+ 9*L->K[    1+i ][j])/16;
    for (i=maxx+2*U->od+1; i<=maxx+4*U->od; i++)
      L->K1[i][j]=L->K[i][j]-(-L->K[abs(3-i)][j]+ 9*L->K[abs(1-i)][j]
                             +9*kval(U,U->maxlev,l,i+1,j)
                             -  kval(U,U->maxlev,l,i+3,j))/16;
    }
}


void coarsenp4x(Stack *U, int l)
```

```
/* 4th order weighting in x direction                          */
/* actually the 4th order approximation is obtained in K       */
/* the stencil is 1/32*(-1,0,9,16,9,0,-1) in x direction       */
{
int i,j;
double q0,q1,q2,q3,q4,q5;
Level *L, *Lc;

L =U->Lk+l;
Lc=U->Lk+l-1;

if ((2*(l/2)-l)!=0) printf("\nwrong coarsening in coarsenp4x\n");

for (j=-2; j<=L->jj+2; j++)
  {
  q0=L->p[  -2][j]; q1=L->p[  -1][j]; q2=L->p[   0][j];
  q3=L->p[   1][j]; q4=L->p[   2][j]; q5=L->p[   3][j];

  Lc->p[   -2][j]= (                   -q1)/32;
  Lc->p[   -1][j]= (      16*q0 +9*q1 -q3)/32;
  Lc->p[    0][j]= (9*q1+ 16*q2 +9*q3 -q5)/32;

  for (i=1; i<=Lc->ii-1; i++)
    Lc->p[i][j]=(-L->p[2*i-3][j]+9*L->p[2*i-1][j]+16*L->p[2*i][j]
                -L->p[2*i+3][j]+9*L->p[2*i+1][j])/32;

  q0=L->p[L->ii+2][j]; q1=L->p[L->ii+1][j]; q2=L->p[L->ii  ][j];
  q3=L->p[L->ii-1][j]; q4=L->p[L->ii-2][j]; q5=L->p[L->ii-3][j];

  Lc->p[Lc->ii  ][j]= (9*q1+ 16*q2 +9*q3 -q5)/32;
  Lc->p[Lc->ii+1][j]= (      16*q0 +9*q1 -q3)/32;
  Lc->p[Lc->ii+2][j]= (                   -q1)/32;
  }
}


void coarsenp4y(Stack *U, int l)
/* 4th order weighting in y direction                          */
/* actually the 6th order approximation is obtained in K       */
/* the stencil is 1/32*(-1,0,9,16,9,0,-1) in y direction       */
{
int i,j;
double q0,q1,q2,q3,q4,q5;
Level *L, *Lc;

L =U->Lk+l;
Lc=U->Lk+l-1;

if ((2*(l/2)-l)==0) printf("\nwrong coarsening in coarsenp4y\n");

for (i=-2; i<=L->ii+2; i++)
  {
  q0=L->p[i][  -2]; q1=L->p[i][  -1]; q2=L->p[i][   0];
  q3=L->p[i][   1]; q4=L->p[i][   2]; q5=L->p[i][   3];
```

```
  Lc->p[i][   -2]= (                    -q1)/32;
  Lc->p[i][   -1]= (      16*q0 +9*q1 -q3)/32;
  Lc->p[i][    0]= (9*q1+ 16*q2 +9*q3 -q5)/32;

  for (j=1; j<=Lc->jj-1; j++)
    Lc->p[i][j]=(-L->p[i][2*j-3]+9*L->p[i][2*j-1]+16*L->p[i][2*j]
                 -L->p[i][2*j+3]+9*L->p[i][2*j+1])/32;

  q0=L->p[i][L->jj+2]; q1=L->p[i][L->jj+1]; q2=L->p[i][L->jj  ];
  q3=L->p[i][L->jj-1]; q4=L->p[i][L->jj-2]; q5=L->p[i][L->jj-3];

  Lc->p[i][Lc->jj  ]= (9*q1+ 16*q2 +9*q3 -q5)/32;
  Lc->p[i][Lc->jj+1]= (      16*q0 +9*q1 -q3)/32;
  Lc->p[i][Lc->jj+2]= (                    -q1)/32;
  }
}


void refine4x(Stack *U, int l)
/* O(4) correction of values of hfi on the fine grid after interpolation */
/* in x direction                                                         */
{
int    i,j,i1,j1,ibe,ien,jbe,jen;
double help;
Level *L, *Lc;

L =U->Lk+l;
Lc=U->Lk+l-1;

if ((2*(l/2)-l)!=0) printf("\nwarning:error in refine4x: l=%d\n",l);

for (i=-2; i<=Lc->ii+2; i++) /* fill also the ghost points */
  {
  ibe=2*i-U->m1; if (ibe<-2)      ibe=-2;      if ((2*(ibe/2)-ibe)==0) ibe++;
  ien=2*i+U->m1; if (ien>L->ii+2) ien=L->ii+2; if ((2*(ien/2)-ien)==0) ien--;
  for (j=-2; j<=L->jj+2; j++)
    {
    jbe=j-U->m2; if (jbe<-2)      jbe=-2;
    jen=j+U->m2; if (jen>L->jj+2) jen=L->jj+2;
    help=0.0;
    for (i1=ibe; i1<=ien; i1 += 2)
      for (j1=jbe; j1<=jen; j1++)
        help += L->p[i1][j1]*L->K1[abs(2*i-i1)][abs(j-j1)];
    Lc->w[i][j] += help;
    }
  }

for (i=-1; i<=Lc->ii+1; i++) /* fill also the ghost points */
  for (j=-2; j<=L->jj+2; j++)
    L->w[2*i][j]=Lc->w[i][j];

for (i=-1; i<=Lc->ii; i++)
  {
```

```
  ibe=2*i+1-U->m1; if (ibe<-2)     ibe=-2;
  ien=2*i+1+U->m1; if (ien>L->ii+2) ien=L->ii+2;
  for (j=-2; j<=L->jj+2; j++)
    {
    jbe=j-U->m2; if (jbe<-2)      jbe=-2;
    jen=j+U->m2; if (jen>L->jj+2) jen=L->jj+2;
    help=0.0;
    for (i1=ibe; i1<=ien; i1++)
      for (j1=jbe; j1<=jen; j1++)
        help += L->p[i1][j1]*L->K1[abs(2*i+1-i1)][abs(j-j1)];
    L->w[2*i+1][j]=(  -Lc->w[i-1][j]+9*Lc->w[i  ][j]
                      -Lc->w[i+2][j]+9*Lc->w[i+1][j])/16+help;
    }
  }
}


void refine4y(Stack *U, int l)
/* O(4) correction of values of hfi on the fine grid after interpolation */
/* in y direction                                                        */
{
int    i,j,i1,j1,ibe,ien,jbe,jen;
double help;
Level *L, *Lc;

L =U->Lk+l;
Lc=U->Lk+l-1;

if ((2*(l/2)-l)==0) printf("\nwarning:error in refine4y: l=%d\n",l);

for (j=-2; j<=Lc->jj+2; j++) /* fill also the ghost points */
  {
  jbe=2*j-U->m1; if (jbe<-2)      jbe=-2;      if ((2*(jbe/2)-jbe)==0) jbe++;
  jen=2*j+U->m1; if (jen>L->jj+2) jen=L->jj+2; if ((2*(jen/2)-jen)==0) jen--;
  for (i=-2; i<=L->ii+2; i++)
    {
    ibe=i-U->m2; if (ibe<-2)      ibe=-2;
    ien=i+U->m2; if (ien>L->ii+2) ien=L->ii+2;
    help=0.0;
    for (j1=jbe; j1<=jen; j1 += 2)
      for (i1=ibe; i1<=ien; i1++)
        help += L->p[i1][j1]*L->K1[abs(i-i1)][abs(2*j-j1)];
    Lc->w[i][j] += help;
    }
  }

for (j=-1; j<=Lc->jj+1; j++) /* fill also the ghost points */
  for (i=-2; i<=L->ii+2; i++)
    L->w[i][2*j]=Lc->w[i][j];

for (j=-1; j<=Lc->jj; j++)
  {
  jbe=2*j+1-U->m1; if (jbe<-2)      jbe=-2;
  jen=2*j+1+U->m1; if (jen>L->jj+2) jen=L->jj+2;
```

```
  for (i=-2; i<=L->ii+2; i++)
    {
    ibe=i-U->m2; if (ibe<-2)      ibe=-2;
    ien=i+U->m2; if (ien>L->ii+2) ien=L->ii+2;
    help=0.0;
    for (i1=ibe; i1<=ien; i1++)
      for (j1=jbe; j1<=jen; j1++)
        help += L->p[i1][j1]*L->K1[abs(i-i1)][abs(2*j+1-j1)];
    L->w[i][2*j+1]=(  -Lc->w[i][j-1]+9*Lc->w[i][j  ]
      -Lc->w[i][j+2]+9*Lc->w[i][j+1])/16+help;
    }
  }
}
```

Appendix K

Program Listing: Sixth Order

```
void sto6k1(Stack *U, int l)
/* stores values of k1 for refine6    */
/* k1 is the value of k minus the 6th  */
/* order coarse grid approximation to k*/
{
int i,j,maxx,maxy;
Level *L;

L =U->Lk+l;

if (4*U->m1<L->ii-1) maxx=4*U->m1+1; else maxx=L->ii;
if (4*U->m1<L->jj-1) maxy=4*U->m1+1; else maxy=L->jj;

if ((2*(l/2)-l)!=0)
  for (i=0; i<=maxx+4*U->od; i++)
    {
    for (j=0; j<=maxy+2*U->od; j++)
      {
      L->K1[i][j]=   L->K[i][j]
        -(  3*L->K[i][abs(5-j)]- 25*L->K[i][abs(3-j)]
         +150*L->K[i][abs(1-j)]+150*L->K[i][abs(1+j)]
         - 25*L->K[i][abs(3+j)]+  3*L->K[i][abs(5+j)])/256;
      }
    for (j=maxy+2*U->od+1; j<=maxy+4*U->od; j++)
      {
      L->K1[i][j]=   L->K[i][j]
        -(  3*L->K[i][abs(5-j)]- 25*L->K[i][abs(3-j)]
         +150*L->K[i][abs(1-j)] +150*kval(U,U->maxlev,l,i,j+1)
         - 25*kval(U,U->maxlev,l,i,j+3)+  3*kval(U,U->maxlev,l,i,j+5))/256;
      }
    }
else
  {
  for (j=0; j<=maxy+4*U->od; j++)
    {
    for (i=0; i<=maxx+2*U->od; i++)
      {
      L->K1[i][j]=  L->K[i][j]-
         (  3*L->K[abs(5-i)][j]-  25*L->K[abs(3-i)][j]
```

```
        +150*L->K[abs(1-i)][j]+ 150*L->K[abs(1+i)][j]
        - 25*L->K[abs(3+i)][j]+   3*L->K[abs(5+i)][j])/256;
     }
    for (i=maxx+2*U->od+1; i<=maxx+4*U->od; i++)
      {
      L->K1[i][j]=  L->K[i][j]-
         (  3*L->K[abs(5-i)][j]-  25*L->K[abs(3-i)][j]
         +150*L->K[abs(1-i)][j] +150*kval(U,U->maxlev,l,i+1,j)
         - 25*kval(U,U->maxlev,l,i+3,j)  +  3*kval(U,U->maxlev,l,i+5,j))/256;
      }
    }
  }
}


void coarsenp6x(Stack *U, int l)
/* 6th order weighting in x direction                            */
/* actually the 6th order approximation is obtained in K         */
/* the stencil is 1/512*(3,0,-25,0,150,256,150,0,-25,0,3) in x   */
/* direction                                                     */

{
int i,j;
double q0,q1,q2,q3,q4,q5,q7,q9;
Level *L, *Lc;

L =U->Lk+l;
Lc=U->Lk+l-1;

if ((2*(l/2)-l)!=0) printf("\nwrong coarsening in coarsenp6x\n");


for (j=-4; j<=L->jj+4; j++)
  {
  q0=L->p[-4][j]; q1 =L->p[-3][j]; q2=L->p[-2][j]; q3 =L->p[-1][j];
  q4=L->p[ 0][j]; q5 =L->p[ 1][j]; q7=L->p[ 3][j]; q9 =L->p[ 5][j];

  Lc->p[   -4][j]=   (                                           3*q1)/512;
  Lc->p[   -3][j]=   (                                    -25*q1 +3*q3)/512;
  Lc->p[   -2][j]=   (                     256*q0 +150*q1 -25*q3 +3*q5)/512;
  Lc->p[   -1][j]=   (          150*q1 +256*q2 +150*q3 -25*q5 +3*q7)/512;
  Lc->p[    0][j]=   (-25*q1 +150*q3 +256*q4 +150*q5 -25*q7 +3*q9)/512;

  for (i=1; i<=Lc->ii-1; i++)
    Lc->p[i][j]=(3*L->p[2*i-5][j]-25*L->p[2*i-3][j]+150*L->p[2*i-1][j]+256*L->p[2*i][j]
                +3*L->p[2*i+5][j]-25*L->p[2*i+3][j]+150*L->p[2*i+1][j])/512;

  q0=L->p[L->ii+4][j]; q1=L->p[L->ii+3][j]; q2=L->p[L->ii+2][j]; q3=L->p[L->ii+1][j];
  q4=L->p[L->ii  ][j]; q5=L->p[L->ii-1][j]; q7=L->p[L->ii-3][j]; q9=L->p[L->ii-5][j];

  Lc->p[Lc->ii  ][j]=(-25*q1 +150*q3 +256*q4 +150*q5- 25*q7 +3*q9)/512;
  Lc->p[Lc->ii+1][j]=(        150*q1 +256*q2 +150*q3- 25*q5 +3*q7)/512;
  Lc->p[Lc->ii+2][j]=(                256*q0 +150*q1- 25*q3 +3*q5)/512;
  Lc->p[Lc->ii+3][j]=(                              - 25*q1 +3*q3)/512;
```

```
  Lc->p[Lc->ii+4][j]=(                                          3*q1)/512;
  }
}


void coarsenp6y(Stack *U, int l)
/*6th order weighting, in y-direction                           */
/*actually the 6th order approximation is obtained in K         */
/*the stencil is 1/512*93,0,-25,0,150,256,150,0,-25,0,3) in y */
/*direction                                                     */
{
int    i,j;
double q0,q1,q2,q3,q4,q5,q7,q9;
Level *L, *Lc;

L =U->Lk+l;
Lc=U->Lk+l-1;

if ((2*(l/2)-l)==0) printf("\nwarning:wrong coarsening in coarsenp6y\n");

for (i=-4; i<=L->ii+4; i++)
  {
  q0=L->p[i][-4]; q1=L->p[i][-3]; q2=L->p[i][-2]; q3=L->p[i][-1];
  q4=L->p[i][ 0]; q5=L->p[i][ 1]; q7=L->p[i][ 3]; q9=L->p[i][ 5];

  Lc->p[i][   -4]=(                                          3*q1)/512;
  Lc->p[i][   -3]=(                                   -25*q1 +3*q3)/512;
  Lc->p[i][   -2]=(                    256*q0 +150*q1 -25*q3 +3*q5)/512;
  Lc->p[i][   -1]=(             150*q1 +256*q2 +150*q3 -25*q5 +3*q7)/512;
  Lc->p[i][    0]=( -25*q1 +150*q3 +256*q4 +150*q5 -25*q7 +3*q9)/512;

  for (j=1; j<=Lc->jj-1; j++)
    {
    Lc->p[i][j]=(3*L->p[i][2*j-5]-25*L->p[i][2*j-3]+150*L->p[i][2*j-1]+256*L->p[i][2*j]
                +3*L->p[i][2*j+5]-25*L->p[i][2*j+3]+150*L->p[i][2*j+1])/512;
    }
  q0=L->p[i][L->jj+4]; q1=L->p[i][L->jj+3]; q2=L->p[i][L->jj+2]; q3=L->p[i][L->jj+1];
  q4=L->p[i][L->jj  ]; q5=L->p[i][L->jj-1]; q7=L->p[i][L->jj-3]; q9=L->p[i][L->jj-5];

  Lc->p[i][Lc->jj  ]=( -25*q1 +150*q3 +256*q4 +150*q5 -25*q7 +3*q9)/512;
  Lc->p[i][Lc->jj+1]=(        150*q1 +256*q2 +150*q3 -25*q5 +3*q7)/512;
  Lc->p[i][Lc->jj+2]=(               256*q0 +150*q1 -25*q3 +3*q5)/512;
  Lc->p[i][Lc->jj+3]=(                              -25*q1 +3*q3)/512;
  Lc->p[i][Lc->jj+4]=(                                     3*q1)/512;
  }
}


void refine6x(Stack *U, int l)
/* O(6) correction of values of hfi on the fine grid after interpolation */
/* in x direction                                                       */
{
int    i,j,i1,j1,ibe,ien,jbe,jen;
double help;
```

```
Level *L, *Lc;

L =U->Lk+l;
Lc=U->Lk+l-1;

if ((2*(l/2)-l)!=0) printf("\nwarning:error in refine6x: l=%d\n",l);

for (i=-4; i<=Lc->ii+4; i++) /* fill also the ghost points */
  {
  ibe=2*i-U->m1; if (ibe<-4)      ibe=-4;      if ((2*(ibe/2)-ibe)==0) ibe++;
  ien=2*i+U->m1; if (ien>L->ii+4) ien=L->ii+4; if ((2*(ien/2)-ien)==0) ien--;
  for (j=-4; j<=L->jj+4; j++)
    {
    jbe=j-U->m2; if (jbe<-4)      jbe=-4;
    jen=j+U->m2; if (jen>L->jj+4) jen=L->jj+4;
    help=0.0;
    for (i1=ibe; i1<=ien; i1 += 2)
      for (j1=jbe; j1<=jen; j1++)
        help += L->p[i1][j1]*L->K1[abs(2*i-i1)][abs(j-j1)];
    Lc->w[i][j] += help;
    }
  }

for (i=-2; i<=Lc->ii+2; i++) /* fill also the ghost points */
  for (j=-4; j<=L->jj+4; j++)
    L->w[2*i][j]=Lc->w[i][j];

for (i=-2; i<=Lc->ii+1; i++)
  {
  ibe=2*i+1-U->m1; if (ibe<-4)     ibe=-4;
  ien=2*i+1+U->m1; if (ien>L->ii+4) ien=L->ii+4;
  for (j=-4; j<=L->jj+4; j++)
    {
    jbe=j-U->m2; if (jbe<-4)      jbe=-4;
    jen=j+U->m2; if (jen>L->jj+4) jen=L->jj+4;
    help=0.0;
    for (i1=ibe; i1<=ien; i1++)
      for (j1=jbe; j1<=jen; j1++)
        help += L->p[i1][j1]*L->K1[abs(2*i+1-i1)][abs(j-j1)];
    L->w[2*i+1][j]=(     3*Lc->w[i-2][j]-  25*Lc->w[i-1][j]
                    +150*Lc->w[i  ][j]+ 150*Lc->w[i+1][j]
                    - 25*Lc->w[i+2][j]+   3*Lc->w[i+3][j])/256+help;
    }
  }
}


void refine6y(Stack *U, int l)
/* O(6) correction of values of WI on the fine grid after interpolation*/
{
int i,j,i1,j1,ibe,ien,jbe,jen;
double help;
Level *L, *Lc;
```

```
L =U->Lk+l;
Lc=U->Lk+l-1;

if ((2*(l/2)-l)==0) printf("\nwarning:error in refine6y: %d\n",l);

for (j=-4; j<=Lc->jj+4; j++)
  {
  jbe=2*j-U->m1  ; if (jbe<-4)      jbe=-4;      if ((2*(jbe/2)-jbe)==0) jbe++;
  jen=2*j+U->m1  ; if (jen>L->jj+4) jen=L->jj+4; if ((2*(jen/2)-jen)==0) jen--;
  for (i=-4; i<=L->ii+4; i++)
    {
    ibe=i-U->m2; if (ibe<-4)      ibe=-4;
    ien=i+U->m2; if (ien>L->ii+4) ien=L->ii+4;
    j1=jbe; help=0.0;
    for (j1=jbe; j1<=jen; j1 += 2)
      for (i1=ibe; i1<=ien; i1++)
        help+=L->p[i1][j1]*L->K1[abs(i-i1)][abs(2*j-j1)];
    Lc->w[i][j] += help;
    }
  }

for (i=-4; i<=L->ii+4; i++)
  for (j=-2; j<=Lc->jj+2; j++)
    L->w[i][2*j]=Lc->w[i][j];

for (j=-2; j<=Lc->jj+1; j++)
  {
  jbe=2*j+1-U->m1; if (jbe<-4)      jbe=-4;
  jen=2*j+1+U->m1; if (jen>L->jj+4) jen=L->jj+4;
  for (i=-4; i<=L->ii+4; i++)
    {
    ibe=i-U->m2; if (ibe<-4)      ibe=-4;
    ien=i+U->m2; if (ien>L->ii+4) ien=L->ii+4;
    help=0.0;
    for (j1=jbe; j1<=jen; j1++)
      for (i1=ibe; i1<=ien; i1++)
        help+=L->p[i1][j1]*L->K1[abs(i-i1)][abs(2*j+1-j1)];
    L->w[i][2*j+1]=
        (  3*Lc->w[i][j-2]- 25*Lc->w[i][j-1]
        +150*Lc->w[i][j  ]+150*Lc->w[i][j+1]
        - 25*Lc->w[i][j+2]+  3*Lc->w[i][j+3])/256+help;
    }
  }
}
```

Appendix L

Program Listing: Eighth Order

```
void sto8k1(Stack *U, int l)
/* stores values of k1 for refine8     */
/* k1 is the value of k minus the 8th  */
/* order coarse grid approximation to k*/
{

int i,j,maxx,maxy,ml;
Level *L;

L =U->Lk+l;
ml=U->maxlev;

if (4*U->m1<L->ii-1) maxx=4*U->m1+1; else maxx=L->ii;
if (4*U->m1<L->jj-1) maxy=4*U->m1+1; else maxy=L->jj;

if ((2*(l/2)-l)!=0)
  for (i=0; i<=maxx+4*U->od; i++)
    {
    for (j=0; j<=maxy+2*U->od; j++)
      L->K1[i][j]=   L->K[i][j]
        -(-5*L->K[i][abs(7-j)]+  49*L->K[i][abs(5-j)]
        -245*L->K[i][abs(3-j)]+1225*L->K[i][abs(1-j)]
        -245*L->K[i][abs(3+j)]+1225*L->K[i][abs(1+j)]
          -5*L->K[i][abs(7+j)]+  49*L->K[i][abs(5+j)])/2048.;
    for (j=maxy+2*U->od+1; j<=maxy+4*U->od; j++)
      L->K1[i][j]=   L->K[i][j]
        -(-5*L->K[i][abs(7-j)] +  49*L->K[i][abs(5-j)]
        -245*L->K[i][abs(3-j)] +1225*L->K[i][abs(1-j)]
        -245*kval(U,ml,l,i,j+3)+1225*kval(U,ml,l,i,j+1)
          -5*kval(U,ml,l,i,j+7)+  49*kval(U,ml,l,i,j+5))/2048.;
     }
  else
  for (j=0; j<=maxy+4*U->od; j++)
    {
    for (i=0; i<=maxx+2*U->od; i++)
      L->K1[i][j]=   L->K[i][j]
        -(-5*L->K[abs(7-i)][j]+  49*L->K[abs(5-i)][j]
        -245*L->K[abs(3-i)][j]+1225*L->K[abs(1-i)][j]
        -245*L->K[abs(3+i)][j]+1225*L->K[abs(1+i)][j]
```

```
        -5*L->K[abs(7+i)][j]+  49*L->K[abs(5+i)][j])/2048.;
  for (i=maxx+2*U->od+1; i<=maxx+4*U->od; i++)
    L->K1[i][j]=   L->K[i][j]
      -(-5*L->K[abs(7-i)][j] +  49*L->K[abs(5-i)][j]
      -245*L->K[abs(3-i)][j] +1225*L->K[abs(1-i)][j]
      -245*kval(U,ml,l,i+3,j)+1225*kval(U,ml,l,i+1,j)
        -5*kval(U,ml,l,i+7,j)+  49*kval(U,ml,l,i+5,j))/2048.;
  }
}


void coarsenp8x(Stack *U, int l)
/* 8th order weighting in x direction                          */
/* actually the 8th order approximation is obtained in K       */
/* the stencil is 1/4096*(-5,0,49,0,-245,0,1225,2048,1225,0,-245,0,49,0,-5) */
/* x - direction                                                       */

{
int i,j;
double q0,q1,q2,q3,q4,q5,q6,q7,q9,q11,q13;
Level *L, *Lc;

L =U->Lk+l;
Lc=U->Lk+l-1;

if ((2*(l/2)-l)!=0) printf("\nwrong coarsening in coarsenp8x\n");

for (j=-6; j<=L->jj+6; j++)
  {
  q0 =L->p[-6][j]; q1 =L->p[-5][j]; q2=L->p[-4][j];
  q3 =L->p[-3][j]; q4 =L->p[-2][j]; q5=L->p[-1][j];
  q6 =L->p[ 0][j]; q7 =L->p[ 1][j]; q9=L->p[ 3][j];
  q11=L->p[ 5][j]; q13=L->p[ 7][j];

  Lc->p[      -6][j]=(                                                      -5*q1 )/4096.;
  Lc->p[      -5][j]=(                                                49*q1 -5*q3 )/4096.;
  Lc->p[      -4][j]=(                                      -245*q1+49*q3 -5*q5 )/4096.;
  Lc->p[      -3][j]=(                          2048*q0+1225*q1-245*q3+49*q5 -5*q7 )/4096.;
  Lc->p[      -2][j]=(                1225*q1+2048*q2+1225*q3-245*q5+49*q7 -5*q9 )/4096.;
  Lc->p[      -1][j]=(      -245*q1+1225*q3+2048*q4+1225*q5-245*q7+49*q9 -5*q11)/4096.;
  Lc->p[       0][j]=(49*q1-245*q3+1225*q5+2048*q6+1225*q7-245*q9+49*q11-5*q13)/4096.;

  for (i=1; i<=Lc->ii-1; i++)
    Lc->p[i][j]=(-5*L->p[2*i-7][j]+  49*L->p[2*i-5][j]
             -245*L->p[2*i-3][j]+1225*L->p[2*i-1][j]+2048*L->p[2*i][j]
             -245*L->p[2*i+3][j]+1225*L->p[2*i+1][j]
               -5*L->p[2*i+7][j]+  49*L->p[2*i+5][j])/4096.;

  q11=L->p[L->ii-5][j]; q13=L->p[L->ii-7][j];
  q6 =L->p[L->ii+0][j]; q7 =L->p[L->ii-1][j]; q9=L->p[L->ii-3][j];
  q3 =L->p[L->ii+3][j]; q4 =L->p[L->ii+2][j]; q5=L->p[L->ii+1][j];
  q0 =L->p[L->ii+6][j]; q1 =L->p[L->ii+5][j]; q2=L->p[L->ii+4][j];

  Lc->p[Lc->ii  ][j]=(49*q1-245*q3+1225*q5+2048*q6+1225*q7-245*q9+49*q11-5*q13)/4096.;
```

```
  Lc->p[Lc->ii+1][j]=(     -245*q1+1225*q3+2048*q4+1225*q5-245*q7+49*q9 -5*q11)/4096.;
  Lc->p[Lc->ii+2][j]=(             1225*q1+2048*q2+1225*q3-245*q5+49*q7 -5*q9 )/4096.;
  Lc->p[Lc->ii+3][j]=(                     2048*q0+1225*q1-245*q3+49*q5 -5*q7 )/4096.;
  Lc->p[Lc->ii+4][j]=(                                    -245*q1+49*q3 -5*q5 )/4096.;
  Lc->p[Lc->ii+5][j]=(                                            49*q1 -5*q3 )/4096.;
  Lc->p[Lc->ii+6][j]=(                                                  -5*q1 )/4096.;
  }
}


void coarsenp8y(Stack *U, int l)
/* 8th order weighting in y direction                               */
/* actually the 8th order approximation is obtained in K            */
/* the stencil is 1/4096*(-5,0,49,0,-245,0,1225,2048,1225,0,-245,0,49,0,-5) */
/* y - direction                                                        */

{
int i,j;
double q0,q1,q2,q3,q4,q5,q6,q7,q9,q11,q13;
Level *L, *Lc;

L =U->Lk+l;
Lc=U->Lk+l-1;

if ((2*(l/2)-l)==0) printf("\nwrong coarsening in coarsenp8y\n");

for (i=-6; i<=L->ii+6; i++)
  {
  q0 =L->p[i][-6]; q1 =L->p[i][-5]; q2=L->p[i][-4];
  q3 =L->p[i][-3]; q4 =L->p[i][-2]; q5=L->p[i][-1];
  q6 =L->p[i][ 0]; q7 =L->p[i][ 1]; q9=L->p[i][ 3];
  q11=L->p[i][ 5]; q13=L->p[i][ 7];

  Lc->p[i][      -6]=(                                                  -5*q1 )/4096.;
  Lc->p[i][      -5]=(                                            49*q1 -5*q3 )/4096.;
  Lc->p[i][      -4]=(                                    -245*q1+49*q3 -5*q5 )/4096.;
  Lc->p[i][      -3]=(                     2048*q0+1225*q1-245*q3+49*q5 -5*q7 )/4096.;
  Lc->p[i][      -2]=(             1225*q1+2048*q2+1225*q3-245*q5+49*q7 -5*q9 )/4096.;
  Lc->p[i][      -1]=(     -245*q1+1225*q3+2048*q4+1225*q5-245*q7+49*q9 -5*q11)/4096.;
  Lc->p[i][       0]=(49*q1-245*q3+1225*q5+2048*q6+1225*q7-245*q9+49*q11-5*q13)/4096.;

  for (j=1; j<=Lc->jj-1; j++)
    Lc->p[i][j]=(-5*L->p[i][2*j-7]+  49*L->p[i][2*j-5]
              -245*L->p[i][2*j-3]+1225*L->p[i][2*j-1]+2048*L->p[i][2*j]
              -245*L->p[i][2*j+3]+1225*L->p[i][2*j+1]
                -5*L->p[i][2*j+7]+  49*L->p[i][2*j+5])/4096.;

  q11=L->p[i][L->jj-5]; q13=L->p[i][L->jj-7];
  q6 =L->p[i][L->jj+0]; q7 =L->p[i][L->jj-1]; q9=L->p[i][L->jj-3];
  q3 =L->p[i][L->jj+3]; q4 =L->p[i][L->jj+2]; q5=L->p[i][L->jj+1];
  q0 =L->p[i][L->jj+6]; q1 =L->p[i][L->jj+5]; q2=L->p[i][L->jj+4];

  Lc->p[i][Lc->jj  ]=(49*q1-245*q3+1225*q5+2048*q6+1225*q7-245*q9+49*q11-5*q13)/4096.;
  Lc->p[i][Lc->jj+1]=(     -245*q1+1225*q3+2048*q4+1225*q5-245*q7+49*q9 -5*q11)/4096.;
```

```
  Lc->p[i][Lc->jj+2]=(           1225*q1+2048*q2+1225*q3-245*q5+49*q7 -5*q9 )/4096.;
  Lc->p[i][Lc->jj+3]=(                   2048*q0+1225*q1-245*q3+49*q5 -5*q7 )/4096.;
  Lc->p[i][Lc->jj+4]=(                                  -245*q1+49*q3 -5*q5 )/4096.;
  Lc->p[i][Lc->jj+5]=(                                          49*q1 -5*q3 )/4096.;
  Lc->p[i][Lc->jj+6]=(                                                -5*q1 )/4096.;
  }
}


void refine8x(Stack *U, int l)
/* O(8) correction of values of hfi on the fine grid after interpolation */
/* in x direction                                                        */
{
int    i,j,i1,j1,ibe,ien,jbe,jen;
double help;
Level *L, *Lc;

L =U->Lk+l;
Lc=U->Lk+l-1;

if ((2*(l/2)-l)!=0) printf("\nwarning:error in refine8x: l=%d\n",l);

for (i=-6; i<=Lc->ii+6; i++) /* fill also the ghost points */
  {
  ibe=2*i-U->m1; if (ibe<-6)      ibe=-6;      if ((2*(ibe/2)-ibe)==0) ibe++;
  ien=2*i+U->m1; if (ien>L->ii+6) ien=L->ii+6; if ((2*(ien/2)-ien)==0) ien--;
  for (j=-6; j<=L->jj+6; j++)
    {
    jbe=j-U->m2; if (jbe<-6)      jbe=-6;
    jen=j+U->m2; if (jen>L->jj+6) jen=L->jj+6;
    help=0.0;
    for (i1=ibe; i1<=ien; i1 += 2)
      for (j1=jbe; j1<=jen; j1++)
        help += L->p[i1][j1]*L->K1[abs(2*i-i1)][abs(j-j1)];
    Lc->w[i][j] += help;
    }
  }

for (i=-3; i<=Lc->ii+3; i++) /* fill also the ghost points */
  for (j=-6; j<=L->jj+6; j++)
    L->w[2*i][j]=Lc->w[i][j];

for (i=-3; i<=Lc->ii+2; i++)
  {
  ibe=2*i+1-U->m1; if (ibe<-6)      ibe=-6;
  ien=2*i+1+U->m1; if (ien>L->ii+6) ien=L->ii+6;
  for (j=-6; j<=L->jj+6; j++)
    {
    jbe=j-U->m2; if (jbe<-6)      jbe=-6;
    jen=j+U->m2; if (jen>L->jj+6) jen=L->jj+6;
    help=0.0;
    for (i1=ibe; i1<=ien; i1++)
      for (j1=jbe; j1<=jen; j1++)
        help += L->p[i1][j1]*L->K1[abs(2*i+1-i1)][abs(j-j1)];
```

```
    L->w[2*i+1][j]=(-5*Lc->w[i-3][j]+  49*Lc->w[i-2][j]
                  -245*Lc->w[i-1][j]+1225*Lc->w[i  ][j]
                  -245*Lc->w[i+2][j]+1225*Lc->w[i+1][j]
                    -5*Lc->w[i+4][j]+  49*Lc->w[i+3][j])/2048.+help;
    }
  }
}


void refine8y(Stack *U, int l)
/* O(8) correction of values of hfi on the fine grid after interpolation */
/* in y direction                                                         */
{
int    i,j,i1,j1,ibe,ien,jbe,jen;
double help;
Level *L, *Lc;

L =U->Lk+l;
Lc=U->Lk+l-1;

if ((2*(l/2)-l)==0) printf("\nwarning:error in refine8y: l=%d\n",l);

for (j=-6; j<=Lc->jj+6; j++) /* fill also the ghost points */
  {
  jbe=2*j-U->m1; if (jbe<-6)      jbe=-6;      if ((2*(jbe/2)-jbe)==0) jbe++;
  jen=2*j+U->m1; if (jen>L->jj+6) jen=L->jj+6; if ((2*(jen/2)-jen)==0) jen--;
  for (i=-6; i<=L->ii+6; i++)
    {
    ibe=i-U->m2; if (ibe<-6)      ibe=-6;
    ien=i+U->m2; if (ien>L->ii+6) ien=L->ii+6;
    help=0.0;
    for (i1=ibe; i1<=ien; i1++)
      for (j1=jbe; j1<=jen; j1 += 2)
        help += L->p[i1][j1]*L->K1[abs(i-i1)][abs(2*j-j1)];
    Lc->w[i][j] += help;
    }
  }

for (j=-3; j<=Lc->jj+3; j++) /* fill also the ghost points */
  for (i=-6; i<=L->ii+6; i++)
    L->w[i][2*j]=Lc->w[i][j];

for (j=-3; j<=Lc->jj+2; j++)
  {
  jbe=2*j+1-U->m1; if (jbe<-6)      jbe=-6;
  jen=2*j+1+U->m1; if (jen>L->jj+6) jen=L->jj+6;
  for (i=-6; i<=L->ii+6; i++)
    {
    ibe=i-U->m2; if (ibe<-6)      ibe=-6;
    ien=i+U->m2; if (ien>L->ii+6) ien=L->ii+6;
    help=0.0;
    for (i1=ibe; i1<=ien; i1++)
      for (j1=jbe; j1<=jen; j1++)
        help += L->p[i1][j1]*L->K1[abs(i-i1)][abs(2*j+1-j1)];
```

```
    L->w[i][2*j+1]=(-5*Lc->w[i][j-3]+  49*Lc->w[i][j-2]
                 -245*Lc->w[i][j-1]+1225*Lc->w[i][j  ]
                 -245*Lc->w[i][j+2]+1225*Lc->w[i][j+1]
                   -5*Lc->w[i][j+4]+  49*Lc->w[i][j+3])/2048.+help;
    }
  }
}
```

Index